INDICA

INDICA

A DEEP NATURAL HISTORY OF THE INDIAN SUBCONTINENT

PRANAY LAL

PENGUIN ALLEN LANE

An imprint of Penguin Random House

ALLEN LANE

USA | Canada | UK | Ireland | Australia
New Zealand | India | South Africa | China | Singapore

Allen Lane is part of the Penguin Random House group of companies
whose addresses can be found at global.penguinrandomhouse.com

Published by Penguin Random House India Pvt. Ltd.
4th Floor, Capital Tower 1, MG Road,
Gurugram 122 002, Haryana, India

Penguin
Random House
India

First published in Allen Lane by Penguin Random House India 2016

10 9 8 7

ISBN 9788184007572

Book design by Tanvi Nathyal with Ahlawat Gunjan
Printed and bound in India by Replika Press Pvt. Ltd.

www.penguin.co.in

For Aria and Avie

Who complete my Circle of Life.

CONTENTS

A PERSONAL NOTE TO THE READER

I love nature. I spent my childhood in different parts of Africa and in small towns of Uttar Pradesh, where my days were spent flipping stones and logs, climbing trees, swimming in ponds, and following bird catchers and snake charmers, and my nights lying on my back, watching the stars and hearing stories from grandparents, uncles and watchmen about their experiences in jungles and with animals. I love the beauty of spectacular mountains just as much as the intricate patterns on the back of a tiny unnamed beetle. But what mesmerizes me, now and as a child, is how nature is constantly at work and how everything is related to each other. This book is a culmination of over twenty years of research, travel, conversations, interviews, and a lifetime awe of nature.

During my journeys criss-crossing India, I have met fascinating people who love nature passionately—qualified scientists, but also ordinary people who take time out of their busy lives to observe nature and record its intricacies and inter-linkages. True scientists are like explorers—they have a fascination for the unknown and an indefatigable quest for answers. They work tirelessly, excavating through layers of rocks and sediment, climbing canopies and descending into caves, or working in laboratories. Every discipline of natural science has its own Marco Polos, Magellans, Drakes, Amundsens, Hillarys and Tenzings who have set off in search for answers in the terra incognita of science. Their works help us understand how our planet and our life came to be and what forces have shaped them. Each piece of rock, every pinch of soil and every little fossil tells a story, and because of the persistence and dedication of the scientists, who decipher these stories and make them available to us, we understand our supremely complex and supremely exciting world better. It is the work of these discoverers that I celebrate in this book.

I have included all that I believe is crucial to understanding the grand story of the formation of India from the time Earth was formed to the arrival of humans; how the position, the contours and the composition of land and oceans changed, how thousands of types of plants and animals evolved and thrived and became extinct and why different places in India look the way they do today. The plan of the book is chronological and there is some overlap in the fifteen chapters to maintain continuity. I have combined scientific depth and historical insights that would interest anybody who has looked out of their window and wondered why there is a river in one place and not another, why rocks in one place look different from those elsewhere or why there is a hill or a forest or a mine in front of them. It should appeal to a wide range of readers, even if they do not have any particular scientific or geographic inclinations. Those not wanting to spend time on details can skip some of the more technical passages and can look at the maps and illustrations through which I have tried to convey the essence of the chapter. For those who would like to dive deeper into their areas of interest, I hope the detailed notes will help in their further exploration.

This story of 4 billion years is one without an end, and the journey to discover our world, our country and us is a work in progress. I welcome you, and thank you for being part of my beautiful voyage.

INDICA

A LONG,
LONG TIME
AGO, BUT
IN A PLACE
NOT TOO
FAR AWAY . . .

1

WHY ON EARTH

Drive out of Bengaluru along National Highway 7 for about 70 kilometres and you will arrive at the 1479-metre-high Nandi Hills. The cooler climate of these lush hills makes them a perfect getaway from the grind of the large, bustling city. From the top of the hills, almost as far as the eye can see, the landscape is dotted with smaller dome-shaped knolls that are mostly bare, with only a few patches of grass and stunted trees growing on them. People who reach the top of the Nandi Hills pay little attention to the ground on which they tread. But do take a moment and look at this grey rock that glistens darkly in the rain because this is no ordinary rock. Belonging to a family of rocks called the Dharwar Craton that was formed about 3.5 billion years ago, it is among the oldest rocks in India and it is literally a part of the very bedrock on which the country stands!

Technically speaking, a 'craton' (Greek for strength) or a 'shield' is an old and stable part of Earth's outer layer (the crust) that has survived the cycles of the merging and rifting of continents. Simply put, a craton is that part of Earth's crust that has remained largely unchanged for billions of years. Most cratons are composed of ancient crystalline

Older rocks, like those in the Nandi Hills near Bengaluru and like this one in Lalbagh Gardens in the heart of the city, became the bedrock around which newer rocks aggregated and peninsular India was assembled, piece by piece.

rocks (appropriately called 'basement rocks') which can extend several hundred kilometres below the surface and are often covered by younger sedimentary rocks. The age of the Dharwar Craton makes India one of the oldest land masses in the world. Only Australia, Greenland and South Africa have rocks that are older than those found here. By comparison, the oldest rocks that lie exposed in Indonesia are no more than 300 million years old. In fact, if we board a train from Bengaluru and travel north to Jammu, we will cover just 2700 kilometres in terms of distance, but will traverse more than 3 billion years of Earth's history, simply by looking out of our window! So how does it come about that some parts of Earth are much older than others? To answer this and

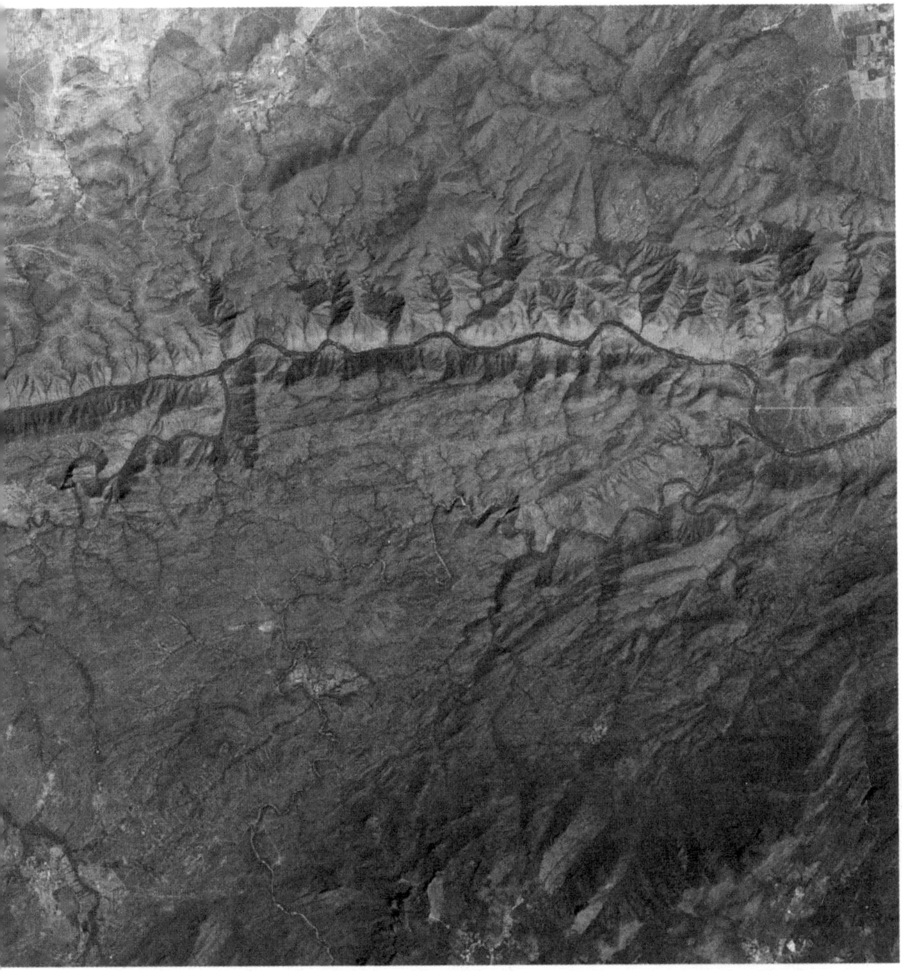

The Moyar Gorge, near Masinagudy in the Mudumalai National Park, about 40 kilometres north of Ooty in Tamil Nadu, is the deepest (260 metres deep), longest (nearly 22 kilometres long) and oldest (about 2.5 billion years old) gorge in peninsular India. South of Karnataka, peninsular India is made of three such rock assemblages or shields which extend from east to west, each of which is separated by a deep ancient gorge. The Moyar-Bhavani gorge is 120 kilometres long and extends from Salem to Coimbatore. Its bedrock on the northern margin is charnockite and its south face is made up of peninsular gneiss.

other questions about the age of the planet, the shape of continents and the composition of life, let's begin at the very beginning. Or *before* the beginning, when our Earth had not even been formed, which was around 5 billion (5,000,000,000 or 5 x 10^9) years ago—a number so enormous that it is almost incomprehensible to most of us. Scientists and thinkers have tried to find ways to make sense of geological time because, once understood, it opens up the mind to the awe-inspiring immensity and magnificence of our world. But we will come to that in a bit.

About 5 billion years ago there was no Earth and no Sun either. For billions of years before our solar system came into existence, generations of stars had lived and died. The birth and death of stars are unimaginably violent events involving the vaporization and subsequent solidification of trillions of tonnes of heavy elements. Stars live, die and explode as a celestial body called a 'supernova'. A supernova is a state of a cosmic body that exists just before an unimaginably massive explosion occurs. Such mammoth fireworks take place because reactions within these bodies go out of control and create so much energy that it cannot be contained inside it and therefore the body, usually a star, bursts. One particular supernova, called the primal supernova, possibly triggered the formation of our solar system. Astrophysicists call it the 'Big Bang nucleosynthesis'.

An artist's representation of the Big Bang nucleosynthesis and the formation of the Sun and the planets. On 11 February 2016, an international team of physicists and astronomers announced that they had heard and recorded, for the first time ever, a faint murmur or chirp which they believe was the sound of two black holes colliding a billion light years away!

Our Sun was surrounded by dust and rocks when it was born. Gradually—and I mean *very* gradually—over the next half a billion years or so, the larger rocks began to attract the smaller rocks and cosmic dust, and these began to clump together under the force of gravity to form early templates of the planets surrounding the Sun, including our *very* own little Earth. Within the belly of each one of these planets lay a molten core and their semi-solid but very hot exteriors were constantly growing. Our Earth was born around 4.567 billion years ago. At first Earth, like all other newly formed planets, was just a hot ball of iron that spun around the Sun. In the early phase, there were twenty infant planets moving randomly around the Sun. The Sun itself was recently formed, therefore, unstable, and its gravitational pull (the force that helps keep all the planets in their orbits) varied, making the movements of these young planets also very erratic.

To begin to make sense of geological time, I am going to borrow an elegant analogy from astrophysicist Nigel Calder. He asked his readers to imagine the 4600-million-year-old Earth to be a 46-year-old woman. Each 'year' of her life is actually a megacentury (100 million years). 'The first seven of those years are wholly lost to the biographer,' Calder says, 'but the deeds of her later childhood are to be seen in old rocks in Greenland and South Africa. Like human memory, the surface of our planet distorts the record, emphasizing more recent events and letting the rest pass into vagueness—or at least into unimpressive joints in worn-down mountain chains. Most of what we recognize on Earth, including all substantial animal life, is the product of the past six years of the lady's life. She flowered, literally in her middle age. Her continents were quite bare of life until she was getting on for 42, and flowering did not appear until she was 45—just one year ago. In the middle of last week, in Africa, some man-like apes turned into ape-like men and, at the weekend, Mother Earth began shivering with the latest series of ice ages. Just over four hours have elapsed since a new species calling itself *Homo sapiens* started chasing the other animals and in the last four hours it has invented agriculture and settled down. A quarter of an hour ago, Moses led his people to safety across a crack in Earth's shell, and about five minutes later Jesus was preaching on a hill farther along the fault line. Just one minute has passed, out of Mother Earth's 46 "years", since man began his industrial revolution, three human lifetimes ago.'

The more mathematically inclined can imagine geological time to be a long piece of string with one inch representing one year. About 6 feet would then represent the average lifespan of a healthy adult in India (about 72 years). The whole of recorded human history would add up to no more than 1 kilometre; the birth of life around 3.8 billion years ago would make the string about 96,520 kilometres long; and the birth of our blue planet 4.5 billion years ago would make the string 1,14,280 kilometres long, capable of going around Earth three times!

Very few rocks from the early years of Earth's life are found now because our newborn Earth was made up mostly of molten iron and other heavy elements. Earth was more like a red hell than a home. However, one crystal which remains unchanged and provides us with a reliable record of the time of the birth of Earth is zircon. Although zircon is only perceived as a cheap substitute for diamonds, scientifically, it is perhaps much more valuable than diamonds because it is the oldest known mineral on Earth, and helps determine the age of rocks and that of Earth itself. The name 'zircon' comes from *zarquan* (Persian *zar*: gold; *qaun*: colour. Traces of iron in its crystals impart its characteristic light golden hue). If a zircon crystal is overexposed to intense UV rays from sunlight, it turns reddish brown, and if the crystal is heated, it turns blue, but zircon's most important property is that it is resistant to weathering. In terms of hardness (on the Mohs scale, where diamonds are hardest at 10 and talc is softest at 1), zircon comes in at about 7.5. This makes zircon harder than glass (5.5) and even steel (6.5). Zircon (or zirconium silicate in its natural form) provides scientists with a reliable way of dating zircon-bearing rocks. A zircon crystal that may only be a hundredth of a millimetre in size gets embedded in granite and when the deep-seated granite is pushed up to the surface and is exposed to the elements, it begins to erode and accumulate in the sands and sediments of riverbeds. Zircon, however, is very durable chemically and not easily altered by physical assault or by the constant flux of rocks forming and deforming around it and, therefore, despite coming up after millions of years, remains virtually unaltered. One particular zircon crystal—a tiny gem just 0.0157 inches long—was discovered in 2001 in the Jack Hills, an

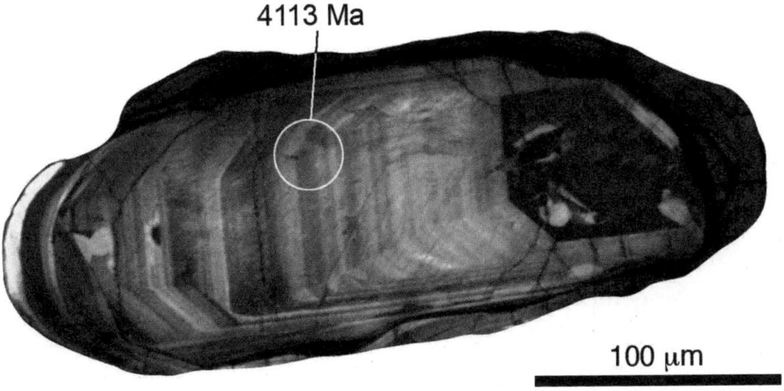

This zircon crystal, about 400 micrometres long is only four times as thick as a human hair. Discovered in a sheep ranch in the Jack Hills in Western Australia, it is believed to be the oldest piece of Earth's crust (about 4.374 billion years old) to have been found and dated yet.

inland range north of Perth in Western Australia. This tiny crystal withstood the violent times after it was created and survived to tell the secret of the age of our planet.

The estimated surface temperature of Earth when it was formed would have been about 1100°C, with waves of liquid metal and molten rock slinking in an endless ocean of lava, and the air would have been made up of a noxious mix of carbon monoxide, nitrogen, methane and hydrogen. A young, errant planet that astronomers call 'Theia' (or Thaea), about the size of Mars, entered Earth's orbit on a collision course. Moving nearly twenty times faster than a bullet, Theia smashed into Earth, causing the largest pyrotechnic event ever to have occurred in our solar system since the creation of the Sun. The tremendous heat generated from the collision melted Earth; only a dense viscous core remained to hold the liquid Earth together. Trillions of tonnes of gases and fluid rock debris blasted into space and Earth was enveloped by siliceous vapour (imagine glass existing as gas!). The collision displaced Earth from its axis of rotation, causing it to tilt at a sharp angle along its equator. This increased the size of the core of Earth and amplified the speed of its spin manifold. A mass of dust and rock circled Earth, looking a little bit like Saturn's red ring. Gradually, over the next 100,000 years, Earth's gravity began to exert itself and attract the rubble of dust and rock back into its orbit. These large rocks eventually coalesced and grew 3200 kilometres wide, about one-fiftieth the size of Earth—this became our Moon. The Moon at this time was the closest it has ever been to Earth: about 64,000 kilometres away as opposed to 350,600 kilometres from Earth's equator as it is at present. The Moon had a gravity of its own and the counteracting effect of the two celestial bodies (Moon and Earth) began to push them apart. Although Earth with its more powerful gravity remained in its place, the Moon began to move slowly away. Like other cosmic events in the life of our planet, the formation of the Moon too was a finely calibrated occurrence. Had it come any closer than 64,000 kilometres when it was newly formed, the Moon would have come crashing back to Earth. At this time, the Moon appeared on Earth's horizon every six hours. An Earth day was only six hours long.

After the molten mayhem, Earth began to segregate into different layers. The innermost shell or the 'core' was composed of complex silicate compounds of iron, magnesium, aluminium and nickel and was

An artist's impression of the collision of Theia with Earth.

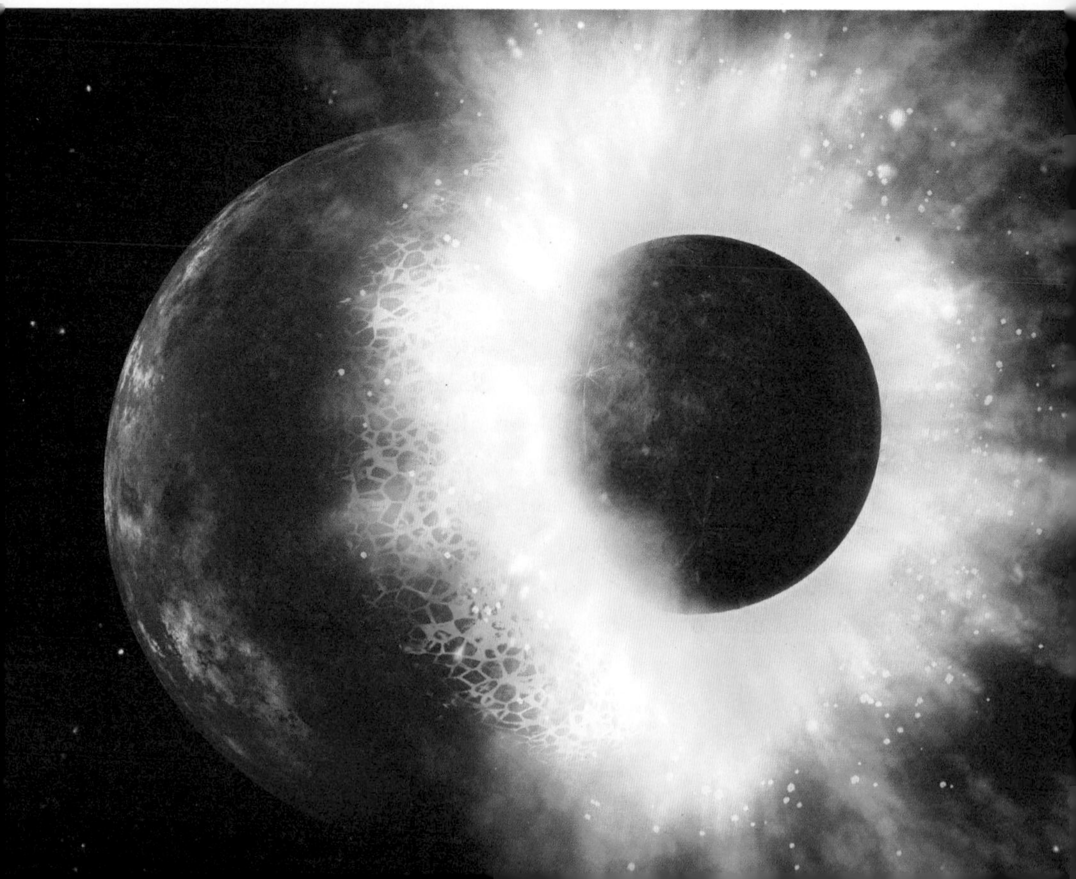

surrounded by a larger but less dense mass of molten iron called the mantle. Covering the mantle was a thin layer of slightly cooler (but still over a 1000°C) crust. Most of Earth's volume is made up of the mantle, the hot filling of rock between the crust and the molten core. In terms of thickness, the core and mantle are roughly equal, but in terms of Earth's volume the core forms only 15 per cent, while the mantle accounts for 84 per cent. The remaining 1 per cent is made up by the crust. While Earth was segregating into layers, the heavy bombardment of meteors continued unabated and the heat from these collisions kept puncturing the thin crust, releasing copious quantities of molten metals. Surrounding the crust was 'air' made up of a viscous amalgam of gaseous iron, silica and manganese mixed with carbon dioxide, hydrogen and helium.

Around 4 billion years ago, Earth's spin slowed and the Moon pulled further away. This increased the duration of day and night on Earth but only by a few hours. The rotation of the Moon too began to slow down as it counteracted Earth's spin. There is little physical evidence from this period which proves that the duration of Earth's day and night have grown longer over time, but there is an elegant example from much later that allows us to deduce what might have happened. About 450 to 400 million years ago—a mere four years ago in the life of our Earth Woman—shallow seas abounded with sponges and corals. Coral colonies, like tree rings, are efficient timekeepers of nature. Each growing coral adds one delicate layer of calcium carbonate every day. If you pick up a coral that has recently been swept off the reefs which dot Pirotan Island in Gujarat, a cross section will reveal clear bands with thin rings inside them. To mark the end of its yearly growth, a coral develops a thick band that separates it from the daily rings of the previous year. When you count the rings within a band under a simple microscope or a powerful hand lens, they add up to exactly 365 days, no less and no more. Travel some 150 kilometres northeast to Kutch or search the sand dunes around Jaisalmer in Rajasthan (which was a shallow sea millions of years ago) and with a little luck you will be able to find, among many different kinds of coral, a tubular coral that looks like a long, dull rod, about a foot or two long. This coral colonized the shallow seas about 170 million years ago. If you travel further northeast for about a thousand kilometres you will reach the high mountains of Spiti. Within the rocks here you can find fossils of coral that

In the first 2 billion years of their existence, the surface of Earth and the Moon was constantly bombarded by meteors. Between 4.1 and 3.9 billion years ago meteor bombardments intensified and gave Earth its water, and many minerals like diamonds and gold.

lived about 380 million year ago. If you make a clean cut across its thickest part (you will need an electrical saw to do this, or a fine sharp-head saw and *really* strong hands) you will find that the rings add up to 410! No matter how you measure it, whether in hours or days, the actual length of a solar year (the time spent in revolving around the Sun) has remained constant. This is because the orbit or the revolution of Earth around the Sun has remained the same since the solar system stabilized. The 410 rings in the 380-million-year-old coral is eloquent testimony to the fact that Earth was spinning *faster* on its axis, making the days shorter. When you divide the number of hours in a solar year by 410 days, you will find that each day was 21 hours long, not 24 hours like it is today. It is these kinds of clues that have enabled astrophysicists to calculate how Earth gradually slowed down and pulled further away from the Moon over time. Their studies also confirm that 3.5 billion years ago the length of a day was just 6 hours long compared to 21 hours 380 million years ago and 24 hours today.

Since the time Earth was formed to about 2.5 billion years ago, frequent collisions with large celestial bodies continued to reset the basic chemistry of our planet. Some of these collisions were so massive that they punctured Earth's crust, releasing huge quantities of metal, especially iron, which spilled out from the core on to the surface. Iron is an extremely versatile element. It reacts easily and combines readily with liquids and gases to create new entities. The furnace-like heat generated by the collisions also separated out other metals which reacted with each other to form new ores. Molecules and atoms that were of similar density or had similar characteristics (such as having the same melting point) bound together to form layers or lumps of ore while some of them became arranged symmetrically to form crystals.

The assault of meteors and asteroids on Earth and the Moon had another very important effect. Inside some of these meteors were tiny crystals of ice and as they entered Earth's atmosphere and hit the ground, the heat from the collision liberated this ice, converting it into water vapour. Another theory suggests that when meteorites succeeded in penetrating the crust and released molten iron on to the surface, it caused carbon monoxide to combine with hydrogen to create water in its vapour form. The surface of the planet was pockmarked with volcanoes that spewed out lava, thus releasing more iron and gases which helped to form

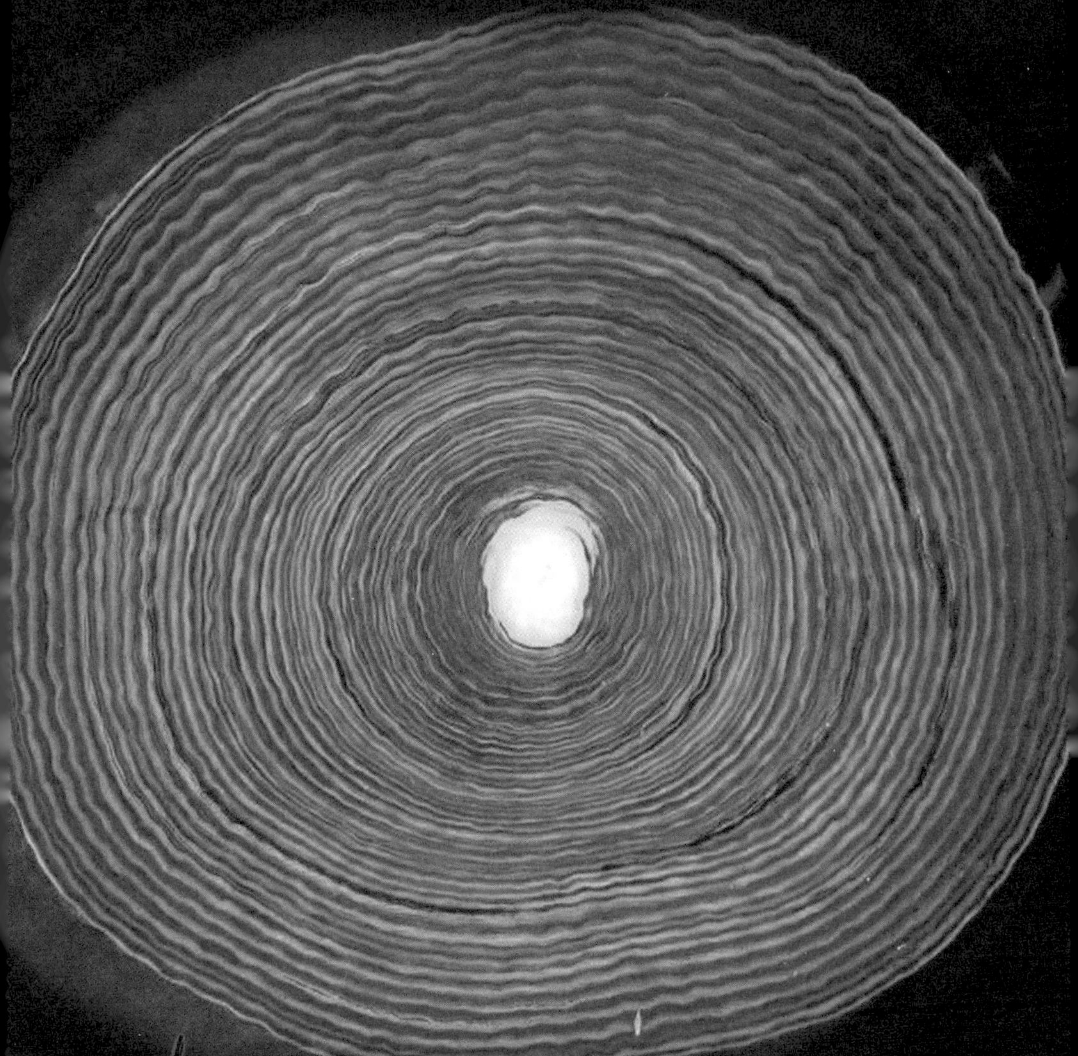

This is a cross section of a 44-year-old deep-sea coral collected off the coast of Newfoundland, seen under ultraviolet light, which illuminates its growth rings. Each day a coral adds a thin layer of calcium carbonate over a previous layer, which is linked to the lunar cycle. Apart from information about the number of days in a year, scientists are able to use coral rings to deduce the temperature, the amount of sunlight and nutrients available in the region, as well as the seasons in which the coral grew.

more water. In either case, as Earth cooled, its surface began to harden and water vapour began to condense to form pools of water. The hardening crust formed basins of continuous rock and over the next few million years or so, enough water had formed to create the first oceans. Incredible as it may seem, every drop of water in every puddle, sea and ocean that we see today is *billions* of years old and was either created by the action of magma on gases, or was carried inside meteors which had travelled millions, perhaps even billions of kilometres to reach Earth. Some scientists contend that the water on Earth could be even older than the Sun since some meteors may have been older than our solar system! The air on Earth however was still noxious and inhospitable at this time, and its temperature was very high—about 75°C.

About 3.2 billion years ago, rocks like the granite which makes up the Nandi Hills and the Dharwar Craton were beginning to form. Granite is formed when molten magma reaches close to the surface of Earth and begins to cool very slowly. The slow cooling hardens the rock and gives it its distinctive large crystals. If you were to take a flight from Bengaluru to Hyderabad (or vice versa), look out for interesting rounded and grey-coloured rocks as you take off and land. Within the city of Hyderabad too you will notice that some rocks seem to have popped up from the ground and some appear to hang precariously, one over the other. When the magma rose to the crust, it formed large domes of granite and as these cooled unevenly, they were exposed to varying levels of heat, and later these domes started to develop horizontal and vertical fissures, causing the rocks to erode gradually in concentric layers, quite like an onion peel. Geologists call these domes 'inselbergs' (German for 'island mountain').

East from Bengaluru, the constant melting and re-melting of rocks formed a variety of granite with an exciting twist: here you find veins of gold embedded in dull grey granite that makes the Kolar Gold Fields. All the gold that exists on our planet today was made in the interiors of massive stars that were billions of kilometres away from Earth. When the lives of these stars ended, they blew up in gigantic explosions. The clouds of gas (chiefly helium) and dust condensed and tiny amounts of gold and other metals from recently extinct and collapsed stars aggregated to form an asteroid. Earth was frequently bombarded by massive asteroids and meteorites (the differences between an asteroid and a meteorite are very subtle, and therefore the two are often used interchangeably). Some of these asteroids were ones that had gold in their bellies, and when they collided with Earth, they spiked the crust and mantle with gold and other metals. Some scientists believe that around 2.5 billion years ago a large asteroid collision took place close to Kolar, 100 kilometres east of Bengaluru. The energy from the collision congealed gold particles which flowed within the granite as veins (geologists call them 'lodes') and streaks. Over millions of years, the action of groundwater passing over these veins concentrated these gold particles which eventually settled in streams and riverbeds or glistened in the sand. Unlike most other metals which are found as ores, gold is found in nature in its shiny, glittery and purest form. Some archæologists believe that the Kolar Gold Fields have been mined since ancient times, perhaps even from the time of the Indus Valley civilization.

Another rock that, like granite, has its origins in the bowels of Earth is basalt. However, unlike granite that simmers and cools slowly within the folds of the crust, basalt comes to the surface and cools rapidly. The quicker a rock cools, the smaller are its crystals. The crystals and minerals in basalt are so fine that they are not visible to the naked eye. Basalt is mostly found on ocean floors, while granite forms the crust of every continent. Basalt is darker and is composed of magnesium and iron, whereas granite is lighter and is made up of silica crystals like feldspar and quartz. When basalt breaks due to normal weathering, it forms vertical columns, while granite splits horizontally.

Another very old rock (although slightly younger than the granite in Dharwar) is gneiss (pronounced 'nice'). Gneiss is formed when older rocks re-melt and then re-form under high pressure which separates the minerals into layers or bands. The rocks that occur between cratons are usually gneiss. This kind of rock lies embedded within the Dharwar Craton, forming distinctive rock formations within it called Closepet granite. Closepet granite (which is actually gneiss) was formed between 3 and 2.5 billion years ago and can be found in the district town of Ramanagaram (as the station is called, or 'Ramanagara' as the locals called it) close to Bengaluru. If the granite rocks of Ramanagara, some of which seem perched precariously one on top of the other, look familiar, it is because they provided the hideout of the archetypal villain Gabbar Singh in the film *Sholay*!

With some of these basic (and humdrum) facts about the oldest rocks in mind, let us go back to Nandi Hills to continue our story of Earth. Nandi Hills are part of the Dharwar Craton, which is the earliest building block of the Indian land mass. The Dharwar rocks lie exposed in a discontinuous arc that begins in northern Karnataka and, after passing through central and eastern Karnataka, ends close to Bellary. They are called the 'Dharwar rocks' after the town of the same name and on a map appear as a V-shape in western Karnataka. Another interesting example of this kind of rock can be seen in Bengaluru as a large grey outcrop of

Ramanagara rocks, which were formed around 2.5 billion years ago, lent an eerie atmosphere to the Bollywood blockbuster, Sholay. *For aficionados of the film, this was the perch of Samba, the henchman of the arch villain, Gabbar Singh!*

peninsular gneiss that sits in the middle of the famous Lalbagh Botanical Gardens. This rock formation is protected by the Geological Survey of India as one of the oldest rocks in the world. Geologists believe that the Dharwar rocks underlie much of peninsular India, especially Karnataka, northern Tamil Nadu, Maharashtra, Andhra Pradesh, eastern Gujarat, western and central Madhya Pradesh, while younger rocks were added over the next billion years in Odisha and Bengal, thus forming the pedestal upon which India sits. The other parts of India are much younger and we will come to their story a billion years from now (or in about fifty pages!). While the cratons and rocks were forming and creating the first land masses, the Moon was proving to be Earth's powerful and compassionate neighbour. About 3 billion years ago, when the Moon was closer to Earth, its gravity exerted a strong pull on the planet.

As Earth cooled and the first oceans formed, its rapid rotation and its proximity to the Moon began to whip up mega hurricanes and storms, some with wind speeds exceeding 500 kilometres per hour. Every night, a tidal wall would march in, overrunning several hundred kilometres of land. The incoming tides would clash with outgoing tides, creating a force that had the power to obliterate any rocks and hills that came in their path. The first rains began at this time and it poured heavily around dawn. These drops of rain contained gases that had been absorbed from the air and were therefore so acidic that when they fell on Earth, they dissolved and softened hard lava rocks. As soon as the rain hit the hot ground, it vaporized and formed clouds only to fall back again as rain. This cycle was repeated over and over again, and the turbid ebb and flow of tides, acid rain and powerful hurricanes continued unabated for the next 100 million years. On the surface of the ocean, tides act like brakes, creating friction between Earth's surface and the space it rotates in. Picture Earth's oceans as a furry tennis ball; the tides then are like the fuzz on the ball that cause friction and act like brakes. Compare this with a smooth dimpled golf ball which offers little friction as it passes through the air and as a result spins and revolves faster and travels much further.

The Sun's constant gravitational pull moved the Moon further away from Earth. As the Moon moved away, Earth's tides began to calm down; their power diminished and the rotation of Earth slowed down. The tides too exerted a gravitational pull of their own, which contributed to pushing the Moon further away because its gravity began to be repelled by the oceans. Even today, our Moon continues to move away from Earth at a speed of about 3.6 centimetres a year, but this distancing is too infinitesimal to have any perceived effect on Earth for a long, long time.

The Moon has also been Earth's shield and has protected it from massive collisions from outer space. The characteristic depressions that you can see on the surface of the Moon even through a pair of simple binoculars are actually craters that range from about 10 to nearly 800 kilometres in width and were formed by the impact of meteors and asteroids. The Moon deflected a large number of asteroids that would otherwise have headed straight towards Earth with possibly cataclysmic effect. But, all said and done, the Moon is just one small body and some meteors did fall on Earth. The impact of many of these collisions can be seen in India as well. The evidence of the earliest meteorite impact in Asia—and the third oldest in the world—was discovered in a small village called Dhala (Lat: 25°3'N, Long: 78°14'E) in Shivpuri district of Madhya Pradesh, about 50 kilometres southwest of Jhansi. This crater is 11 kilometres wide and spreads over 64 square kilometres, making it the largest meteorite crater in South and South East Asia. What is unique about the Dhala crater (it was named after the place where it was discovered, as is standard practice) is that calling it a 'crater' is a bit of a misnomer. There is no large depression here as one might expect in a crater; instead, a plateau-like structure rises from the otherwise flat terrain that surrounds it. It is, in fact, one of the few complex impact structures in the world and one of the best preserved. An impact structure is created when the heat from the meteorite impact causes the molten rock surface to rebound instead of becoming depressed.

Open Google Earth and type the following latitude and longitude: 25°17'59" and 78°8'3". From your bird's-eye view, you should be able to see the Dhala crater shaped like the smudged pugmark of a tiger, standing out because its colour differs from the surrounding terrain. You can reach Dhala by road from Shivpuri or from Jhansi. Travel south from both these places and find your way to State Highway 19 until you reach a small settlement called Dhala that lies at the base of a flat-topped mountain. If you alight at Dhala and walk towards the hill, you will come to a mound made up of maroon rock, interspersed with smaller orange-brown rocks. This is 'impact breccia' or fragments of angular rocks embedded inside more rock. As you climb upwards you begin treading on smooth and slippery rock at first and then come to loose brown rocks that are easier to climb. Further up, the slope gradually becomes steeper and you will be forced to follow the narrow footpaths used by goatherds that will take you to the top of the nearly 400-metre-high flat-topped hill. The climb should take you about 15 minutes. Once there, you will notice that the top of the hill is made of a single flat rock. The Dhala impact created a complex structure that has the characteristic central uplift of melted rock which popped out when a massive meteor crashed into it. The rebound of the molten rock was so massive and exaggerated that it resulted in the formation of a plateau, 11 kilometres in diameter, which rises out of nowhere in an otherwise flat terrain. The meteor that actually transformed the land in Dhala has long since vanished but it has left telltale signs of existence in the form of glass fragments (formed by the shock melting of the silica and quartz or 'breccia') which are embedded in rocks, and these provide clues to the cataclysmic impact of the meteorite.

Researchers have estimated that the Dhala crater was created between 2.5 and 1.6 billion years ago and the collision could have had some impact on the direction and tempo with which evolution progressed. But to the 2867 residents of the village of Dhala (and about 5000 cattle and goats), the flat top of the mountain is just a place that provides meagre fodder for their goats and a little fuel wood. They are not aware of the astronomic and geologic significance of this structure because Dhala was first reported in a conference only as recently as 2005 and detailed research to understand what actually occurred here nearly 2 billion years ago, was published only in late 2007.

A better-known meteorite structure—this time an actual crater—lies in Lonar in Buldhana district of central Maharashtra. Unlike Dhala, Lonar is a popular destination, overflowing with tourist traffic and in dire need of protection. You need to travel some 700 kilometres east of Mumbai in order to see this, India's most famous, crater. The Lonar crater is 2 kilometres wide and 170 metres deep. At its centre lies a placid blue lake, dotted with temples on the periphery. A plaque at the rim of the crater states that it is *'the only natural hypervelocity impact crater in the basaltic rock in the world'.* Some scientists believe that part of the meteor still lies embedded some 600 metres below the south-eastern rim of the crater. If you travel down the depression, closer to the bottom of the solid rock and soil horizons, there is a layer of soil that is richly embedded with tiny glass fragments. The Lonar meteorite fell nearly 50,000 years ago and it is possible that early human settlers in the Indian subcontinent actually saw the falling of the meteorite.

The tiger-paw-shaped Dhala crater in Shivpuri is an extremely rare meteorite impact structure. Instead of a crater, the meteor penetrated the crust and magma oozed out to form a plateau. The maroon-brown plateau stands out dramatically on the otherwise flat terrain. (courtesy Google Earth)

Another more recent crater lies in Luna (Lat: 23°58'N, Long: 69°32'E) in the *banni* plains near Bhuj, Gujarat. The impact occurred on the soft mudflats of Kutch; some earth scientists estimate that the meteorite crashed on Earth around the time when the settlements of the Indus Valley civilization were growing into cities. Archæologists have even found human remains around the crater's lake.

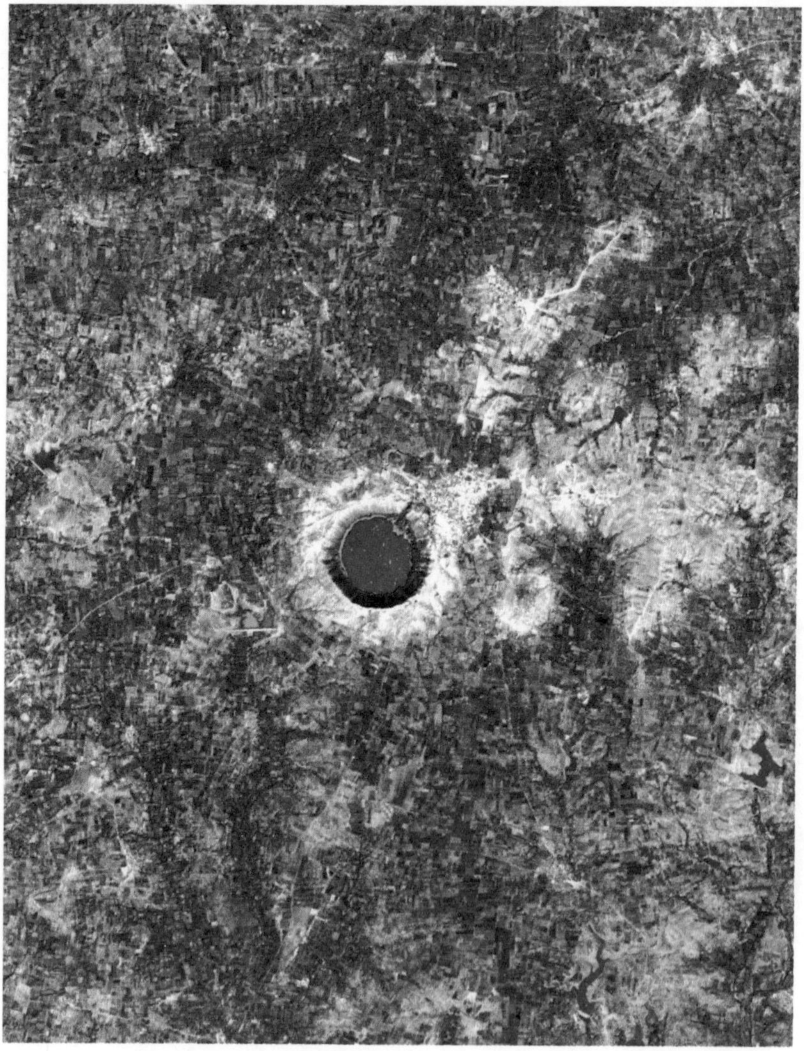

The Lonar crater is a deep depression in the ground made by a meteor collision 50,000 years or so ago. A circular lake lies at the centre of the crater, which is saline and is fed by small seasonal streams. Notice another small green lake to the right, close to the town, which was formed from a small portion of the meteorite, which disintegrated and fell here.

There are three other craters in India that were probably created by meteors but we do not know this with absolute certainty. About 120 kilometres south-west of Delhi is the Ramgarh crater (Lat: 25°32'N, Long: 76°62'E) in Baran district of Rajasthan with the town of Ramgarh situated at its centre. Although it is still being debated whether this is a crater or merely a large, exposed volcanic vent, minute glass shards embedded within its layers suggest that it is more likely to be an impact crater. Another candidate crater lies offshore in the Arabian Sea between Mumbai and the Gulf of Cambay and can be seen from space. Called the Shiva crater, its discoverers claim to have found a circular and unusually deep depression on an otherwise flat seabed and believe that the massive meteor responsible for it fell around 65 million years ago, creating enough destruction to end the reign of the dinosaurs. However, whether or not it is indeed a meteor impact site continues to be debated. In 2009, some geologists suggested that Srinagar's famous Dal Lake was actually a depression created by a meteor. There are a few other craters that have been reported in India but these remain controversial because their age, origin and impact are yet to be confirmed.

Coming back to the story of the Moon: As it moved away from Earth, it rapidly began to cool. Because the Moon is smaller in size and weight, it lost its internal heat relatively quickly and developed a solid core. The larger Earth, on the other hand, developed a thin, shell-like crust with a massive, hot molten core which helped it retain its energy. The Moon's gravity too waned and soon it could no longer hold on to its hydrogen-rich atmosphere which simply dissipated into space. Some of it, perhaps, was absorbed by Earth. It took 1.2 billion years for the Moon to change from being a fiery red-hot ball to a cold, lifeless sphere. But the bombardment of metcors on the Moon has continued unabated, the craters giving it its characteristic pockmarks.

Apart from protecting our planet from meteors, the Moon also ensured that Earth was stable on its axis and did not wobble (the way a slowing top does). Had there been no Moon, Earth would have wobbled wildly, creating chaos in our atmosphere, climate, oceans and, therefore, life. The Moon's constant gravitational pull ensures that Earth revolves at a relatively steady speed and maintains its inclination at the axis. But this will not last forever because, remember, the Moon is distancing itself from Earth at the rate of 3.6 centimetres per year. Expect chaos to set in when it moves far away enough, several million or perhaps even a billion years from now!

Human societies instinctively value the celestial silver presence of our Moon. The idea that the Moon is important for us is age-old. Nearly 3500 years ago, people in Brahmagiri (Lat: 14°48'N, Long: 76°49'E), a village in Chitradurga district in Karnataka, shaped rocks into tall rock columns called 'megaliths' and arranged them in a large circle. Over 250 such arrangements have been found in the area around Brahmagiri. Such megaliths have been found in other parts of India as well, for instance in Chaibasa (Lat: 22°57′ N, Long: 85°82′ E) in Jharkhand, and slightly more recent ones in Arossim in Goa and in parts of Manipur. Historians suggest that the collection of rock columns arranged in a semi-circle in Brahmagiri was an observatory that recorded the movement of the Moon and the Sun. Considering that there were no machines at that time to lift or move these massive boulders—some of which weigh more than 2 tonnes—this was truly an extraordinary achievement. What is even more amazing is the scientific accuracy with which these arrangements of rocks are able to predict the lunar cycle. Every eighteen and a half years, the Moon dips down on the horizon and stays there for a couple of weeks, a phenomenon known as the 'lunar standstill'. Some smaller rocks in the circle point towards a hill on the east and measure the exact height of the Moon on each night. Large rocks placed to mark the rise and fall of the Moon and a circle of stones indicates the point from where important lunar events can be witnessed when they occur. The builders of these monoliths would have been able to use their observatory to accurately predict lunar events every year.

Despite the Moon's protection, some extra-terrestrial material continues to reach Earth even today. Every year 300,000 tonnes of cosmic debris and dust finds its way to the surface of Earth. This spreads all over the land, oceans and seas. If this fine debris were to fall all at once at a single place, it would have the potential to cause immense devastation. Astrophysicists watch out for meteors and also use radio frequencies to record interferences caused by waves to determine how much cosmic dust is reaching the surface of Earth. So the next time you

MENHIRS AT POKARIA.

Rocks have been erected on flattened ground in India for more than 3000 years. They have been used as sight-lines to mark the rise and set of the Moon and the Sun on the horizon. Some columns or menhirs (French, maen: stone and -hir: long, or standing stone), have been used as sepulchres and commemorate important events. These menhirs from Pokaria near Chaibasa in Jharkhand were drawn by the British explorer Valentine Ball in 1868.

hear a whooping sound on your radio, it could well be because of cosmic interference in the radio signal and not a glitch in your radio or the relaying station.

While meteors and cosmic dust pose great dangers, they did provide many of the constituent elements that made our Earth, and, crucially, they also brought water. The Moon too, for its part, not only protected Earth from cataclysmic collisions, but also had a stabilizing effect that helped jump-start life. Its presence has continued to help protect life on Earth from decimation. But how did life begin in the first place? It wasn't until 3.8 billion years ago—38 years ago in the life of our Earth Woman, when she was barely 8 years old—that Earth cooled down enough and water began to form pools; when the first infinitesimal forms of life appeared. We must now revisit the rocks near Bengaluru—and then venture further along the ancient cratons—in our search for the earliest signs of life in India.

2

BREATH OF LIFE

The first billion years of the period I am going to talk about now (from 1.8 billion years ago to about 800 million years ago) have been termed the Boring Billion (or The Barren Billion, The Dullest Time on Earth, Dullsville, and even Earth's Middle Ages) by scientists. True, some parts of this story are not particularly scintillating. However, this period is critical to our understanding of how life evolved on Earth, particularly to the story of India and therefore I will dive into it, head first—the 'head' in this case being life itself.

That life exists on Earth is a spectacular coincidence. None of the 250 or so other planets that are known to science appear to support life. Of course, this does not preclude the possibility that there may be planets, unknown to us, where life exists. For the present, however, no other planet seems to have the right ingredients to sustain life as we understand it. Some planets are too close to their sun and therefore too hot. Others are so light that they would float on water, and some others

Most hills in the Dharwar Craton that are in excess of 4000 feet around Bengaluru and Mysuru are made up of Closepet granite. Savanadurga (4024 feet, left) though not as high as the Nandi Hills (4851 feet) is the single largest block of rock in Asia. Closepet granite hills like Savanadurga have a characteristically rounded shape with very smooth and steep sides. Again for movie buffs this was the location of the elephant procession going up the hill in David Lean's A Passage to India.

are full of poisonous gases or composed of elements that are incapable of supporting life. Earth, though, is at the perfect distance from the Sun to avoid getting incinerated. Its gravity is just enough to keep its atmosphere from dissipating into outer space, and its favourable atmospheric pressure just right to prevent life forms from being crushed. And all these processes needed to occur in the right order for life to become possible.

Life arrived on Earth at Hell's gate, about 3.8 to 3.5 billion years ago. The planet was so hot at the time that it is difficult to imagine this rather inhospitable environment facilitating the appearance of life. In fact,

The surface of Earth was hot and inhospitable 3.8 billion years ago. Water vapour condensed under a thick toxic layer of gases to create massive clouds which fell as acidic rain, etching the cratons to create the first seas.

the process of the first emergence of life is so complicated that scientists have struggled long and hard to decipher this enigma.

Some scientists believe that the first simplest replicable proteins, which are the basis of life, may have been formed in the depths of oceans. Deep sea volcanic vents called 'black smokers' (so called because they are black and emit hot carbon dioxide and methane) released ionically charged bubbles into the mildly acidic (due to a high concentration of carbon dioxide) water of the ocean. Compared to the alkaline bubbles that were emerging from the vents, this acidic water was richer in the subatomic particle called 'proton'. Protons carry an electric charge which enables chemical reactions. Deep in the ancient sea, these alkaline vents surrounded by acidic water were forming a proton gradient (protons flowing from areas of high to low concentration) thereby creating an electrically-charged environment. Here, in the presence of catalysts like iron, trace metals and elements like copper, molybdenum, strontium and selenium, gases from these vents produced nitrogenous compounds called 'amino acids' which are the building blocks of simple proteins. These ingredients and the exchange of miniscule amounts of energy may have worked in tandem to jump-start life. Even today, all living creatures have a similar electrically charged proton gradient within their cells that produces the energy required to undertake even the simplest activities. Deep sea vents still exist within the deepest trenches of oceans and we know very little about the organisms that live there. If all life on Earth were to end one day, it is possible that these deep-sea vents will give rise to new life, starting from scratch, perhaps taking a different course and creating life forms that are entirely different from those we have seen or which exist today.

Life has never been created in a laboratory, so our understanding of the processes that created the first life is, at best, mostly speculative. There are other theories proposed by scientists on how life might have emerged. A second popular theory suggests that the first life perhaps originated on beaches where water met land. The ebb and flow of the sea created shallow pools where chemicals from land and sea could mix and match. These shallow pools were rich in minerals and free elements, especially carbon, phosphorous, nitrogen and sulphur, which, in the presence of the Sun's ultraviolet (UV) rays, synthesized organic molecules like sugars and proteins that are the building blocks of life. But the trials for life must have had several failed attempts to make organic molecules that could self-assemble and sustain life as a cell.

Scientific experiments have shown that a thin layer of a simple fat known as a 'lipid bilayer' would probably have been formed first, which folded into a sac containing water and minerals. Although flimsy and delicate, the lipid bilayer would have provided an envelope for nutrients to be held within. The minerals in the sac had the ability to bind with simple sugars (nucleosides) and amino acids (the building blocks of proteins and enzymes). The key to the successful creation of life lay in the assembly of three parts—an outer layer or membrane, simple sugars and genetic material that could replicate on its own. This genetic material was perhaps a protein itself or the building block of proteins: ribonucleic acid (RNA). The outer membrane and RNA needed to be compatible with each other and coordinate their replication. To become functional, these parts also needed a proton gradient to jump-start the cell to life. It was only the successful reproduction of these cells for generation after generation that made the life of the earliest cells possible on Earth.

This hot spring near Leh, Ladakh, is home to both the earliest sulphur anaerobes and the oxygen-producing blue-green bacteria, which grow along the margins of the pool where it is cooler. These tough life forms have survived every extinction event since they evolved nearly 3 billion years ago. Should all life on Earth be wiped out, chances are that organisms quite like these will resume the process of evolution, forming, in all probability, entirely different types of creatures than those we know!

To see under what conditions early life could have come into being, we need to visit a mineral spring in Ladakh or in northern Himachal Pradesh or eastern Arunachal Pradesh. These springs emerge from heated rocks that lie below young mountain folds, and trickle out from the foot of a mountain. As they bubble out from cracks and fissures of granite, they form small shallow sulphurous pools of effervescent water. Looming over these pools is a plume of vapour and gases, almost like their own personal low clouds. Inside the water is a spectacular mosaic of algal and bacterial mats of different hues of green, yellow and orange. This velvety, slightly slimy layer is probably what the ancestors of all of us would have looked like! In the past, there were many hot springs in India but most have disappeared due to overexploitation. The famous Sahastradhara spring near Dehradun, which was once a popular picnic spot, is now reduced to a mere trickle, with very little life to show. But these blue-green mats of bacteria and algae are probably more complex than what first emerged in such pools 3.8 billion years ago.

Even after the first life forms had emerged, the development of life on Earth in all probability went through a number of fits and starts. The first breath of life was a mixture of methane and hydrogen sulphide. The earliest cells inhaled these noxious gases to create sugars and amino acids and exhaled sulphur dioxide or carbon dioxide along with water. These life forms are called 'anaerobes'. Mark Roth of the Fred Hutchinson Cancer Research Center in Seattle conducted a very interesting experiment with hydrogen sulphide (H_2S) to show how life may have emerged and survived in cycles. In high doses, hydrogen sulphide is lethal to organisms but, in 2010, Roth showed that when organisms are exposed to a sub-lethal dose of the gas, their bodies go into a coma-like condition. The movement of muscles ceases, the sensation in the nerves diminishes and vital functions like the beating of the heart and breathing are greatly reduced. Such effects are seen in virtually every organism, ranging from microbes to worms and mice. A higher dose of H_2S would kill the animal, and a slightly-lower-than-lethal dose would only make it lethargic. As soon as the dose of H_2S is reduced considerably, even after the animal has gone into coma, the effect of the gas wears off and the animal wakes up from its slumber. This experiment proves that it is possible to exist in a state between death and life, and the state in which the organism remains can be manipulated by the level of H_2S (and perhaps others gases too). When conditions were adverse, some life forms probably remained at rest, only to wake up when conditions were favourable once again. Roth's experiment has proven to be critical to our understanding of how life may have emerged and how life forms may have survived changing conditions.

But what *is* life? Scientists have long struggled to define it but most agree that life has three essential properties. First, living systems have the ability to self-assemble, which is against nature's tendency towards chaos, disorder and destruction. Second, life is a self-sustaining chemical system possessing the ability to evolve with and adapt to changing environments. Third, life is an interlinked chemical process designed to transfer energy from one organism to another. This means that life does not exist in isolation but depends on some source of energy. Virtually all life depends on other life forms to survive, multiply and evolve. Different life forms come together and shape their environment and that of others, making survival—the game of life and death—possible.

When the first life forms came into being, our Earth Woman was 11 years old. And, as you have seen, these first forms of life were neither complex nor elaborate. But it was these organic bubbles of jelly that gave rise to the earliest living functional cells, the Last Universal Common Ancestor, LUCA, of everything alive today. Every cell of every living thing today works on the same simple operating system as LUCA, only the degree of sophistication has changed. The maestro of the orchestra of life is the genetic material which directs the proteins, sugars and minerals. It receives signals about the external environment, decides the speed and timing of the reactions and communicates this information to all cells. It breaks down sugars, clears waste, repairs the membrane and reproduces. The first genetic material that powered life and was used by the simplest cells was RNA. About 200 million years later, more sophisticated genetic material called DNA (deoxyribonucleic acid) evolved and in turn gave rise to more complex microbes. DNA is the record keeper of all protein design (including RNA, which in some ways is a simpler form of DNA), and is the genetic (or the hereditary) material of all cellular organisms.

All cells of all organisms contain some RNA but only a few, very simple viruses (like HIV that causes AIDS, Ebola, SARS, influenza, polio and measles, to name a few) are completely RNA-driven and do not possess any DNA. These RNA-driven viruses infect a host cell, disassemble its DNA and produce mini-me viruses, causing disease and, in some instances, even the death of the host. Unlike RNA, DNA has the capacity to store more genetic information, is a more stable molecule and can reproduce with fewer mistakes. However, the relative instability of RNA means that RNA-based organisms mutate quickly (in order to evade the immune response of more complex organisms with advanced immune systems) while DNA-based organisms like us cannot keep pace. This explains why certain viruses (like HIV), although simple in structure, are most difficult to beat!

The evolution of DNA allowed more complex organisms to come into existence. The RNA within a cell of a DNA-based organism was demoted to its current role as a messenger and transcriber of DNA codes. In all higher organisms, DNA is located in the cell nucleus as a double-stranded helix, a tightly coiled, staircase-shaped molecule, which, thanks to crime shows and paternity tests, everyone is familiar with. Nevertheless, allow me to pause the story of life for a moment and tell you about the discovery of this genetic material. In 1869, Swiss physiological chemist Friedrich Miescher set out to isolate and characterize the protein components of leukocytes (white blood cells). For this, Miescher washed pus-soaked bandages of patients and filtered the leukocytes to extract proteins from them. During this process of separation, Miescher discovered one particular kind of sticky protein, high in phosphorous content, that did not break down easily under the action of alcohol and the weak reagents he had used to separate the other proteins. This key protein that he called 'nuclein' is what we now know as DNA. More than 50 years passed before the significance of Miescher's discovery was appreciated by the scientific community. Several scientists in the early twentieth century were able to isolate the components that made up this resistant gooey protein but no one knew what it looked like or how it worked. In 1952, the double helix structure of DNA was first discovered by James Watson and Francis Crick, using a single X-ray diffraction image (labelled 'Photo 51') taken by Rosalind Franklin (who, according to some, was the rightful discoverer of the structure of DNA) and Raymond Gosling. In 1962, after Franklin's death, Watson, Crick and Maurice Wilkins jointly received the Nobel Prize in Physiology or Medicine. In subsequent decades, scientists like Har Gobind Khorana and his collaborators deciphered the genetic code, a feat for which they were awarded the Nobel Prize in 1968, and their discovery led to the birth of the discipline of molecular biology.

DNA is found in every living cell of the human body except in the red blood cells and it is possible for us to extract DNA at home. For this we will use Miescher's technique without using pus-soaked bandages! And you don't need a sophisticated lab to do this. All you need to see your own DNA is three cups, some water, rubbing alcohol, a washing detergent and a little bit of patience. In the first cup, mix half a glass of water with one tablespoon of table salt. Label this 'cup 1'. In the second cup, mix three tablespoons of water and one tablespoon of liquid dish-washing soap, and label this 'cup 2'. Take one teaspoon of the salt-water solution from the first cup and swirl it in your mouth for 30 seconds. Spit the salt-water from your mouth into the third cup and label this 'cup 3'. Add one teaspoon of the soap solution from cup 2 into the contents of cup 3, and gently swirl the mix for about a minute. Next, take three tablespoons of rubbing alcohol (the kind that doctors use before giving you an injection) and gently pour it, drop by drop, down the side of cup 3 so that it floats on the surface of the contents without mixing. Wait for about a minute, and you will see a cloud of bubbles forming below the layer of alcohol. This is your DNA. If you can collect a lot of saliva, cup 3 will have *lots* of DNA to show. You can use a plastic straw or stirrer to make a small blob of your DNA. It looks a bit like candyfloss!

Genetic material like DNA, RNA and specialized proteins are unique to individuals. Scientists have studied the genetic differences between today's living things as 'molecular clocks' to estimate that LUCA emerged about 3.8 billion years ago. When the first life form emerged in or around water, Earth was still very hot and being on it would have felt a bit like being inside a hot chemistry lab with the air smelling of rotten eggs and sewer gas. The first cells were simpler than the simplest organisms alive today, but they were little superheroes. They tolerated the heat being produced by Earth, survived the harsh UV rays of the Sun, used oxides of carbon or sulphur or methane as a source of energy and produced oxygen as a by-product. Interestingly, in doing so, these first cells sacrificed their own well-being like true superheroes to help life develop on Earth because oxygen was toxic to most of these early organisms as it disrupted their vital chemical pathways.

There was one more abundant available source of energy—sunlight, but it took another 400 million years (3.4 billion years ago) before life discovered the means of harnessing the power of the Sun's rays through a process called 'photosynthesis'. For a long time, scientists could find no physical evidence of the first appearance of oxygen in the atmosphere, but in the summer of 2014 one of the earliest soils on Earth was discovered in the village of Madrangijori (Lat: 21°69'N, Long: 85°53'E), 6 kilometres north of Keonjhar in central Odisha. The study of the Keonjhar paleosols (ancient soil horizons formed during the earliest geological periods) by Indian and Irish scientists reveals that this quartzy soil layer was formed more than 3 billion years ago and provides important chemical evidence of the processes that took place on Earth as oxygen was making its first appearance.

Around this time a new type of organism called 'aerobes' emerged, and began to exploit oxygen as a source of energy. The most important among the aerobes was cyanobacteria. At one time, cyanobacteria used to be called 'blue-green algae' but they are, in fact, bacteria, not algae—the difference being that algae have a nucleus that contains the information carrying genetic material, while the genetic material of cyanobacteria is distributed all over the protoplasmic jelly inside their cell, rather than in specialized compartments like the nucleus. Cyanobacteria breathed in carbon dioxide, used the Sun's energy to fix the carbon to make sugar through photosynthesis that, in turn, released large amounts of oxygen in the air. And it was these humble aerobes—millions and millions of them—which, over the next billion years, orchestrated the appearance of free oxygen (O_2) in the planet's atmosphere. By the time the first oxygen-breathing life forms came into existence, a full 1.4 billion years had elapsed, and our Earth Woman was 14 years old. One of the earliest adaptations of single-celled microbes was to form colonies at places where water met land and in shallow seas. Most successful among these were the stromatolites (from Greek, *stroma*: stratum; and *-lithos*: rock). Stromatolites are colonies of billions and billions of cyanobacteria that grow over or around calcium carbonate (like limestone), containing traces of phosphates that are present in shallow water. A variety of single-celled cyanobacteria become arranged in layers and grow one over the other to form a colony. Most stromatolites are layered, although column-like or cabbage- shaped structures are also common. The oldest stromatolite fossils were found in Australia and South Africa and are dated between 3.5 and 3.2 billion years old. Along with free-living aerobes, they too began adding oxygen to the atmosphere and started to become agents of change.

The fossils of these pioneering colony-forming life forms can be seen in several places in the iron-rich rocks of the Chitradurga region of Karnataka. The beautiful red, orange and black striated rocks that you

A 30-foot statue of Vishnu, the Preserver, reclines peacefully beside a pool in the Bandhavgarh National Park, Madhya Pradesh. The green cover on the pool is cyanobacteria and the top few inches of the water are dominated by air-breathing bacteria. It was communities of cyanobacteria that produced oxygen first and made complex life possible. Even today, it is these aerobes that produce 60 per cent of the world's oxygen—another reason to protect our ponds and lakes. As long as these organisms survive, Vishnu can rest in peace.

see here are actually layers of stromatolites that are 2.8 to 2.6 billion years old. If you go north of the town of Shimoga in Karnataka, you can see thin layers of stromatolite embedded within yellow limestone. Similar stromatolites that are arranged layer upon layer in soft limestone can be found around Chakrata near Dehradun, although these are much younger. In Sonbhadra, near Benaras, stromatolites found embedded within rocks appear like clumps of cauliflower. In the Pin Valley of Himachal Pradesh, Saraikela in Jharkhand and the area around Jodhpur, Rajasthan, stromatolites are set deep within hard rock that is chiefly composed of iron. If you travel about 25 kilometres south-east of Udaipur, Rajasthan, you will come to the village of Jhamarkotra. The rocks here are rich in phosphate and if you look at these large flat rocks, you will notice a mottled grey layer above them. The villagers call these rocks *magar-machh bhata* or crocodile rocks. Geologists have dated these to be between 1.6 billion to 700 million years old. So if you were to draw a line along the stromatolite fossil trail you would be able to roughly mark out the margins of the Indian land mass at this time. Stromatolites have been long dead in India and most of the world, except in two places: a

Stromatolites are peculiar structures produced by some of the earliest oxygen exhalers that lived over vast spans of geologic time and were perhaps the most successful and dominant life forms of their times. A single type of blue-green bacterium or several types of aerobes co-operated to create successful stromatolite colonies. Stromatolites are diverse, with quirky and attractive designs. Fossilized stromatolites are found in sedimentary rocks across India.

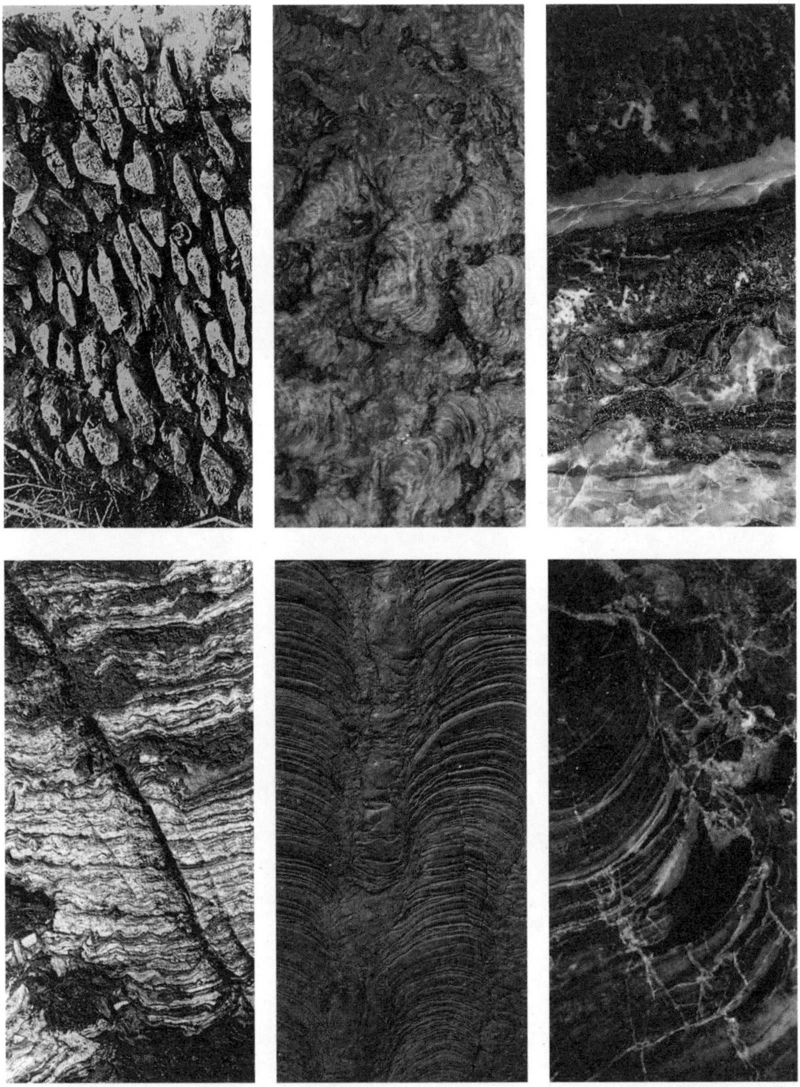

Clockwise from top left: The crocodile-skin stromatolite is from the phosphate mines of Jhamarkotra near Udaipur; the polished pink stromatolite is from Dharwar, and can be viewed in Shakti Sthal, New Delhi; the dark and coarse wave-like stromatolite is from West Mayurbhanj, Odisha; the orange, brown, white and black layers that appear like the semi-precious stone agate but is actually a colony of four different organisms is from the Vindhyan foothills near Mirzapur, Uttar Pradesh; the grey and white concentric circles and lines are 2.6-billion-year-old stromatolites from Chandil in Jharkhand; the thin black layers of stromatolites from Chitradurga, Karnataka, are perhaps 3.1 billion years old and among the oldest confirmed life forms to be discovered in India; the cabbage-shaped stromatolite is from Sonbhadra, Uttar Pradesh (opposite).

The alternate layers of red oxidized iron and grey oxygen-deprived silica-rich iron are deposited one over the other which make banded iron formations (or BIFs). BIFs can occur as massive rocks and help geochemists determine levels of oxygen at various times on Earth. BIFs formed in three major cycles and all three can be seen in India. The earliest episode of the rise of oxygen occurred between 2.7 to 2.4 billion years ago; the second around 1.9 billion years ago; and the third (the Snowball Event) occurred 750 million years ago. The BIF featured above is from Sundergarh, Odisha, and is about 2.4 billion years old. It commemorates the resting place of former prime minister Indira Gandhi—perhaps a fitting geological tribute to the iron lady of India!

lagoon in the Bahamas and Hamelin Pool in Western Australia where they are still growing *very* slowly, adding about five centimetres every hundred years. Scientists think that the conditions here exclude grazers and hence replicate early conditions which enabled these stromatolite colonies to survive. Could these relict colonies play a role in the future if our climate were to change?

The earliest stromatolites emerged around 3.5 billion years ago and produced oxygen for nearly 1 billion years before it could make a significant change to the composition of air. All the oxygen that was being produced was immediately converted into oxides by the highly reactive iron present on Earth (like the iron-rich soil we see in Keonjhar in Odisha). As oxygen levels began to build up in the soil, a process of mass rusting turned the soil red, like the soil you see in the iron belt in Karnataka and Goa. Even the dissolved oxygen reacted under water with the iron present at the bottom of the seas, leaving little oxygen in the air. Ferrous iron (Fe+2) is water-soluble and became concentrated at the bottom of the oceans and seas. However, with the gradual build-up of oxygen in the atmosphere and in the seas, another oxide of iron, ferric iron (Fe+3), began to be formed. Both ferric iron and ferrous iron began to settle as successive bands at the bottom of iron-rich seas and lakes as oxygen levels fluctuated. Once deposited, the layers hardened one above the other and gave the appearance of a layered cake—thin strawberry -jam-coloured striations of highly oxidized iron (ferric oxide, Fe_2O_3) and dark chocolate lines of less oxidized iron (ferrous oxide, FeO). These rocks are known as the banded iron formations (or BIFs). Travel about 200 kilometres north-east of Bengaluru to Chitradurga district and you will see the spectacular beauty of BIFs. Travel a little west, and the BIFs here assume new colours and form intermittent broad red and brown ribbons of iron and manganese oxide found in Sandur in the Bellary district, and those found around Shimoga have a layer of grey silica (called 'quartzite'). These large sedimentary rocks (about 1–2 metres high) were formed under the sea nearly 2.5 billion years ago and took millions of years to come to the surface.

Gradually, all possible surfaces that could absorb oxygen (or 'sinks') became saturated and excess oxygen began to build up in the atmosphere, causing its level (O_2) to rise dramatically. This is called the Great Oxygenation Event (GOE) and it took place around 2.3 billion years ago. As more and more oxygen was produced by cyanobacteria and aerobes, some of this oxygen reached the upper regions of the atmosphere and was converted into ozone (O_3) under the action of UV rays. Around 1.85 billion years ago, the ozone layer began to absorb much of the Sun's UV rays, allowing only a fraction to reach the surface of Earth and this slowly began to cool the hotcase-like Earth, but it was not until 850 million years ago that the ozone layer was fully formed. As oxygen levels increased, the battle for domination between the aerobes and anaerobes ended with the aerobes colonizing almost the entire surface of the planet and shallow waters, forcing the anaerobes to retreat to places like hot springs or deep caves that had little or no oxygen. The once fiery red and then purple Earth turned into a nurturing blue-green gem. Despite the passage of billions of years, it is relatively easy to see the blue-green bacteria that transformed Earth and changed the course of evolution. It is the green slime that covers ponds and streams, and rice fields and irrigation canals and sewage drains. Still, Earth's atmosphere

The 2.2-billion-year-old mudstones and sandstones around Talya and a few outcrops around Chitradurga in Karnataka are the earliest evidence of glaciation in the Indian subcontinent. Large oval rocks (called dropstones) like this quartz rock were carried by moving glaciers and eventually deposited in a shallow sea along with fine sand and silt. The 500-metre-thick rock formation in Talya overlies the 3.4-billion-year-old granite rocks. Dating studies suggest that the glacial period lasted about 120,000 years. Similar formations in South Africa and Namibia (about 2.3 billion years ago) pre-date Talya, suggesting that this glacial event was a part of the second episode of glaciation.

was not stable. So much oxygen was added to it over the next billion years or so that free oxygen began to react with atmospheric methane, a greenhouse gas that makes the air very warm, and converted it to carbon dioxide (which traps less heat than methane) and water, causing Earth to cool further. The climate cooled to such a degree that it began to freeze at the poles. Thus began Earth's first Ice Age, and we can see telltale signs of this glacial episode that occurred after the Great Oxygenation Event not far from where we found stromatolites and banded iron formations. If you travel west from Chitradurga on State Highway 13 towards Shimoga for 32 kilometres (or 77 kilometres east of Shimoga), you will arrive at a village called Talya (Lat: 14°01'N, Long: 76°27'E). Here, on the slopes near the village, you will notice rounded stones embedded within layers of red-brown sandstone. Ancient icebergs dropped them over shallow marine sediments and are about 2.2 billion years old.

During the first Ice Age, not only did atmospheric methane decline but volcanic activity too ceased. With so much more oxygen being added to the atmosphere, the climate began cooling considerably. When volcanic activity resumed on Earth, the ice caps on the poles melted and Earth once again began to warm up. It was almost as if the planet was in the hands of an errant child who was playing with the thermostat, alternately making it too hot or too cold! Despite the changing conditions, aerobes continued to dominate Earth, and in another 1 billion years (about 1.85 to 0.85 billion years ago) the oxygen concentration in the atmosphere began to increase again. On the surface of Earth at this time, land had not been divided by oceans and seas as it is today. All the land lay clumped together as a single land mass, a 'supercontinent' that Earth scientists call 'Rodinia' (after the Russian word for 'motherland'). Gradually, rifts began to appear and tear apart the supercontinent and, between 850 and 800 million years ago, these rifts widened, allowing water to fill up and create new shorelines. When new shorelines are created, more carbon-carrying sediments get buried in deep seas—a process called 'carbon burial'. As carbon levels (chiefly methane and carbon dioxide) began to reduce in the atmosphere due to carbon burial, the air began to cool, setting in motion another episode of global cooling.

A second and even more intense glacial interval called the 'Snowball Earth' began about 850 million years ago and lasted intermittently for nearly 165 million years (until 635 million years ago). During this Ice Age, Earth was covered from Pole to Pole and looked like a snowball. Around 750 million years ago, there were perhaps a few isolated places along the equator which were free of ice (so Earth resembled a slush-ball and not a snowball, according to some scientists). During this time, the land mass that comprised India, Madagascar and Australia was attached to East Antarctica and lay south of the equator between 20° and 40°N, receiving the full glare of the Sun.

This glacial event was not continuous and was broken by a few brief warmer periods. Although the number, timing and duration of these glacial episodes are yet to be fully understood, there is consensus that the most intense glaciations occurred between 760 and 700 million years ago. A second intensive glacial cycle happened around 635 million years ago, and this was followed by the final glacial event about 582 million years ago. The calculation of the duration and timing of these events is based on studies done on cap carbonate. Cap carbonates are distinctive carbonate rocks that are formed due to complex geochemical processes and mark the return of warm conditions at the end of a glacial episode. In between each of the glacial strata lie layers of small marine fossils that tell us about the kind of life that existed in each of the warmer periods and it is by studying these layers of marine fossils that we know that the last event of glaciation was very important for life. The freezing during this period was less intense than the previous events and as the ice receded (about 560 to 551 million years ago), first complex multicellular organisms emerged in the sea. But let's come back to multicellular organisms in a bit.

Another place where you can see evidence of successive glaciation events is in the Lesser Himalayas—starting from Solan in Himachal Pradesh in the west, running through Mussoorie and tapering off eastwards in Nainital. Here you will find layers of white and grey limestone, and between these lie differently coloured shale (a rock made from the deposition of fine clay). To see these, you will need to travel by road from Chandigarh to Shimla. Just where the hills begin to rise, near Solan, stop and look for these greyish-green rocks which are visible in road cuts. They consist of delicate layers arranged like reams of paper. Carefully pluck out a few of these sheets and you may see the distinct imprint of a marine creature.

Since we are talking about impressions and fossils of life forms, let me digress a little and elaborate upon this very important element on which a large part of our understanding of life in the past, and the past itself, hinges. 'Fossil' is a commonly used term but what exactly is a fossil? The word fossil comes from the Latin *fossilis* which

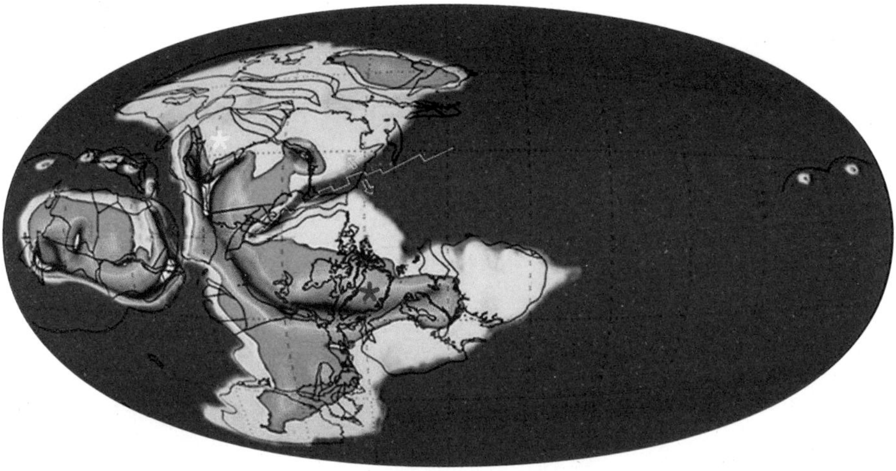

Earth, around 750 million years ago, closer to the end of Big Freeze would have looked a bit like this. India (marked with a yellow star) was indistinguishable from Antarctica and Australia which were smothered under a thick layer of ice. Notice that Greenland (marked in a red star) lay south of the equator and was largely arid while India was mostly under ice.

means 'dug up'. Fossils are the remains or outlines of an animal or plant preserved in rock or sediments. They can be massive, like the bones of a dinosaur or tree stumps, or microscopic like those of pollen grains and plankton. Fossils can also be impressions and imprints such as the burrows and tracks of various organisms that lived in or moved through a particular area. Occasionally, under certain special conditions, whole communities of soft-bodied forms end up being preserved, like in the case of small worms and jellyfish (which can be found in the Vindhyan range or in the sandstone from around Jodhpur). Perhaps the most abundant fossils of all are the microscopic shells of marine organisms. The bottom of all oceans and seas from ancient times to the present day accumulates layer upon layer of microfossils (microscopic shelled animals) and, over millions of years, they become exposed as chalk deposits which can be hundreds of metres thick, like those found in the chalk cliffs of Dover in

England. This shows how productive the oceans are and how resistant these delicate microskeletons are to decay over time! Other abundant fossils include the tough-walled spores of ferns and the pollen of conifers, palms and flowering plants. Apart from providing answers to scientific questions, some fossils also have high economic value. Coal, lignite (a less decomposed form of coal), natural gas and petroleum are products of plants and microscopic organisms that died millions of years ago, and are therefore called 'fossil fuels'.

The probability that any organism, after its death, will be preserved in rock is very small—probably a million to one. Once an organism dies, it decays due to other organisms, those inside its body and those that feed on it from the outside. This decay, the physical and chemical breakdown of a dead organism, begins within a few hours of its death. Only if a life form is buried under fine sediment or is frozen immediately after its death is this breakdown process arrested. Fossils, very simply, are organisms that have turned to stone. But unlike under Medusa's stare, these organisms turned to stone after dying, rather than when they were still alive.

There were two other mini Ice Ages which had an important effect on life and have left fascinating signs in India. Between 299 and 290 million years ago, Earth witnessed a short, intense glacial event, the evidence of which lies in a very unlikely place. Take State Highway 114 from Jodhpur towards Jaisalmer, and before the road forks towards Pokhran, turn right on to State Highway 28. Stay on this road for about 40 kilometres, until you reach Phalodi. Just north and north-east of Phalodi the landscape changes slightly. You will notice a flat valley—what geographers call a 'peneplain'—created in geological history by the erosion caused by water. Only, in this case, instead of water, the plains were sculpted by ice. All over this area you will find rounded red, orange and purple pebbles, many of which have disintegrated under the hot and cold of the desert and from the corrosive blasts of desert sand. Among these you will also find the brown volcanic rocks that form the columns of Jodhpur (called 'rhyolite'), red sandstone and orange siltstones. This entire area extends to about 80 square kilometres but is best viewed near the villages of Kheerwa (Lat: 27°29'N, Long: 72°32'E) to Nokha (Lat: 27°61'N, Long: 72°66'E), and Baap (Lat: 27°22'N, Long: 72°20'E) to Phalodi,

A great way to learn about what lies beneath is by observing rock layers like this 100-foot rock-cut near Narsinghpur in Madhya Pradesh. These rocks are characteristically pink because iron, a highly mobile and reactive element, leached into new rocks as sediments that were deposited in marine and shallow lagoon environment. The bottom layer is pink granite which is about 1.6 billion years old, the source of this iron. Above it lie alternating layers of sandstone, shale, and calcium-rich rocks (limestone and pink dolostone) which yield fossils of some of the earliest multicellular organisms, more than 1 billion years old.

through which State Highway 15 passes. Here you will find heaps of stone and large boulders (some of which are about 1 metre high or wide). These large boulders bear what look like nail scratch marks but are actually marks caused by the grating action of glaciers over stone. Surrounding these glacial rocks are gravel beds made up of myriad small stones that have been rounded by the action of ice and water and within these are simple shells and the remains of other marine creatures. During this time, terrestrial life had begun to establish itself and early amphibians were claiming their place on land, but more on this a little later. The levels of atmospheric oxygen were the highest ever during the period—at about 35 per cent compared to 21 per cent today.

The same Ice Age that affected Rajasthan persisted a little longer (from between 306 to 270 million years ago) on the eastern margin of the subcontinent. Clues to what happened during this Ice Age lie in the coal and steel belt of India, in the state of Jharkhand. The rocks that hold

Some of the best evidence of the last glacial event (about 270 to 260 million years ago) in the Indian subcontinent is in Dudhinala, which lies along the Ranchi–Hazaribagh highway in Jharkhand. Here, up to several horizontal feet of rocks and boulders lie embedded between alternating layers of sedimentary rock. But soon you may not be able to see this at all. The stones and rocks are being indiscriminately quarried and the ancient rivers that lead into the Damodar and Barakar rivers have been reduced to effluent drains.

these secrets lie in a 12-kilometre-wide, 200-kilometre-long strip called the 'Talchir formation' by geologists. The formation begins north-east of Ranchi, the capital of Jharkhand, and continues further eastwards towards the coal capital of India, Dhanbad. If you travel from Ranchi towards Hazaribagh by road, just before you reach the mofussil coal depot town of Kuju (Lat: 23°73'N, Long: 85°51'E), look out for a section of hill that was cut through to build the new highway. Here you will see the Talchir boulder bed that is characterized by rounded grey and white boulders and smooth rocks that lie embedded within a layer of hardened sandstone and look quite distinct from the rest of the hills. The white and grey circular rocks are made up of granite and are surrounded by ochre-coloured sand. Referred to as 'dropstones' by geologists, these circular rocks were transported here by glaciers and melting ice sheets. Glaciers carry with them any impediment they encounter in their path—boulders, rock debris, fine sand and sediments. The rocks grind and grate under the weight of the glacier and melting water and, over a long distance, become smooth and rounded. Once the glacier reaches a lake or sea, the ice melts and the boulders are 'dropped' along with slurry of sediment. The boulders descend to the bottom of the lake bed and the sediments slowly settle around them. Because there is very little disturbance in this placid environment, over several tens of thousands of years, more and more sediments are deposited on top, and gradually it begins to harden into a layer of solid sedimentary rock. This is the process by which the Talchir dropstones (and the Talya dropstones during the first intense Ice Age) got embedded in sand and eventually appeared as rocks. If you travel about 12 kilometres north-west of Kuju

This nondescript cross section of rocks depicts three billion years of geological history. The dull grey rocks are the basement granitic rock of the Singhbhum Craton. Above these are the glacial deposits (notice the embedded dropstones) which are 300 to 270 million years old. The topmost layer is the more recent sedimentary rock.

along National Highway 33, which connects Ranchi to Hazaribagh, you will come to a small riverbed called Dudhinala (Lat: 23°74'N, Long: 85°49'E). Walk about 2 kilometres west along the river's course and you will see layer upon layer of gravel and sediment beds stacked one over the other. Count these and you will find seven such layers. This suggests that there were at least seven clear glacial deposits and that glaciation took place in quick succession before a milder climate set in.

This brings us to the question of how life on Earth survived the millions of years of freezing that should have wiped out all life. This is because cyanobacteria never became extinct. The explanation of how this sensitive bacteria could survive for millions of years under a blanket of ice, without sunlight, comes from freezing lakes in Antarctica that have not melted in several million years. Here, despite the thick cover of ice, some sunlight still manages to reach the bottom. Water that has no impurities, upon freezing, produces a virtually clear, spotless layer of ice. Like a thick piece of quartz or glass, this ice permits enough rarefied sunlight to penetrate into the freezing water to allow the bacteria to survive.

If you are wondering why life was able to survive in frozen water but not on land, it is because water or, perhaps more accurately, its solid state, ice, possesses a special property. Water in its solid state floats in its own liquid form. Water molecules are loosely held together by weak chemical bonds that are constantly forming and breaking and therefore remain liquid. As the temperature drops close to 0°C, these bonds begin to hold fast and form hexagonal lattice-like structures called 'ice crystals'. On freezing, the space between the lattices in each ice molecule actually increases, unlike in other solids where the space compresses. Therefore despite the molecule being tied-up and constrained, the space between the individual molecules makes the solid hollow and lighter than its liquid form. Apart from being one of the few naturally occurring solids that are lighter than their liquid form, ice also solidifies from top to bottom. Even when the top of the water freezes, organisms are able to live in the liquid, albeit very cold water, at the bottom. On land, on the contrary, freezing takes place from bottom to top and therefore organisms would be unable to survive an Ice Age on land. These are simple explanations but there may be more to the mystery of why some life survived under frozen sheets of ice because, as the great chemist Felix Franks put it, 'Of all known liquids water is probably the most studied and least understood.' Water abounds in mysteries, so much so that the unexplained expansion of water during freezing was called 'anomalous expansion', until it was satisfactorily explained in the 1960s. One puzzle that continues to remain unexplained is why hot water freezes faster than cold water in sub-zero conditions. This phenomenon was first discovered by a Tanzanian high-school student named Erasto B. Mpemba (and is named the Mpemba effect) in 1963. Mpemba discovered in a classroom experiment that hot ice-cream mix froze faster than colder mix. His teachers chided him and students mocked his insistence until his teacher repeated the experiment and obtained the same result!

So how were the Ice Ages relevant in shaping the course of evolution? Here's how. After millions of years of freezing, it took a few thousand years for the ice to melt. The exposed land began to react with the melting water. The intermittent volcanic eruptions that had taken place during the Big Freeze had brought new minerals like sulphur to the surface and these combined with different minerals to create still newer ones (like gypsum). The role of minerals is often ignored in the story of life but they were crucial in the creation and sustenance of new life. It is quite fascinating to look at the evolution of minerals in the light of how life transformed it, and how life in turn was transformed by it.

Scientists believe that in the very beginning (about 4.5 to 4 billion years ago) there were merely twelve minerals in the swirling dust clouds that eventually formed the solar system. When the Sun heated the newly formed planets, the heat rearranged these minerals and created new ones, and these perhaps numbered around sixty. Minerals like diamond were created when the planets were forged out of cataclysmic explosions or implosions deep within the mantle. Over the next 500 million years, as geochemical processes within Earth commenced, the number of minerals increased to around 500. About 4 billion years ago, reactive metals like iron, nickel, aluminium and copper continued to remain pure in their natural state and glistened on the surface. As land and oceans formed, processes beneath Earth's crust triggered new chemical reactions which increased the number of minerals to about 1500. When the earliest single-celled organisms appeared about 3.8 billion years ago, things began to change. We have already seen that photosynthesis by blue-green bacteria, other single-celled microbes and the first primitive plants increased atmospheric oxygen dramatically. Although this intense oxygenation that occurred between 2.5 to 1.9 billion years ago was toxic to many organisms (and perhaps caused the extinction of many early life forms), it also led to the formation of new minerals which created niches for life forms to evolve differently. In an oxygen-rich environment, the chemical processes of oxidation (that is either the addition of an oxygen atom or the removal of a hydrogen atom) and weathering created several new varieties of metal-rich minerals. Approximately half of the 4300 known mineral species were created as a direct result of oxidation and weathering. While the origins of life may have been dependent on certain minerals, other minerals too were formed only because life enabled their creation. Two copper minerals—the beautiful blue azurite and the intense green malachite—often found lumped together, were formed in oxygen-rich environments. They cannot occur in an environment that is devoid of oxygen. It is sometimes even possible for scientists to estimate when certain minerals

Microbial mats emit gases from below the soil surface which take the form of doughnut-shaped domes, like this one from the Muth Valley in Spiti, Himachal Pradesh. Another place which is relatively accessible and where signs of early life can be seen is the Vindhyan range that extends from Chittorgarh, Rajasthan, in the west, and goes in a slightly upward arc before ending near Sasaram, Bihar. The Vindhyas contain an amazing variety of rocks, stromatolites, microbial mats and worm trackways.

were formed based on how much oxygen is required for their formation and at what time Earth contained that much oxygen in its atmosphere. Around 2 to 1.7 billion years ago microscopic organisms and the earliest colonizing creatures began combining calcium with carbonates, silicates and phosphates that were freely available in water and began to build-up. Gradually some marine creatures began to form large reef-like complexes. This started to litter the sea floor and the vast accumulation of debris from of these marine organisms became a source of minerals like calcite and aragonite. Much of this was deposited in the shallow seas which once covered the region from central Rajasthan to eastern Madhya Pradesh. As the Aravallis rose, the energy released converted the calcium-rich minerals into marble. We see these many-billion-year-old rocks as the marble found in Makrana and the white gorge of Bedaghat which the river Narmada cuts through near Jabalpur. Early marine plants also played their part by decaying and providing the compounds for organic minerals that make up rocks like coal or black shale. When exposed to water, air and varying temperatures, these compounds created many more varieties of mineral and ores.

The first plants and animals that emerged after the Big Freeze began to rework the soil and sediments. Burrowing creatures and microbial mats played an influential role in changing the texture of sediments, mixing and creating new types soils and minerals. Scientists call this process 'bioturbation'. In places where there was sufficient moisture, intense microbial activity began. A succession of microbes grew one on top of another, and one layer replaced another microbial mat. This led to a lot of organic matter being buried and eventually so much of it was accumulated under the soil that it began to decompose and release gases from below. These gases formed doughnut-shaped domes that can be seen alongside microbial mats and worm tracks in the Muth Valley and along the Pin river in Spiti, Himachal Pradesh.

Although some fossilized burrows of worms can be found along the Vindhyan range, in parts of southern Dharwar and the sandstones of western Rajasthan we see an explosion in the number and variety of burrows from 565 to 542 million years ago, which suggests that the varieties of burrowing creatures had increased manifold during this period. When plants first established themselves on land 445 million years ago, they produced acids around their roots that converted minerals of volcanic origin, such as mica, feldspar and pyroxene, into clay minerals which ultimately formed rich soils. (This is one reason why volcanic islands like the Andaman Islands, Hawaii and Indonesia are so verdant). The action of heat, rain and the first burrowing animals and early plants helped soils to attain the form in which we know them today—the substrate upon which all life depends.

About 700 million years ago, more sophisticated oxygen-producers like the photosynthetic microbes, multicellular organisms and early plants began to dominate and the stromatolites' role in producing oxygen diminished. However, their successful strategy of aggregating together and living as colonies was adopted by the first multicellular animals like sponges as well. Every cell in a sponge is exactly similar to the other and they feed simply by letting water pass through their bodies and filtering out small organisms. If any part breaks or any cell disintegrates, they add more cells around them and form another independent colony.

But why did single-celled creatures turn more complex? The answer to this lies in their genetic make-up. For nearly 3 billion years (from 3.8 billion to 900 million years ago) single-celled organisms ruled the land and seas. The addition of more oxygen and the creation of new minerals triggered the evolution of new metabolic pathways in organisms. It was during this period that the powerhouse of every complex cell, the mitochondria, the largest proton pump producing energy inside a cell, evolved—which, in turn, led to a rapid increase in the number of multicellular animals, a process biologists call 'radiation'. The most accepted theory of how the mitochondrion was created is that one powerful cell engulfed and 'enslaved' another cell to provide it with energy and in exchange it received nutrients. Their mitochondria powered these cells to evolve rapidly into newer varieties. Two major forces drove evolution at this time: random mutations of genetic material which could become stable species on their own over a few generations (known as 'vertical evolution') and the mixture of genetic material between newly evolving types (known as 'horizontal evolution'). Imagine this vertical and horizontal process as the twisting of a Rubik's cube. With each twist and turn, new combinations of life forms are created; each slightly different from the other, quite like how the minerals were created. With the mitochondria in place inside their cells, many microbes began to move, chase, capture and digest food more efficiently. Some creatures began to team up to survive in these changing times. This created the first multicellular forms of life among organisms. While the first Ice Age was manipulated by cyanobacteria, which helped jump-start multicellular life, it was the intermittent episodes of the second (the Snowball event) that triggered an explosion of life.

The period roughly between 1.8 and 0.8 billion years has been referred to as the 'boring or the barren billion' and even the 'dullest time in Earth's history', pointing to the lack of any major geological event that could have driven biological progress or spurred evolution. Until recently most scientists believed that the relative stability of the landmasses caused evolution to come to a standstill during this period. However, some scientists have proposed that this period of stability may have nurtured the emergence of microorganisms which developed new photochemical pathways and evolved into truly self-reliant organisms, the autotrophs (*auto*: self; *troph*: feeder), which later evolved into more complex organisms. The 'boring' period, in fact, was rather interesting as it provides clues on the relationship between tectonics, climate and evolution.

This 565-million-year-old rock bears a trackway called a Cruziana *and is made by an early multicellular animal which lived in shallow seas. It is relatively easy to find these in the old sandstones of western Rajasthan. This rock is from Dulmera near Nagaur, Rajasthan.*

The first multicellular creatures were a simple aggregation of a few types of cells (as opposed to complex fungi and plants that have between ten to fifty types of different cells, and animals which evolved much later and possess a few hundred to a few thousand different cell types). As multicellular life in the seas became more complex, organisms developed a distinct head, mouth, body and digestive system, and were able to attach themselves to sea floors and feed on floating organic matter. A few could even move and swim. These soft-bodied animals were the ancestors of

The 400-kilometre-long Ken river originates in the Kaimur range, the eastern most mountains of the Vindhyas. It cuts through an amazing variety of formations, like this one near Raneh in Chhatarpur district of Madhya Pradesh where you can see several types of rocks in one place. The white-grey rock in the foreground is a cooked calcium-rich rock called dolomite which is about 120 million years old. It lies over a layer of slaty shale, which is about 550 million years old and bears impressions of sea creatures. The pink rock is quartzite, which overlies maroon sandstone that is about 90 to 65 million years old. The grey rock at the bottom right of the picture is 600- to 400-million-year-old granite and below it is the deep grey volcanic basalt which is about 2 billion years old.

molluscs and other burrowing animals that live on the sea floor today. The impressions of large multicellular jellyfish-shaped and yet relatively simple animals can be seen in the sandstones around Bikaner, Jodhpur and Nagaur in Rajasthan. You can also find these in old rocks that are exposed in the lower hills between Mussoorie and Nainital (called the 'Krol formation'). Gradually, a few multicellular life forms began to diversify their cells and some began to form tissues which could perform more sophisticated tasks. This period is called the Ediacaran (580 to 542 million years ago, just as the Snowball Earth was beginning to wane), and is named after the first fossil impressions that were found in hills of this name outside Adelaide in Australia. In India, we find similar fossils in Sursagar (26°15.77'N; 73°0.14'E) which is seven kilometres northwest of Jodhpur. The Ediacaran period ended with a minor Ice Age and, after the world thawed, it created another opportunity for the evolution of newer types of organisms with greater complexity.

Around 570 million years ago, a few early enterprising organisms developed a new reproductive strategy—sex! Although the oldest evidence of sexual reproduction is from a fossil of microbe found 1.2 billion old rocks from Canada, sex as a reproductive strategy did not become the driving force for evolution until about this time. Sex opened up a plethora of possibilities. It enabled a greater exchange of genetic material, creating opportunities for its mixing and causing the emergence of newer varieties of complex multicellular organisms at great speed. The occasional errors in translation would result in individuals that were different from their parents and lead to the creation of new life forms. Animal life erupted and diversified into a kaleidoscope of forms in what palæontologists call the 'Cambrian Explosion'. As a rule of thumb, life forms that use more oxygen are capable of producing more complex and larger proteins. This often translates into the development of new structures which help these life forms exploit their environment in novel ways. One such important evolutionary leap was movement. Among the first creatures to develop this talent were the *Cnidarians* (or coelenterates). Most members of this family (like the sea anemone and coral) are attached to rocks on the sea floor while others like comb jellies and jelly fish are free-moving. They evolved to become thin, pin-shaped worm-like creatures with no arms or legs that wriggled on the bottom of the sea floor. Delicate hair-like trails created by such worms were preserved in the soft sand and can be seen on rocks in places like central Karnataka, along the Vindhyan range and in western Rajasthan.

This organism had three important characteristics that enabled it to move successfully. The first was the presence of a head that allowed it to direct itself, the second was a body plan, and the third was bilateral symmetry. It is these three characteristics that would also enable future animals to move successfully.

Bilateral symmetry was a critical development and was triggered by a set of genes called the *Hox* gene (from *Homeobox*, shortened from Greek *homeotic*, meaning 'similar') and it is essential for even the most evolved animals to have it in their DNA. Without the *Hox* gene no animal more complex than a worm can live, reproduce or survive. The symmetry of the first worm-like creature was governed by a symmetrical nervous system, which comprised a knot of nerves at one end that extended along the length of the body. This knot of nerves was a precursor to the earliest brain. The first worm-like creature had another simple but remarkable sensory organ—a primitive eye. The eye was capable of differentiating between light and darkness and therefore could move towards or away from a source of light. Gradually, some worms also began to develop thousands of microscopic hair-like structures called 'cilia' under their bellies, which would beat in a wave-like motion that helped them move more efficiently in water. In some worms, these cilia fused together to form a thin membrane that helped them glide through water. When minerals made from calcium, silicon and phosphorus were released due to the melting of the glaciers, complex multicellular animals used these to develop a new hardened external skeleton made of a complex carbohydrate called 'chitin'. To help them move, they also developed jointed limbs, and their bodies were divided into segments or sections. These segmented worms that had evolved from soft-bodied marine ancestors became the first arthropods (from Greek *arthro*, joint; *-podos*, foot), the ancestors to insects (Latin for 'cut into section'), the myriapods (or myriad feet, like millipedes and centipedes), arachnids (Greek for spider; also includes scorpions and ticks among others) and the crustaceans (like crabs and lobsters). It was from this period that the design of almost all higher organisms—from sea slugs to anemones, from centipedes to crabs—was decided. All higher organisms (with very few exceptions) from this time on would have a mouth on their head, a thorax, an abdomen and limbs. This again was the handiwork of the master gene, *Hox*.

Nature works as a tinkerer. It innovates with the materials it has in its surroundings, and works towards achieving maximum results with minimal effort. The *Hox* genes turns on and off during growth. Special proteins that bind to these genes act as switches controlling them. Small variations in the switching off and on mechanism can cause different regions, segments and appendages to assume dramatically different forms, which is one reason why we see such a fantastic variety of insects!

This is the crucial bit. We know that all information of life is contained in its DNA (the master molecule or the hardware) and its genes (the software). DNA gets packed in together with other genetic material in what is called a chromosome. Genetic material like DNA exists so that these creatures can go forth and proliferate. In an average human, adult or child, there are more than 3.2 billion letters of coding. If you were to take out just one strand of DNA from a single cell it would stretch as long as six yards, about the length of the average sari. Now think of the several billion trillion combinations of genetic code, each with the capability of doing something different inside or outside your body! It is these combinations of genes working in tandem or individually which make each life form unique with several possibilities, unlike, say, a fruitfly (with a little less than 16,000 genes). Of course, we cannot switch on or switch off our genes on our own volition, and that's where the master gene *Hox* comes into play. Until about a decade ago, scientists believed that we humans were made up of at least 100,000 genes. Geneticists inferred this by studying fruitflies and other lab animals but they were wrong. In 2004, the Human Genome Project left the most chauvinistic among us crestfallen when it reported that we were made of less than 30,000 genes, the same as mice, and the same number of chromosomes as the grass in your lawn, fewer chromosomes than even a salamander swimming lazily in a frigid pond, a couple of hundred times less than the rose bush or the fern that is growing in your garden. But here is the good news. Our genes have several repeated sections or duplications which, some scientists believe, act like a kind of backup in case some genes get damaged. Creatures with long but distinct genes with no repeats perhaps may not recover from an environmental assault (say a dose of radiation) and may therefore be more prone to becoming extinct.

Coming back to our story of arthropods in the seas, predation probably provided a strong incentive for animals to get bigger and become more complicated. Many animals began developing new body plans including mouthparts, armours and arms. It was an ecological arms race in size and complexity: bigger predators had an advantage in catching prey, while larger prey could more easily avoid being eaten. The need to escape or repel predators in turn probably inspired the first scales, spines and body armour, as well as some of the bizarre body plans.

During the Cambrian Explosion (or radiation as some scientists prefer to call it) the sea was teeming with crustaceans and insects with heavy armour and weapons and the increasing threat from predators in water forced some creatures to seek refuge outside, making the arrival of the first animals on land imminent.

The first animals that ventured on land came primarily to lay eggs on the margins of beaches, out of reach of sea-dwelling predators. Many sea creatures like sea scorpions and crabs, adopted a unique strategy. To ensure that their progeny survived hostile seas, males and females of a species came up to the surface to mate. The males probably returned to deep waters after the act to fight their sea wars while the females stayed back in the shallow waters close to the margin and deposited their fertilized eggs in the moist sand around the water's edge before returning to the sea. The larvae and the young meanwhile remained in the spawning grounds till they matured, venturing into deeper water only after they had reached adulthood.

As more oxygen became available in the air, animals were able to stay longer on land and venture further out. About 430 million years ago, arthropods spent more time out on land not only to breed but increasingly to feed as well. They evolved lungs which are pressed like pages of a book (called 'book lungs') while others developed tubes (or trachea) which connected the body cavity to the external environment to enable their blood (called hæmolymph) to rapidly exchange oxygen on land as well. Many evolved complex mouthparts to feed on plants or on each other. Perhaps the most bizarre and interesting creature to adopt this strategy was a type of an armoured arthropod called the trilobite that flourished on the seafloor. Fossils of trilobite show that they had three lobes (which gives them their name). They also had a head that often had prominent eyes, and a trunk with segments that could flex in their mid-body followed by a tail in which the segments were fused. Trilobites found in India resembled a modern woodlouse; some were as small but a few were a hundred times bigger and, of course, completely aquatic. To see fossils of this sea creature in India you need to go to very unlikely places that today are far away from the seas: 5500 metres above sea level in Zanskar in Himachal Pradesh, or in abandoned limestone quarries near the Mussoorie Hills. The rocks in these places were under water until relatively recently. Trilobites appeared rather suddenly in the fossil record about 525 million years ago, and rapidly became abundant and varied. More than 20,000 distinct species of trilobites have been catalogued, and these vary in size, shape and complexity. In Utah, USA, some trilobites that have been unearthed were as large as a Frisbee, many more were the size of modern crabs, and yet others were smaller than

During their 270-million-year reign in the seas, trilobites had captured virtually every niche on the sea floor—that of predator, scavenger, bottom feeders and filter feeder. Some scuttled along the sea floor, a few darted off in short bursts, and others cruised at various depths. The last trilobites died shortly before the Great Dying, about 251 million years ago. This fossil is from the abandoned limestone quarries that lie between Dehradun and Mussoorie.

a pea. The ones we find around Mussoorie are about the size of a grain of daal (lentil) to chana (horsegram). Trilobites that occupied every part of the sea, from its deep floor to the shallow seas near the beaches, also varied in their shapes. Some were spiny with bizarre appendages while others, like those found in India, were relatively simple and, as I have mentioned before, almost like the woodlice found under flower pots or in humid garden soil. Many had complex eyes which they used to catch smaller prey, others were blind and perhaps lived at the bottom of the

The enigmatic velvet worm is potentially among the first creatures to have colonized land. It was last documented in India in Arunachal Pradesh in 1911. The elegant creature pictured here is from the forest floor of Malaysia.

seas and therefore did not need eyes in the deep gloom. Trilobites eventually died out about 270 and 252 million years ago, during the greatest extinction ever known which wiped out nearly 95 per cent of all species in the seas. There has been no evidence found to suggest that trilobites ever took to fresh water. Had they done so perhaps they or their descendants would have been alive today.

Some other creatures who evolved alongside the trilobites are still with us. Travel to the Muth Valley in Spiti, Himachal Pradesh, and you will see signs of the arrival of life on land in India. There are fascinating impressions of tracks of a many-legged arthropod, perhaps a centipede, and several other creatures that criss-crossed the land and left telltale signs of their passage. The size of the tracks of these pioneers on land suggests that these creatures were very large, perhaps as large as a man's cycle or even a small-sized car.

We know that animals started venturing on land about 440 million years ago but we do not know which was the first to come ashore and take to the land. It is likely that several animals tested life out of water before some claimed their share of land. An organism that has been found in rotting wood and leaf litter in the undergrowth of the beautiful, moist and dense tree fern forests of northern Arunachal Pradesh is similar to or perhaps even the same as one of the earliest animals that made land their home. It is a predator like none other in its size. It resembles an elongated caterpillar, adorned in a colourful orange, pink or red velvet

The earliest evidence of landfall in India is this trackway made by a centipede-like animal that can be seen on a rock face in Muth Valley, Spiti. The 9-inch-wide trackway (the coin is 1 inch wide) suggests that this creature was more than a metre long. So the next time you see what looks like a tyre-mark impression on a rock it could be the footprint of an ancient creature!

coat and is aptly called the 'velvet worm'. It is about 2–5 inches long with fourteen to twenty pairs of legs. This hunter uses a lasso of sticky goo that it fires from special glands on its forked head to capture insects or small worms. The enzymes in the goo dissolve the prey so that it is half-digested even before it is eaten. This former sea-dweller gives birth to live young. Molecular studies of velvet worms reveal that they share a common ancestor with arthropods and are even older than trilobites, sea scorpions and horseshoe crabs. Sadly these beautiful creatures have hardly been studied in India and we don't know how many different species exist in the country. The only record of the existence of the velvet worm in India comes from the documents of an expedition made in 1911–12 in the Dihang Valley that lies at an altitude of 350 to 750 metres in Arunachal Pradesh. Since then, as far as I know, no serious attempts have been made to look for and study this amazing little creature.

Another potential pioneer on land could have been the humble earthworm. One particular species found in India was definitely among the early colonizers. It is a relatively large earthworm, *Drawida grandis*, which can grow up to about fifty centimetres long (or about the diameter of a family-sized pizza) and is found in deep burrows in the forests around Mysuru. What makes this earthworm unique is that, unlike other earthworms, it lacks dorsal pores or openings on its back to regulate the release of water. This absence of dorsal pores is seen in primitive marine worms. Since *Drawida* lives exclusively on land, the absence of its pores suggests it made the transition from sea to land and was possibly among the first creatures to make land its home.

Leaving the water and moving to land was a giant leap for all animal life on Earth, and over the course of the next several million years (542 to 475 million years ago), many animals moved back and forth between land and water and almost all major evolution took place at the water's edge.

Even before the first arthropods had escaped the increasingly hostile water and ventured on to land, early plants had already made landfall. During the initial years, plant life, like animals before them, had remained in water, but scientists have found signs of early plant life on land in places where water met land. Most scientists agree that algal mats from the sea floor were among the first to venture out and survive

For a short while in Earth's history, fungi dominated over plants and created the first forests on land. Some fungi were slender with columns that rose about 2–3 metres high, while others were squat and shaped like cabbages.

alternating dry and wet conditions on land. Plants adopted a similar strategy of spreading along the ground like the algae which protected them from the harsh sun and desiccation. They developed hair-like roots that began to penetrate deeper into the soil to seek nutrients. Some tissues shot up vertically and developed into the first stems and leaves. These first plants that colonized the margins of water almost 480 million years ago were no more than a few centimetres tall. They lacked sophisticated vascular tissue to conduct water and minerals, and the transfer of nutrients took place through osmosis where nutrients moved from cell to cell, crossing the thin membranes of the cell wall due to differences in pressure. These plants had no leaves, flowers or buds, and they reproduced using vegetative means, through spores, or simply broke into smaller pieces each of which would take root and become another individual plant.

While plants were still inconspicuous on land and animals had not even begun to venture out of the water, one unlikely organism began to dominate the landscape. These were enormous fungi like *Prototaxites* which had massive fruiting bodies that stood more than 8 metres tall. It was the largest land organism around 570 to 530 million years ago which, along with other fungi, formed fungal forests along margins of water, and towered over the low, carpet-like mossy vegetation. In India, a large disc-shaped organism called *Aspidella*, which lived around 560 million years ago, has been found in Jodhpur. These fungi had hair-like structures at their base which helped them anchor themselves firmly in moist soil. However, not much is known about the lifecycle and behaviour of *Aspidella* and other fungi like it.

There's a lot more to fungi than meets the eye. They are enigmatic, almost misfits in the order of life. Taxonomists (scientists who classify all things living or those that have lived in the past, and have a penchant for naming things in dead languages like Latin or ancient Greek!) therefore accorded them a place of their own, an independent kingdom as it were, and called them *fungi* (Latin for mushroom). Fungi are not considered plants since they lack chlorophyll, nor are they animals; yet they have relatively complex and different cell structures from both plants and animals. They are a bit like bacteria and mostly rely on dead matter to survive and reproduce. They can grow in the absence of oxygen, and they help decompose leaf litter and feed on dead and decaying organic matter on the forest floor. They also produce strong acids that have the ability to dissolve minerals from rocks like iron, silica and magnesium and speed up the process of soil-making. Some fungi that had colonized the land at the time when plants appeared also developed the ability to cause disease in plants, insects and eventually animals that evolved later. Others penetrated the soil and developed close partnerships with roots of plants creating new chemical cycles, many of which are still only partly understood by science. Strangely, the cell walls of fungi contain chitin, the identical chemical that makes the external skeleton of insects and crustaceans. Again, science has few answers as to why such completely diverse creatures like fungi and insects should share this unique chemical compound. However, we do know with some degree of certainty that fungi originated in the sea and were among the earliest creatures to appear on land.

When plants started arriving on land around 440 to 420 million years ago, fungi began a massive turf battle with them. Plants had also evolved in water and had begun to come to the edges of the land to seek more sunlight and to be closer to the rich nutrients that were being washed in from land. The first plants (like *Cooksonia*) were moss-like in stature and grew in the shadow of the fungi on land. They were at first out-competed by fungi and algae. However, over the next 10 million years or so, plants evolved the ability to produce something wonderful that helped them beat the fungi. This was lignin (Latin for wood), a molecule that was both strong and flexible; a material that could support a lot of weight, yet bend in the wind without breaking. The first plants which synthesized lignin helped their shoots to grow erect and supported the transportation and storage of nutrients and water more efficiently. This gave them an edge over the fungi and soon they began to beat back the fungi. However, the newly dominant plants also competed with one another for sunlight and nutrients. Around 380 million years ago, plants developed a new structure, the leaf, which helped capture more carbon and release more oxygen in the atmosphere. Plants were also more suited to survive drier conditions, while fungi withered away in dry and cold environments. At first, shrub-like forests of primitive plants made up of horsetails and ferns evolved. Most plants had developed

Nearly 300 million years ago, Mamal in Kashmir would have looked like this tree-fern forest of Arunachal Pradesh. The Mamal forest was dominated by moss-like lycopod trees and living in their shadows were horsetails, seed ferns and the earliest conifers. Giant tree-ferns could grow as high as 30 feet but were still dwarfed by the 100-foot-tall asparagus-like poles of the lycopods (or the clubmosses). As the climate became cooler

and drier, the clubmosses shrank in size and retreated to the forest floor where they now exist as the microscopic velvet that lines the trunks of trees or covers moist forest floors. Ferns continue to live in habitats ranging from tropics to cooler sub-polar latitudes, and come in a variety of shapes, colours and sizes. There are floating ferns, ferns that climb like creepers, ferns that live on trees, and ferns that themselves are trees.

true roots and leaves, and some plants gained fibrous bulk which gave them the strength to penetrate deeper into the soil and grow tall and become more like the trees we see today. But lignin also had a downside: it is hard to chew, swallow and digest, and, even in that environment, it was slow to degrade. Termites and other organic munchers of today had not yet evolved and it would be nearly 100 million years before lignin would face any real threat. The humble fungi and bacteria evolved a set of new enzymes to digest lignin, but it took them another 20 million years to do this. In the meantime, shrubs and the first trees kept growing and many of these got buried and remained preserved in the silt and sediments. Tree fossils from varying periods of time can be seen all across India—from near Jaisalmer in Rajasthan to Bankura in West Bengal to Ariyalur in southern Tamil Nadu. Though tree fossils closely resemble the rocks around them in colour, if you look closely, you will be able to see the bark and sometimes even the tree rings on the cross section.

One of the earliest surviving plants from these times in India can be commonly seen along the roadside in the hills around Darjeeling and Sikkim. This is the tree fern, a fern which resembles a palm and grows in the understorey of the moist hill forests of northeast India. Tree ferns emerged around 320 million years ago and their stems were made of true wood.

Just like in the sea, where several sea creatures exploded on to the scene almost simultaneously, the arrival of plants was also pretty explosive and caused a green revolution on an otherwise brown and mostly bare land. Another interesting plant from these times is the cycad which emerged around 350 million years ago. Cycads are easily confused with palms but they are different. They have no flowers, and their male and female reproductive parts are located on separate plants, which bear primitive-looking berries. Cycads are found across moist regions of India and in recent years a handful of new species of these primitive plants have been discovered in forest fragments of Odisha, Andhra Pradesh and Karnataka.

Oxygen was the key to the emergence of complex forms of life both at the time of the Great Oxygenation Event and post the Snowball Event. It was the abundance of oxygen that led to the proliferation of life but the increase in the levels of oxygen meant a decrease in carbon dioxide.

The 'greening' of the continents acted as a sink for carbon dioxide, causing a reduction of its concentration in the air, gradually cooling the climate. The furious growth of plants and trees raised the oxygen level. The level of oxygen in the atmosphere when the first trees evolved was about 50 per cent higher than it is today (about 21 per cent). This allowed early insects that had made land their home, to grow much larger than they are today. Dragonflies of that time were as big as eagles and millipedes were the size of gharials! The early insects co-evolved as land plants diversified, thus forging a relationship between plants and insects that endures till this day.

The drama and violence of Earth's processes is central to the creation of life because every time Earth experiences big changes, new opportunities for life open up. Even small disturbances can cause extinctions of species and sometimes lead to the collapse of entire ecosystems. It is hard to believe that since life began nearly 3.8 billion years ago, 99 per cent of all living things that have ever lived on Earth have become extinct. What we see today is just a miniscule 1 per cent of all life that has ever lived on Earth. In fact, it is almost a miracle that life continues to exist at all!

For the next few million years, the water's edge would continue to be the real scene of action and would set the scene for the next big stride in the evolution of life—the rise of the first fish which would become ancestors to all animals with backbones.

3

FINS, FLIPPERS AND FEET

Walk along Marina Beach in Chennai at dawn or at dusk where the tides meet land and you may spot a rather curious creature hidden in the sand. It is the size of your fingernail (about 3-5 mm), looks like a thin worm and has a translucent body that shimmers in the sand. To catch it, you will need to scoop out a handful of sand and let it pass through your fingers until all that remains in your hand is this wriggling creature. Put your delicate catch in a small bottle cap with a little water, and look at it under a simple hand lens and you will see that this organism has a central nerve (the notochord) running down its back. This is the *Branchiostoma* (from Greek, *branchio*: gills; *-stoma*: mouth) or the lancelet (because of its knife-like shape), one of the earliest fish to have evolved. It lives in shallow sand along beaches and feeds on floating organic matter and worms that are even tinier than itself. It breathes through its translucent skin, and although it has thin slits along its pharynx which give the appearance of gills, these slits have no capillaries, and therefore serve no function. *Branchiostoma,* or something that closely resembled it, is believed to be the common ancestor of virtually all vertebrates—organisms with backbones—that exist today and the *Branchiostoma* itself is likely to have remained unchanged since it first emerged 530 million years ago!

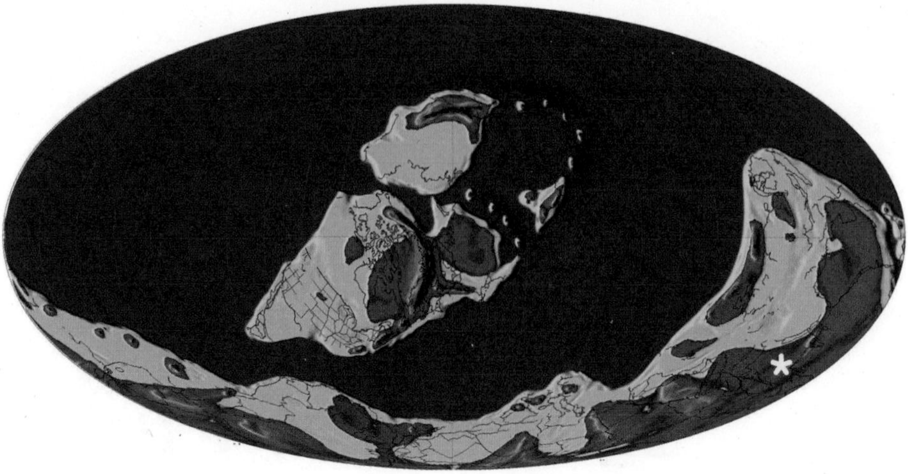

Earth, 420 million years ago; India is roughly where the yellow star is.

The story of fish began with a fish like the *Branchiostoma* struggling to survive among armoured crustaceans. Cutting a long story short or compressing the saga of the nearly 500 million years of evolution of fish into a few paragraphs, this is broadly how it would all have happened: a few hundred thousand years after the first fish evolved, triggered by the action of a new set of genes, some organisms developed a hard rigid outer coat to encase the notochord. This became the 'backbone', and the organisms came to be known as 'vertebrates'. The switching on and off of these genes led to the creation of new appendages that advanced the functioning of existing organs.

Branchiostoma don't have eyes or a true brain but what they do have in surprising abundance is a photo-pigment called 'melanopsin' which responds to stimulus from light. So if you put these creatures in a bowl with sand and water and switch on a light, they will swim and float to the surface and back, and if you gradually reduce the light they will stay embedded in the sandy gravel. What function could such receptors have? Unlike worms which only move away from strong light sources, these creatures clearly have a more sophisticated response to even slight changes in the amount of light, which they use to regulate body functions. Recently scientists have found that melanopsin is a critical chemical which informs the part of the brain called the hypothalamus to adjust to different times of the day. We call this the 'circadian rhythm' or the body clock, and as vertebrates evolved, this rhythm became more and more complex. So the next time you experience jet lag, remember that the earliest vertebrates were probably dealing with the same problem, and were the first to synthesize this remarkable chemical which helps our bodies adjust and respond to changing light and heat (and many other stimuli).

Branchiostoma have a single set of *Hox* gene but as vertebrates evolved, more genes were added which made them more complex. Among the first signs of complexity was the development of new external tissues and appendages like scales and fins which helped fish move faster in water. Like modern-day lampreys and hagfish that live in deep seas (like in the trenches of the Indian Ocean), the early fish had no jaws. Around 480 million years ago or so, fish acquired jaws and, over time, teeth grew on them, enabling them to eat other fish. In order to defend themselves, their scales became thick and in some fish the scales began to fuse together to form tough plates and armour. Within a few million years (a very short period in the evolutionary and geological scale) a bewildering

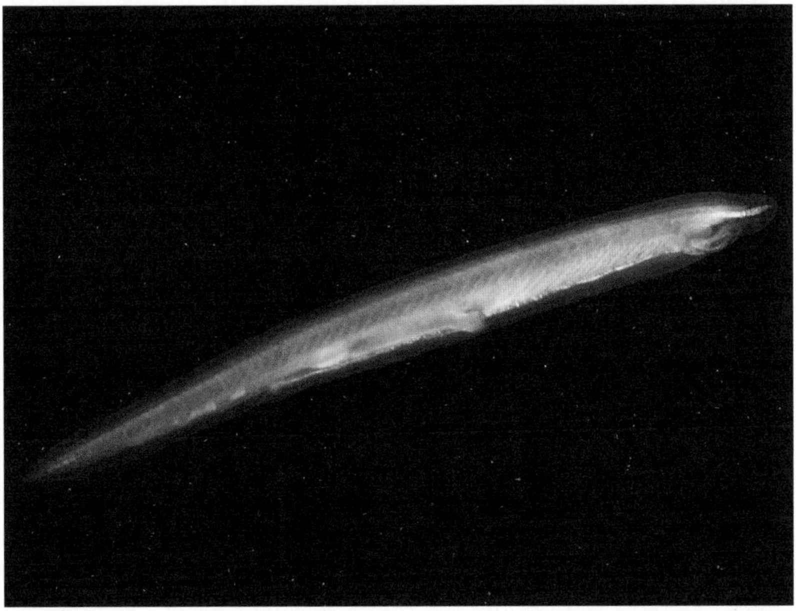

The Branchiostoma, *which is about the size of your thumb nail, is what the first animal that developed the earliest spinal cord might have looked like. This fragile animal has probably survived millions of years of devastation on land and in seas, but can be found only on beaches which have not been entirely run over by humans in India.*

variety of vicious fish had evolved, each outcompeting and out-eating the other. Some of these fish were so large and so fierce that they could rip apart present-day sharks with ease. As more and more fish became carnivorous, every species of fish had to protect itself from constantly evolving predators. Somewhere between 430 and 360 million years ago, when our Earth Woman turned 42 years old, heavy armour and large jaws started to become a liability for fish. Heavy armour, in particular, restricted their movement and the next logical step in terms of the evolution of fish was to move the hard armour (the exoskeleton) from the outside to the inside (first as cartilage and later as bones). This made fish agile and fast and enabled them to hunt and escape easily. Soon, having armour and a mean set of teeth became a thing of the past.

By this time, plants had already established themselves on land and had begun to add oxygen to the air. The organic run-off from land began to enter the seas, which attracted diverse microbes along the edges of water. They, in turn, gradually enriched the depths of the seas. The fish that had discarded their primitive armour and teeth began to develop specialized feeding habits and breathing mechanisms to take advantage of new sources of food.

It was around 440 million years ago that four main lineages of bony fish began to evolve and about 419 million years ago, these lineages diverged into four distinct families. The first family is of jawless fish that have survived virtually unchanged since then and continue to live in the depths of the oceans, feeding on decaying organic matter. These include lampreys and hagfish that live on the dissolving carcasses of animals (which is why they don't need jaws) found in the deep trenches of all seas and oceans. The second is the family of cartilaginous fish like sharks that had evolved before bones developed in nature and who became the undisputed predators-in-chief once armoured fish became extinct. The other two groups emerged when bones began to develop in the bodies of fish. The more dominant between these are the ray-finned fish. Some of the most popularly consumed fish in India—like rohu, hilsa carp and perch—belong to this category. You will notice that these fish have fins made of a web of skin supported by numerous thin radiating bones (called rays) which make them light and fast swimmers, enabling them to out-manoeuvre other fish. Their speed and agility helped the ray-fins and sharks outcompete and outlive the last remaining armoured predatory fish and they have dominated the water ever since. Ray-fins have used their fins to steer, mate, defend and even fly! Flying, obviously, is not a trait associated with fish, but some fish have developed the ability to skim and leap over long distances on the surface of water, making it appear as if they are flying over water. Flying fish are ocean-living surface fish and they glide over long distances using elongated pectoral fins and winged gliders. When a shoal takes to the air and glides on the surface of water, the fish together look like shimmering streaks of silver. Three common species of small, four-winged flying fish are found in India: *Cheilopogon*, which is commonly seen in the fish markets of coastal Maharashtra; and *Hirundichthys* and *Cypselurus*, which are found along the coasts of Tamil Nadu (locally called 'parava' or 'kola meen') and southern Kerala. These fish are fried and eaten as a popular snack in the toddy shops of Kerala and the small eateries that dot the highway along coastal Tamil Nadu.

The fourth family of fish is the lobe-finned fish which are now nearly extinct. They played a crucial role in advancing life on land. Their

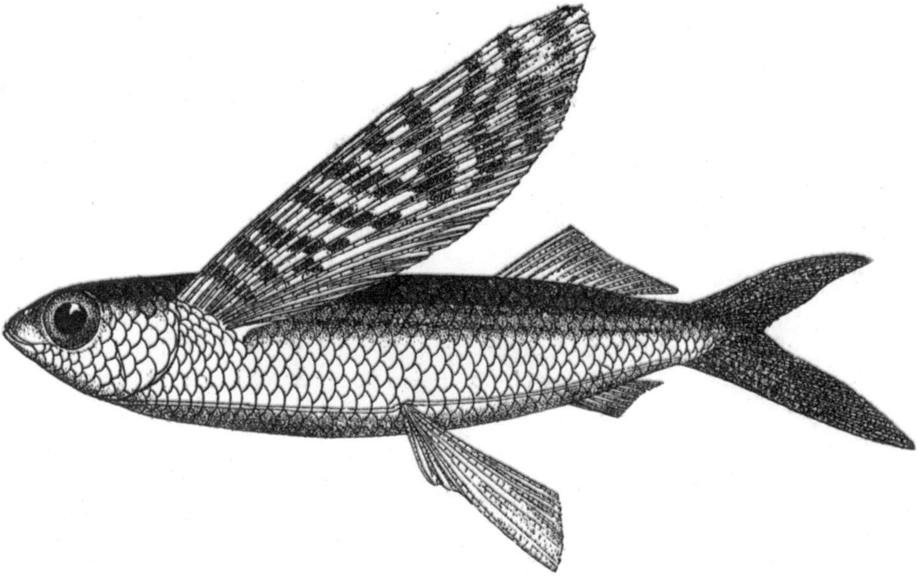

*This beautiful 1876 illustration of a flying fish (*Cypselurus*) that lives along the southern coasts of India demonstrates elegantly how fins became a versatile appendage for the fish. Flying fish use fins to skim for a few metres above the water surface so that they can escape predators. Schools of flying fish look like tinsels of silver over water.*

distant descendants which survive today include modern-day lungfish or *Dipnoi* (from Greek, *di*: two; *-pneu*: breathe, as it breathes through a modified air bladder and gills) which live in Africa, Australia and South America. A few lobe-finned fish like the coelacanth have remained unchanged since their emergence and live in deep oceans. This group of fish has bones at the base of their fins and it is believed that they correspond to the one upper and two lower limb bones that led to the development of all four-legged animals (called tetrapods, and comprising all amphibians, reptiles, birds and mammals).

Fish were the first vertebrates to evolve and the only ones to continue to live in the place of their origin. They are arguably the most successful vertebrates on the planet. Half of all living vertebrates today are fish,

Lungfish, as seen in this illustration of an Australian one, were among the first vertebrates ever to develop the ability to breathe air through their mouths, fill their lungs and survive long, dry spells that could last as long as a few years. Lungfish have died out in India but are still found in Africa, South America and Australia.

and of them the ray-finned fish, with nearly 15,000 species, make up half of all living fish. In other words, ray-finned fish dominate the world of fish just as beetles do in the insect world!

Somewhere around 400 and 380 million years ago, nearly 120 million years or so after the first fish emerged (or about six to eight years ago in the life of our 46-year-old Earth Woman), a fish similar to the modern-day lungfish began to venture to the edge of water. The development of bones and strong pectoral fins which acted like limbs aided this relocation, allowing for a variety of movements and providing the mechanical support necessary for life on land. Bones also became the main site for blood cell production and helped in the organization of the muscles and blood vessels which helped in transporting oxygen to all organs of the body, and provided a layout for nerves.

The examination of fossils of early tetrapods and the discovery of new fossils from around 360 million years ago suggest that some fish evolved four well-developed legs even before they took the first steps on land (which happened nearly 40 million years later, close to around 320 million years ago). The design of their legs and the strength in them show that these limbs would not have been able to support their body weight on land. However, they would have been able to help them raise their heads, enabling them to breathe in oxygen-poor water. A second clue to the early tetrapods' entirely aquatic existence comes from their jaws. Their jaws suggest that they were carnivorous. It would have been quite impossible for full-grown adults, some of whom were as large as present-day scooters and auto rickshaws, to survive by feeding exclusively on the small insects and arthropods that were found on land at that time. These small insects, however, would have been an adequate diet for the young of these tetrapods and it is likely that the young lived on land (so as to not compete for survival with the larger and fiercer predators that were found in water) but went back to water on attaining adulthood. Since early fish took to land much after they had developed legs, they used this interval to develop stronger legs and lungs that prepared them to take to life on land.

The ability to breathe out of water was critical for the fish and fish-like creatures with evolutionary ambitions to begin life on land. Most fish breathe through gills, but a few could also breathe in through their mouth—in these cases the air directly entered a specialized sac, the forerunner of modern lungs. The ability to breathe through the mouth and the evolution of air sacs perhaps occurred as a survival strategy when oxygen levels in the sea dropped because of increased volcanic activity. These fish would come up to the surface for a gulp of air, fill their air sacs and descend into the depths of the seas. One survivor from this time is the lungfish which, as I have mentioned earlier, lives now only in eastern Australia, central Africa and South America. *Ceratodus*, ancestors of the modern lungfish, were found in India too but became extinct about 90 million years ago and their fossils have been found all over the country, from Kashmir to Kutch to Kanchipuram. Modern lungfish live in slow-moving rivers and muddy pools and come to the surface to take gulps of air. When water runs out, they aestivate or undergo a summer sleep (the opposite of hibernate) in dry riverbeds where they lie inert in their moist burrows, waiting for the first rains to release them from their mud prisons. Given that lungfish have the ability to breathe outside water, it is plausible that ancestors were the first to set fin on land. But modern lungfish do not ambulate in search of water when their pools dry out or go to land in search of food. So in effect, lungfish fulfil only one criterion of living on land: they can breathe air on land, as opposed to only in water. Our quest for a possible candidate to first venture out of water seeks something that is not only capable of breathing out of water but also possesses the capacity to move on land. This ability has been observed in another fish, the walking catfish.

Walking catfish is abundant in India; so abundant, in fact, that not much has been written about its unique behaviour that sets it apart from most other fish. The walking catfish thrives in isolated ponds, paddy fields, mudflats and the slow-moving canals of northern, eastern and peninsular India, places that hold stagnant water and are low in oxygen. Because of its ability to live in oxygen-depleted water, it is caught and sold live in fish markets across India. Like the lungfish, the walking catfish is capable of surviving for a short while out of water. When there is plenty of water, it stays in the muddy bottom, coming to the surface for a quick gulp of air and returning to the depths once again.

This fish does not need to be an agile swimmer because it seldom pursues a prey. Its diet consists of worms, insect nymphs and decaying vegetation usually found relatively easily at the bottom of water. It has eyes on the side of its skull and whiskers that help it navigate its way in muddy water, a trait which helps it on land as well. However, what sets it apart from lungfish, and most other fish, is its ability to *walk* in search of water. When the water dries up, instead of building itself a mud burrow, this fish does something quite amazing: it drags itself out of the mud and 'walks' to find a new water body. Using its robust front fins (called pectoral spines), it flexes its supple body and, slithering to and fro, affects an awkward but effective flip-flop way of moving on land. This fish has additional physiological adaptations that allow it to stay out of water for extended periods (sometimes up to a couple of hours and even longer if they remain moist). In addition to air sacs, this fish has gills that are structurally hardened to prevent them from collapsing under pressure on land. Under the hardened gills lies a dense mat of hair-like structures which captures oxygen more efficiently than other types of gills. The gills rapidly diffuse oxygen from the air through a rich network of capillaries into the blood stream. Carbon dioxide and oxygen diffuse in opposite directions across the gill surface, making the process of exchanging gases quicker and more efficient.

Some small ray-fins, like gobies which evolved in India, also swallow air bubbles, and never come to the surface because it makes them vulnerable to predation. Gobies are common in aquariums, so if you ever notice a fish consuming air bubbles in a fish tank, remember it is doing so to breathe and to maintain its buoyancy. Most fish have a large gas bladder which helps them rise and descend rapidly in water. In the catfish this organ is much smaller because it does not need such agility, and the extra space is used to accommodate enlarged air sacs which aid its breathing. The catfish depends on its highly specialized gills to survive out of water. Once out, the walking catfish gulps air through its mouth which enters its gills and bladder and then diffuses into its bloodstream. The catfish undertakes its overland journey usually at night when it is cooler and when it is better able to sense (or smell) water in a distant pond or a paddy field. There are also fewer predators at night. The walking catfish is quite a sight!

However, we cannot say with certainty which the first vertebrate to venture on land was because there is no conclusive proof of which fish could have undertaken this momentous journey. A few other fish (ray-finned, not lobe-finned), like the *Polypterus* that is found in Africa and India and can breathe and walk out of water, and the waterfall-climbing cave fish (*Cryptotora thamicola*) which has developed a pelvis and can climb on moist cave walls, are also potential candidates, but not quite. They are good prototypes of how the early tetrapods could walk and survive, but they do not possess the muscular limbs that enable walking on dry land.

The walking catfish is a tough survivor. These hardy fish were accidentally introduced in southern USA where they have multiplied, and hordes of them can often be seen 'walking' across highways and farms, in their quest for new water!

Once fish ventured on to land, they had to adapt to life in this new habitat. For this, they needed to reshape their triangular skulls into a broad, flat skull with elevated eyes. Over time, they started getting comfortable with short sojourns on land so long as they stayed moist. Yet they did not completely leave water or become amphibious. They were perhaps a little like the modern mudskipper which is found aplenty in the mudflats of the Sunderbans in West Bengal and in the estuarine forests that dot the coasts of India. Short periods of exposure to surface air did not pose any danger of desiccating the fish that ventured on land,

and a little sunning on mudbanks actually proved to be very useful. A warmed-up body helped the fish digest food faster than when it was in water; and this allowed the offspring of such fish to grow faster, mature sooner and reproduce more successfully than other fish. Many living amphibians and reptiles like alligators and crocodiles still bask while they digest their meals. The early land-venturing fish developed stronger bones to keep them propped up while they lay in the sun, or else they would have died of suffocation from the collapsing weight of their lungs. A small modification in the bone structure also helped these fish. Fossil records have revealed that fish that basked gradually lost the bony connection between the skull and neck. This made their neck flexible and allowed the head to move from side to side (which was also useful for catching fast prey in shallow waters), and chew and swallow better. In addition, they had wide mouths that enabled them to gulp large quantities of air and they stayed with their mouths facing the sky when basking in the shallows. These amphibious fish developed robust and fleshy fins in front and had muscular and mobile elbows which helped them in station-holding (staying in one spot) while in water as well as raising their heads out of water. These fish also used their front fins like crutches to help prop up their bodies and wriggle forward on land with their tails swishing sideways to further assist in this movement.

Apart from having to keep their skin moist, the first animals that left water also had to fend off threats from arthropods who had arrived on land earlier (about 430 to 410 million years ago) and had grown so large in the oxygen-rich atmosphere that they probably feasted on the first vertebrates to venture on to land. But since insects and crustaceans had also arrived on land 60 million years earlier, there were some crunchy feasts awaiting the vertebrates along the water's edge too. When the first vertebrates began to venture along the edges of the water, the forests had already begun raising the levels of oxygen and moisture in the air. Several small pools had formed on the forest floor and this aided the movement of these fish further inland. The gradual advance of these amphibious fish to land was also inevitable because large lakes and seas had become too hostile for their survival, just like for insects before them.

The landfall of amphibians marked the beginning of terrestrial life for vertebrates. This illustration imagines how the first vertebrates (tetrapods) would have walked on land. They needed to pull themselves out of water, slither over soft mudbanks and walk efficiently on hard ground. Their gait would have been similar to that of modern-day geckos or salamanders. The oldest trackways are dated to be around 400-million-year-old and were discovered in Poland, but no fossil of the creature that made it has been found yet. The oldest tetrapod fossil found until now is only about 375 million years old.

The next evolutionary leap forward took place about 10 million years later: the development of true amphibians (Greek for 'both lives'), who evolved from an ancestor of the lobe-finned fish (like the lungfish). Within a period of just about 2 million years (363 million years ago, give or take a million), amphibians had completely evolved and were markedly distinct from their fish-like ancestors. We know this from fossils records where ankles, jaws and skulls of amphibians are preserved. Amphibians can live both on land and in water, but need the latter for reproduction, where they spend a large portion of their early development. Moisture is essential to them, as their skin desiccates if they do not return to water frequently.

The first amphibians were of modest size (between 0.5 to 1 metre in length) compared to the amphibious fish. Amphibians were the first true four-limbed creatures to evolve and they were, in all probability, the common ancestors of all tetrapods that live on land today. So there is a fish and a frog in all of us! Terrestrial tetrapods have a spine and a pelvis attached to their limbs that helps them move and distribute their weight evenly on land. This is true for virtually every tetrapod. Even in animals that appear limbless, there is a spine and a pelvis under the skin—made of interlocking spurs and bones, which enables movement. In snakes too you find special muscles that emerge from the sides of vestigial pelvises; these are remnants of limbs. In virtually all living tetrapods, limbs develop as arms and legs, or flippers and feet, all of which are attached to an interlocking joint that enables their movement, and each limb ends in digits like fingers or talons. In hoofed animals, these fingers are fused but when dissected or viewed through an X-ray machine, the digits are clearly visible.

From 360 to 290 million years ago amphibians grew tremendously in size and became the dominant predators on land. To see how large these ancestors of frogs and newts actually grew, we need to travel to an unlikely place, the rolling hills around Srinagar in Kashmir. Fossils excavated from the Mamal range in Kashmir provide the earliest evidence of landfall by amphibians in India, but before we describe this we will go back to see what was happening here a few million years earlier. The hills of the Mamal range lie next to the 12-square-kilometre wide Nishatbagh formation that surrounds the famous Nishatbagh

Gardens in Srinagar. The Nishatbagh formation is older than the Mamal formation and is rich in plant fossils and a few isolated fossils of the earliest insects from India (mostly cockroaches and extinct winged flies) have also been found here. Sedimentologists (geologists who study how sediments are deposited in water and on land) have confirmed after studying the layers of rocks that the Nishatbagh formation was once an island with meandering streams that opened into a large lagoon to its north and east, and was surrounded by several smaller islands of varying sizes. About 295 million years ago, this land was largely covered by forests of cycads, horsetails, giant ferns (called 'pteridosperms'—plants with winged or air-dispersed seeds), lycopods (trees that looked like scaly telephone poles) and tree ferns. There were also some members of the moss family that grew nearly 30 metres tall. From space these islands would have looked like dense evergreen rainforests. Around 270 million years ago, there was a massive underwater volcanic eruption that produced so much lava that it smothered the islands and caused them to sink under its weight. You can see this volcanic lava even today as a 2000-metre layer of basalt which makes up the base of the snow-covered Pir Panjal mountains (also called the Panjal traps) that lie west of Srinagar. The Panjal traps have layers of grey limestone that are raised on a pedestal of dark grey basalt. The volcanic activity lasted a short while—perhaps a few thousand years—and life resumed once again when these islands were raised up from under the sea. You needn't undertake an arduous climb or a cumbersome journey to find signs of the succession of life that followed. All you have to do is drive to the adjoining hills east of the Nishatbagh Gardens. Here, at the edge of the Nishatbagh formation, you will find slightly younger rock beds (about 268 million years old) that were overlain by sediments of the sea and inland streams which are now covered with 200-metre-high orange-red sedimentary rocks called the 'Mamal formation'.

The Mamal range extends from Anantnag to Budgam, and ends near Srinagar. It is named after a tiny village (Lat: 34°01'N, Long: 75°31'E) famous for the oldest Shiva temple in Kashmir in Pahalgam district, where rocks from this period are best exposed. The 12-kilometre wide and nearly 45-kilometre-long Mamal formation is rich in fossils of plants that had recolonized the islands, and as we go from the lower

(older) to upper (younger aged) rocks we find a diverse variety of plants and early amphibians which start small and gradually grow larger. The hills of Mamal during the time of the amphibian landfall were also islands and were not a part of the mainland Indian Peninsula (geographers

Nishatbagh in Kashmir or Dhanbad in Jharkhand would have been covered by such fantastic forests around 270 to 250 million years ago. The forests were dominated by cycads, horsetails, giant ferns and primitive palms, and their frequent submergence due to massive flash floods led to the formation of coal.

call them 'extra-peninsular islands'). Palæontologists have carried out many excavations and have catalogued the plant and animal fossils found in this region in their effort to reconstruct the life that once existed here. The plant fossils of Mamal are different from the ones found in Nishatbagh. Tall trees, primarily *Glossopteris*, dominated the times and within this rich and humid wetland vegetation lived a large ray-finned fish called *Amblypterus* (*ambly*: to walk, *-pterus*: with wings or fins) that hunted in the shallows and mudflats but could not venture on to land. Another ray-fin found here which was entirely aquatic was the *Gardinerichthys* (or Gardiner's fish). A compact (about 2–4 feet long) but menacing amphibian with an oversized skull called *Kashmirosaurus* (or Kashmir's lizard) was the dominant carnivore on the water's edge. *Kashmirosaurus* fed on fish and other amphibians. Its arch rival was another carnivore, the *Lysipterigium*, which also lived along the banks. *Lysipterigium* were precursors of the families that would become the first true amphibians and from whom modern amphibians would evolve. These heavily built animals had scales, claws, bony plates and a massive head (compared to their body). They were a little smaller than modern-day crocodiles but very much alike in characteristics. Many smaller amphibians and fish that feasted on the insects, and on each other, also lived along the water's edge.

With the evolution of amphibians, although vertebrates had begun to move to land, their colonization on land would be considered complete

Predatory fish like the Amblypterus *terrorized the waters around Kashmir about 260 million years ago. They hunted in pairs like barracudas and had a tough armour which was made of large scales, perhaps to fight off others of their kind.*

only when they began to reproduce without having to be near water. It took another 15 million years after the emergence of the first amphibians for this to happen and for the appearance of the ancestors of reptiles who were able to lay their eggs on land away from water. Reptiles evolved from land-living amphibians about 315 million years ago and had tough skin with scales on it, reminiscent of the fish. Unlike the wet skin of amphibians, the reptiles' scaly skin allowed them to live on land even when away from water for long periods of time, giving them a greater advantage over amphibians in the rapidly drying conditions on Earth.

Between 300 and 280 million years ago, the land started to become arid and the newly evolved reptiles, who were better able to adapt to the rapidly drying conditions, began to dominate over amphibians. Only a few species of the limbless amphibians called *Caecilian* (Latin: blind ones, pronounced like 'Sicilian'), a worm-like burrowing creature, had scales. This is an adaptation that this amphibian acquired for living out of water. But caecilians and reptiles are not similar in any other way and molecular studies suggest that caecilians were not the ancestors of reptiles. It was perhaps a latent ancestral gene that came into play, enabling reptiles to grow thicker scales made of a protein called 'keratin' (Greek for 'horn-like') which later evolved into claws in reptiles and reptile-like mammals. Gradually, specialized cells were formed under the scaly skin of reptiles that helped them regulate temperature more effectively than amphibians. They developed other special abilities like camouflage under nervous control, as seen in chameleons and garden lizards. Reptiles also developed cone-shaped teeth and tapered skulls with eyes on either side, a feature that they later passed on to mammals.

Although the first amphibian landfall occurred in Kashmir, evidence of the evolution of early reptiles can be seen in the Kundaram formation in Andhra Pradesh. The rocks and sediments of Kundaram formed along the ancient river valleys of eastern Andhra Pradesh which were carved out by the Godavari and Mahanadi rivers. These rocks are slightly younger in age (280 to 251 million years, a period called Late Permian) than those found in the Mamal formation of Kashmir but bear within them a variety of fish, amphibian and reptilian fossils. Fossils of the earliest mammal-like reptiles have also been found in the sediments and rocks here. The dominant reptiles found in India had mammal-like features and belonged to a lineage called the 'dicynodonts' (or two dog teeth). Many of these dicynodonts have been found all over the world, from Madagascar to eastern and southern Africa, and Brazil to Andhra Pradesh in south-eastern India. Their distribution across continents shows that these land masses were in close proximity during this time. In fact, they were all joined together in a supercontinent called Gondwana that formed after the break-up of Rodinia (between 800 and 750 million years ago). Landlocked regions reduce opportunities to create new species (a process called 'speciation') and therefore the same species were found across large land masses. Among these were the earliest dicynodonts—specialized herbivores that had evolved a secondary palate which helped them breathe and chew at the same time. Dominant among the dicynodonts in peninsular India were *Endothiodon* and *Wadiasaurus*. These were massive cow-sized reptiles that grazed and lazed in the luxuriant fern forests of the region. Their bodies were bulky and their limbs short and strong. They had very short snouts and no teeth except for two tusk-like upper canines, which they used for display and fighting, and for shearing and grinding the tough vegetation.

Gradually, reptiles began to develop a fondness for meat and acquired more dominant mammal-like features, especially teeth, the arrangement of the skull bones, the eye sockets and the ear. The pairing of backbones and ribs, and the arrangement of girdles and hips which give mammals a more erect and less sprawling posture were also found in these reptiles. The *Cistecephalus* (or box head), a stubby look alike of the modern-day gecko that scurried in the undergrowth of forests, was among the first animals to show mammal-like traits: it had incisors, canines and cheek teeth, just like modern mammals do. Those teeth also tell us that it was an omnivore. Its legs seem to have been adapted for running and swimming, a little like the water monitor. It was between 1 and 2 feet long and lived like a mole. It ate fleshy roots and nosed around for crunchy insects and worms when it burrowed in the soil. It had a strong neck and a thick wedge-shaped skull that was flat at the top to help it burrow. It used its stout front legs and sharp toes to break the ground and its thick and muscular rear feet to push the soil behind it as it dug.

A typical day in central Andhra Pradesh, 260 million years ago. A colony of Cistecephalus *(left foreground) is terrorized by a mammal-like reptile while reptilian grazers (in the background) go about feeding on water reeds and succulent plants.*

Around this time, tectonic activity on Earth began in fits and starts, and Gondwana began to stir. Some tracts of land were subsumed, causing ingress of the sea while others were raised into folds. This gradually changed climatic patterns in the region. At first, heavy rains occurred in short spells which carried away with them the forest litter and the gigantic tree fern of those times, the *Gangamopteris* (Greek *Gangamon*: small round oyster net; Latin *-pteris*: leaf). *Gangamopteris* was the dominant tree in the Nishatbagh formation but the upper layers of the rocks show that another genus of ferns, *Glossopteris* (Greek for tongue-shaped leaf) was beginning to displace it. Many palæobotanists believe that *Glossopteris* has been the most abundant plant ever to have existed. A small patch of forest could be home to several species of *Glossopteris*, and palæobotanists have recorded at least 70 species from India alone. After a short period of normalcy, heavy rains began again, drowning and sweeping away lush forests of *Glossopteris* and thick tropical litter into lagoons and depressions, and smothering large swathes of giant tree fern forest into the debris. Ferns, cycads and horsetails dominated these forests and there were no flowering plants. Massive floods became more frequent and more violent. No sooner had the forest grown than it would be drowned again, and a fresh stand would spring up in a very short time, only to be swept off again. A flood could submerge a 20-metre-high forest within a couple of days. This occurred over and over again for nearly a million years. Vegetation accumulated in stagnant water which decomposed into an organic mass called peat. As more and more sediments were deposited, often several times in one season, the weight of the fresh organic matter squashed the previous debris, and microbes rapidly decomposed the vegetable matter by raising the temperature of the peat. Soon, the numerous peat bogs were belching out copious amounts of organic gases like methane and sulphurous fumes. The decomposition reduced the weight of the peat and converted it into the light, shiny and brittle carbon deposit we call coal. Geologists believe that around 300 to 270 million years ago, a network of rivers and large deltas as well as forests and swamps existed where present-day Bihar, Jharkhand, eastern Odisha and West Bengal stand and this place became home to the Eastern Coalfields. It took a 10- to 12-cubic-metre layer of submerged *Glossopteris* rainforest to make 1 cubic metre (roughly 1 tonne) of coal. The depth of eastern India's coal mines exceeds 4 kilometres and they are spread across 4000 square kilometres. It would take 200 times all the existing forests of the world to create the coal reserves that exist in India alone!

In comparison, the submergence of the *Gangamopteris* forests—the predecessor of the *Glossopteris*—formed only a thin layer of coal mixed with sediment like the ones we find in Kashmir. The *Gangamopteris* layer can also be seen under undisturbed soil sections in the coal seams along the Barakar river, the main tributary of the Damodar. The Barakar originates in Hazaribagh in Jharkhand and meanders through the mound-shaped Parasvnath Hills in Giridih district of Jharkhand which is famous for its Jain and Hindu temples. Winding through collieries, thermal power stations and depleted industries, it traverses 225 kilometres before meeting the Damodar in Burdwan district of West Bengal. The several layers that make the Barakar formation contain within them a diverse collection of plant fossils from 280 to 230 million years ago.

But despite the deluge washing away forests and its creatures ever so often, life never had it better. The arrival of vertebrates on land led to the proliferation of many orders that exist today—amphibians, reptiles, mammals (all except birds), all of whom emerged from the first animals that left water. However, the good times did not last. The vertebrate revolution ended abruptly as Earth began to witness the greatest extinction ever known. The decimation of the *Glossopteris* forests was the first sign of a larger catastrophe that awaited all life. About 96 per cent of all life was obliterated forever around 251 million years ago. Scientists simply call it The Great Dying, and life would need every trick it had learnt so far to survive the greatest catastrophe it had faced.

The exact cause of the extinction remains undetermined, but it is believed that around 252 million years ago the volcanoes in Siberia—they can be seen as flat-topped plateaux in the Perm region of Siberia today (the term 'Permian extinction', as the Great Dying is more commonly called, comes from here)—erupted, causing a massive outpouring of lava that started it all. However, several subsequent events may also have triggered the near-complete decimation of life. I will talk about the three theories, among several others, that seem most plausible to me. The first, of course, is the theory that the Siberian volcanoes themselves caused the carnage. But some climate experts and geologists were not convinced because, in the 1980s, scientists studying rocks in Greenland discovered rocks from the Permian period that showed a spike in the levels of Carbon 12 (C12). This was far greater than could be expected from the release of carbon dioxide during the normal decaying process that leads to the formation of coal, or from the accumulation of peat, or even due to the action of volcanoes.

So what could explain the rise in C12 levels? Scientists thought that there had to be another source of such a large spike. The answer came from rather unexpected quarters. In the 1990s, geologists exploring for oil discovered reservoirs of methane hydrate all along continental margins of the Gulf of Mexico. When a large amount of organic matter gets quickly buried in deep seas, the methane produced reacts with cold sea water and solidifies under pressure to form large quantities of ice-like methane hydrate. But methane hydrate is a gas at temperatures between

27 to 30°C and therefore when exposed to warm water or air, it releases massive amounts of C12 carbon as methane. Even in deep oceans, it takes a very small increase in temperature to release this potent greenhouse gas. When a small amount of this white crystalline solid, quite like dry ice, came in contact with warm water, say when a volcanic or a hydrothermal vent ruptured through a seam of methane hydrate, it would have released massive quantities of C12. So a second theory attributing the anomalous spike in C12 to the action of methane hydrate seems plausible.

A third theory, proposed in June 2014, has suggested that a methane-producing microbe (*Methanosarcina*) was the chief perpetrator of the rise in atmospheric methane that caused the Great Dying. Scientists who support this theory say that the Siberian volcanoes released the necessary metal which catalysed the essential enzymes required by *Methanosarcina* to digest the rotting plant debris, and this produced copious amounts of methane.

Some scientists believe that the extinction took place not as a single event but in episodes that were perhaps spread over tens to several hundred thousand years. Whatever the real reason for the Great Dying may have been, what we know as a fact is that around 251 million years ago nearly 96 per cent of all species that existed died. It was the 4 per cent survivors who took over the land and water, and started all over again.

How else should we remember these times? Apart from the selfish reason of the survivors who became the ancestors of all mammals, this period remains significant to our modern existence for other reasons as well. This period gave us coal, a cheap source of generating power and electricity. In the early nineteenth century, coal-fired steam engines powered factories that made the Industrial Revolution possible. The Industrial Revolution led to ideas and innovations that built machines and established industries, launched ships, organized labour, effectively reshaped society and rewrote the rules of trade, economics and politics. It also drove countries to war. Coal became the power behind the industrial progress of Britain and France, enabling them to conquer, exploit and rule over less-industrialized but resource-rich tropical countries like India. In 1836, the East India Company appointed a 'Committee for the Investigation of the Coal and Mineral Resources of India' (or the 'Coal

Committee') whose chief object was to assess the availability of coal and the possibility of its immediate exploitation in order to power steamships on the Ganga and its tributaries. At first, a handful of geologists and officers from the British army and surgeons assisted in such surveys. But in 1848 a permanent corps of geologists and surveyors came to work for the East India Company. Following this, the Company instituted the Geological Survey of India in 1851. In 1860, 280,000 tonnes of coal was being extracted in India, and by 1868, more than 4,92,700 tonnes was produced. During the First and Second World Wars, coal from India was used to make arms and supplies for the British and their allies. The production of coal in India has doubled roughly every twenty years. India has in excess of 61 billion tonnes of coal in reserve, and, in 2014, nearly 585 million tonnes were extracted. Coal remains the world's largest source of energy and the largest source of human-induced emission of carbon dioxide, a mild but insidious greenhouse gas.

In effect, all things come around. The making of coal took place when amphibians and fish ruled the world, and it is this coal that is now proving to be their nemesis! But this time round it is not volcanoes or the climate that is killing them; it is highly evolved humans and their use of coal. At the time of the Great Dying, our Earth Woman was 44 years old, and life on land and water had nearly been wiped out. Only some plants and a handful of lucky animals survived in the barren and desolate landscape. It would take almost 10 million years (about a month in the life of our Earth Woman) for these survivors to revive the planet and for life to begin anew.

4

REVIVAL

If you could have seen Earth from space 250 million years ago, after the Great Dying had killed most life on the planet, the land would have seemed dusty brown because of the near obliteration of all greenery, and the seas would have looked inky dark. The blue-green gem-like Earth had become barren and desolate. There were very few wispy clouds in the sky and rain was only an occasional event. The blame lay entirely with the gas we call methane.

Excessive methane in the atmosphere had destroyed the ozone layer in the stratosphere—our natural sunscreen which protects life from deadly UV rays—and it would take a lot of oxygen to heal it eventually. Methane is a potent greenhouse gas because it traps the inward radiation from the Sun. Gram for gram, methane is twenty-five times more potent than carbon dioxide in trapping the Sun's heat. The smell of methane and hydrogen sulphide would have been stifling, pervasive and noxious, and with heavy air trapping the heat, Earth would have felt like a casserole.

The Nigahi mines in the Singrauli region of Madhya Pradesh make the northwestern margin of the massive Permian coal deposits of India. When forests of central and eastern part of Greater India were swept away by thousands of years of successive flooding, they got squashed and decomposed to form layers of coal deposits. It took 10- to 12-cubic-metre layer of fallen rainforest to make one cubic metre (or roughly one tonne) of coal. By that measure, it would take 200 times all the existing forests of the world to create the coal reserves that exist in India alone!

The supercontinent Rodinia which lay in the southern hemisphere had begun to break up into smaller fragments between 800 and 750 million years ago. Over the next 450 million years these fragments began to aggregate again to form another supercontinent called 'Pangæa' or 'Pangea' (Greek for 'all earth'). India lay wedged between Africa, Madagascar, Antarctica and Australia and although rocky outcrops like the Aravallis were already discernible, the land mass was contiguous and, of course, it would have been difficult to say where today's countries began and ended. We know all this from the discovery of uniform fossils of plants and animals in what are now far apart countries and from some unmistakable commonalities in the physical features between countries.

Pangæa was shaped like a thickened letter C. The top curve of the C consisted of land masses that would later rearrange themselves into modern Europe and Asia. North and South America formed the curved back of the C with Africa nestled inside the curve, while India, Australia and Antarctica made up the bottom of the curve. The huge area contained within the belly of the C was the Tethys Sea, though of course there was no one around then to call it anything! The northern hemisphere and most of the rest of the southern hemisphere were covered by a massive ocean we call 'Panthalassa'. Panthalassa surrounded Pangæa. If you girdled the planet through its midriff with an imaginary line like the equator, it would have bisected the Mediterranean between Spain and Morocco and cut through North America diagonally, starting very close to where Boston is today. North of this line was 'Laurasia' (a continental mass made up of present-day Europe, Asia minus India, and all of North America). In the south was the land mass we know as 'Gondwana' and together they formed the supercontinent Pangæa from around 300 to 200 million years ago. Laurasia began to separate from Gondwana around 250 million years ago and they were completely detached from one another around 200 to 180 million years ago.

Since all land was connected until about 200 million years ago, some cosmopolitan herbivores like *Lystrosaurus* (or shovel lizard), a survivor from the Great Dying, could have walked west from Sydney to reach Kolkata in a matter of days, and would have been able to cross over to the centre of Antarctica or into Cape Town from Chennai in about three

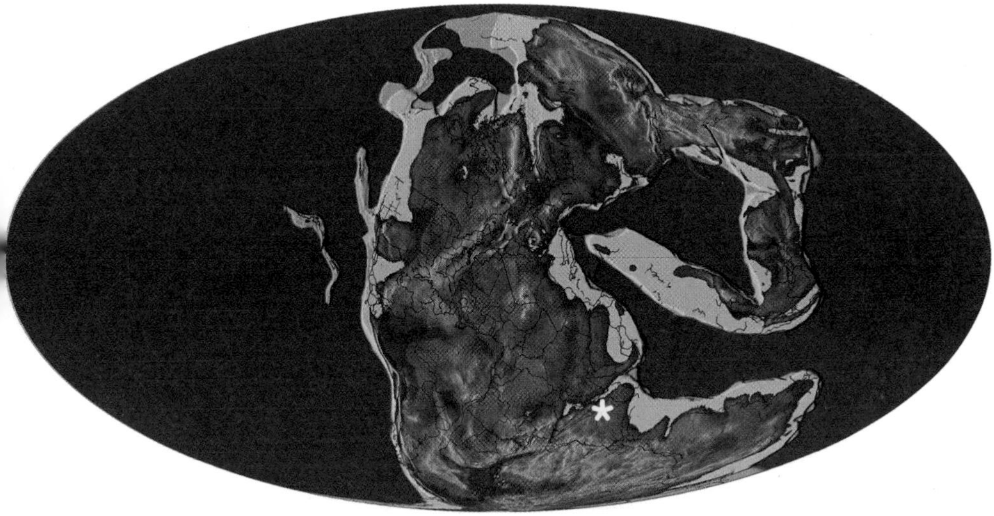

About 251 million years ago, all land was fused together and looked like the open-mouthed Pac-Man. The yellow asterisk shows the position of India. Notice that India was tightly squeezed between Australia in the east, Antarctica in the south and a narrow channel separated Africa from the western margin of India.

weeks. Before the Great Dying, all these parts were covered with thick forests which grazers like *Lystrosaurus* had colonized.

During the Great Dying, the most widespread death occurred in the seas. The enigmatic trilobites, many species of clams, sea urchins and coral along the margins of seas disappeared and many families of fish were wiped out, never to be seen again. *Glossopteris* forests and the giant insects and arthropods that crept about within them as well as many large amphibians and reptiles were also extinguished forever. At about 60°C, the heat on land was stifling. Water too was piping hot with temperatures around 40°C. For 5 million years (around 252 to 247 million years ago), life on Earth went through what was possibly its most difficult phase. But some plants and animals survived and hung on tenaciously for dear life. Oxygen was at its lowest (at about 10 to 12 per cent of all gases in the atmosphere) and all creatures on land and sea had to either find some means of quickly adapting to living in these extreme conditions or perish. There was so much heat in the air that though it would have rained in the tropics, much like today, as there were no forests, these rains would not have brought down the temperature. Only a few hardy ferns withstood the heat. Some animals avoided being singed by moving to higher latitudes and a few survived by living underground. In the oceans, a few species of shellfish, crabs and fish survived the extinction.

These conditions spelt doom for the aerobes, but for sulphur- and methane-loving microbes it was a time of revival. As we have seen in other periods of hardship (and will see in three more episodes before the present), adversity causes life forms to take up new designs—any design—that will help them survive. The handful of survivors—molluscs, fish, amphibians, reptiles, mammal-like reptiles—all came up with unique innovations that helped them to breathe in the suffocating air. Several molluscs and corals were replaced by others, and new types of ammonites and nautiluses emerged to take over different depths of the seas. Some reptiles too returned to the sea. But why return to water where it was more difficult to breathe? There is currently no definitive answer to this, but it has been seen again and again that in times of crisis, many animals leave land and return to water. One possible reason could be that although the greatest devastation during the Great Dying occurred in water, recovery too began along the margins of water, perhaps in small pools that were spared the effects of harmful gases. These pools were home to microbes and small invertebrates, a handful of ray-finned fish and sharks, as well as amphibians and reptiles. Also, within the piles of rotting plant debris and decaying vegetation were seeds of the gymnosperms (*gymnos*: naked, *-sperma*: seed) that took root and slowly resumed the greening of the land. Nearly 9 million years after the Great Dying, the climate was still hot and dry with only a few intermittent spurts of rain but, bit by bit, life began to painfully limp back. Over the next 1 million years or so, plants and plankton began to replenish oxygen in the atmosphere, which neutralized the greenhouse gases and started to mend the ozone hole. Earth began to heal itself and the persisting acrid smell of methane and hydrogen sulphide vanished from the air. On the largely arid land, after the near disappearance of *Glossopteris*, another fern called *Dicroidium* took over, while other newer varieties of ferns, gingkoes, palm-like cycads, conifers and horsetails began to jostle for space. These plants were better adapted to living in arid conditions because their leaves were coated with a thick waxy cuticle or were reduced to needles and spikes like most modern-day conifers. We know about how early plants struggled to revive land from the gas holocaust from plant fossils found in the Panchet formation of Jharkhand and West Bengal. The Panchet formation is exposed for about 8 kilometres downstream, along the banks of the Damodar river, and the village of Deoli (Lat: 23°36'N, Long: 86°51'E), in particular, offers exciting possibilities of seeing imprints of ferns and fossils of reptiles and mammal-like reptiles. It took nearly 15 million years before the forests regained their former glory. This is more than 200 times the time that our own species, *Homo sapiens*, has been in existence!

The entire story, that of the survivors who inherited the land and seas, especially the fish which transmogrified into amphibians and then became the earliest reptiles and reptile-like mammals, can be completely reconstructed through the fossils found in one region in India. To locate this region, open a good atlas with a physical map of India and draw an imaginary triangle with the apex just above Kolkata. Move your finger west, crossing the Damodar towards the Satpura Hills, down along the Godavari river, which flows east of Hyderabad, and trace this imaginary line back to Kolkata. You will find that this triangle is bisected by the Son and Mahanadi rivers. The area west of the Mahanadi where the mighty Godavari is joined by a smaller and lesser-known Pranhita river is of particular interest and is called the Pranhita-Godavari Valley (or the P-G Valley).

The Godavari is the second-longest river in India after the Ganga. It originates just 80 kilometres away from the Arabian Sea in the Western Ghats of central Maharashtra and then traverses 1465 kilometres across the breadth of the Indian peninsula to the Bay of Bengal in the east. The Pranhita-Godavari have been on course for over 300 million years, and these two rivers have created a wide valley with small, undulating hillocks which have the most complete geological sequences of the Gondwana.

The P-G Valley is truly remarkable because it holds within it the entire spectrum of vertebrates—from primitive fish and early amphibians to the first dinosaurs and early mammals. Recently, even hand tools made by early humans have been discovered in Yammanpalle village in Adilabad that lies within the valley. The valley itself is not continuous and is made up of five distinct formations that are exposed parallel to the rivers. We don't necessarily find formations placed one over the other. Often we find that one formation is exposed and is well preserved while others may be denuded at one place or completely absent; and it is rare to find all formations intact in a single place.

On a geological map, the formations of the P-G Valley appear like a patchwork quilt of different colours with irregular stripes, triangles and rhombuses. These stacked-up layers of rock and soil are rich in fossils and are named after villages where key discoveries were first made. The oldest layer among these is the 'Kundaram formation' which contains fossils of animals that became extinct during the Great Dying. It is one of the few places in the world where fossils of reptiles that existed *before* the rise of the dinosaurs have been found.

Reptiles evolved from amphibians into the first true land vertebrates and became central to the evolution of *all* vertebrates from there on. These pioneers gave rise to the ancestors of mammals, dinosaurs and birds. The first reptiles appeared around 310 million years ago and have existed in some form or another ever since, evolving continuously, with ancestors of snakes being the last to evolve about 180 million years ago. Most reptiles we see today emerged around 70 million years ago and constitute only a small fraction of the many diverse and bizarre kinds of reptiles that existed before them. There are roughly 6000 species of reptiles still living—twice as many as living species of mammals. And among reptiles, lizards dominate, just as beetles and ray-finned fish do in their respective classes.

Just above the Kundaram formation sits the 'Kamthi formation'. These rocks and sediments were formed within a narrow band of time (about 251 to 247 million years ago) and bear important fossils of the survivors of the Great Dying. There is an interesting story associated with the Kamthi formation. In the early months of 1968, palæontologists from the Geological Survey of India were digging around the villages of Adilabad district where the discovery of isolated bones held the promise of finding more complete fossils from the Triassic period (252 to 201 million years ago), the age which followed the Great Dying. But despite many days of digging under an oppressive sun, palæontologists found few fossils of any relevance or importance. As the team was about to give up and move on to another site, their local guide casually informed them that a few years ago, people from the nearby village of Kamthi had discovered two stones which they thought looked like a local goddess. He said that the villagers had consecrated these rocks and erected a small temple where they held an annual local fair. The scientists decided to visit the temple and were amazed to discover that the 'stones' were actually the beautifully preserved and complete fossils of the primitive reptile, *Lystrosaurus*. They were able to convince the villagers to give them one of their goddesses and with that the search for fossils in the quarries around Kamthi resumed with redoubled vigour!

Above the Kamthi formation lies the 'Yerrapalli formation' (247 to 237 million years ago) and the rocks here tell us that this area was once a lowland forest that was broken by streams and lakes, which created opportunities for the revival of fish and amphibians after the Great Dying. The soils reveal that there were distinct dry and wet seasons and the presence of three types of extinct lungfish (*Ceratodus*) suggests that dry seasons were long and that the rains fell in short, intense spells, much like in places where lungfish live today.

To get a sense of some of the most remarkable animals that evolved after the Great Dying, we need to visit the village of Maleri in Adilabad district in Telangana, and its adjoining areas. The Maleri formation holds fossils from 15 million years (237 to 222 million years ago) after the Great Dying and the relative ease with which one can find these in exposed beds suggests that the forests during this period had fully recovered. If you happen to be in this region, look out for a fine-grained whitish-coloured sandstone rich in calcium. Interspersed among these layers of rocks is a red-coloured rock made from soft clay called 'shale', within which it is common to find fossilized wood from these times. Farmers in the region around the formation can point you to strangely shaped rocks. These include fossil tree trunks and some odd-shaped rocks that are the skull or hips of the remarkable reptile *Lystrosaurus* whose fossils have been found in rock sequences from before and after the Great Dying, indicating that this reptile somehow survived the Great Dying.

Lystrosaurus was a cow-sized, podgy, stump-tailed, beaked reptile that waded in shallow lakes and streams, and fed on succulent plants. We do not know for certain why or how *Lystrosaurus* survived. Some scientists believe that a few parts of Earth remained less affected by methane, and that a handful of survivors were thus able to recolonize

Drama at dawn, 220 million years ago. A herd of cosmopolitan grazers, Lystrosaurus *(left) startle a small mammal-like reptile,* Thrinaxodon *(in the foreground, right) which has ambushed and killed an amphibian at the water's edge. In the background, a menacing reptile,* Chasmosaurus, *lurks in the still waters, ready to snap at an unwary visitor.*

the land once the forests returned. Herbivores had to adapt to changing vegetation during the Great Dying in order to survive and *Lystrosaurus*'s survival strategy included reduction in the size of its beak and the development of stronger jaws to chew the tough foliage of those times. *Lystrosaurus* was capable of traversing long distances in search of palatable vegetation and a suitable environment. They became abundant and dominated the unclaimed landscape immediately after the catastrophe and its fossils have been have been found on all Gondwana (except South America and Australia) and this (along with fossils of *Glossopteris*) is a reliable indicator that all southern continents were once conjoined in a single, giant land mass.

A few other herbivores had also emerged around this time. *Hyperodapedon* (meaning many 'basement-shaped teeth', also known as *Paradapedon*) was adapted to graze on succulent vegetation that

The gharial-like Parasuchus *used surprise to capture its prey. It attacked any creature that came to the water's edge like the herbivorous* Hyperodapedon, *which too was armed with sharp curved teeth and was capable of inflicting a painful bite.*

grew along lake shores. It was stocky, large (about 1 to 1.5 metres in length) with a barrel-shaped body and a broad head that ended in a narrow beak. A smaller herbivore, *Exaeretodon* (Greek for 'exceptional tooth' denoting its exaggerated molars), was a podgy animal about 1 metre long, with teeth that had a specialized grinding action to chew the tough vegetation of the relatively arid areas it lived in.

Hyperodapedon and *Exaeretodon* were mammal-like reptiles because they had deciduous teeth (teeth that are shed in infancy and are replaced by permanent teeth when the animals become juveniles), an important characteristic of mammals, including us humans. This meant that babies could not chew, and perhaps the mother chewed the cud for their infants and regurgitated the vegetation for them to eat.

The tenacity of mammal-like reptiles and other quadrupeds is an encouraging sign for those betting on mammals and their future,

especially from the perspective of design. Most mammals on land started to walk with a more erect posture unlike their ancestors who walked with a sprawl. The design of the shoulder, the pelvic girdle and the limbs of early animals was such that they had a sprawling gait, very much like that of most living amphibians and reptiles who walk (if at all) with their legs splayed wide apart. Physiologically, this is an inefficient 'design' because such animals can either breathe or run, but cannot do both at the same time. Having a more erect stance improves the functioning of the lungs and alleviates pressure on the abdominal cavity. Walking on two feet (the hindlimbs), which evolved in the earliest dinosaurs, made them very efficient runners.

The fossils within the Maleri formation show how reptiles were gradually becoming mammal-like. Another survivor of the Great Dying found in the Maleri formation was a carnivorous mammal-like reptile called *Thrinaxodon* (meaning three-pronged teeth). *Thrinaxodon* lived in burrows and made surprise attacks on small animals in the undergrowth. Around 246 million years ago, when methane slowly filled the skies, it imposed gradual physiological changes on some animals (a process that evolutionary scientists call 'selective pressure'). As the crises escalated, these physiological changes enabled animals like *Thrinaxodon* to cope better than others. This small mammal-like reptile enjoyed an innate advantage over large reptiles. Life underground had prepared *Thrinaxodon* to live in a low-oxygen environment. It had developed a bony secondary palate, a larger unobstructed airway powered by a muscular diaphragm, and fewer thoracic ribs which made breathing more efficient as oxygen levels decreased even further. *Thrinaxodon* was possibly the ancestor of the creature that eventually evolved into the earliest mammals. Its fossils have been found in rock layers formed before and just after the Great Dying, but it perished when early dinosaurs arose. Apart from the fossils of reptiles and mammal-like reptiles, the Maleri formation also has fossils of flat-headed amphibians (about 1–4 feet long) and fish that lived along the edges of lakes and streams. Another mammal-like reptile that is believed to be closely related to the ancestors of mammals was *Deccanodon*. It was discovered in the Maleri formation but not much is known about the behaviour of this aggressive, dog-sized predator which perhaps hunted small amphibians and reptiles.

Two important lessons emerge from the revival of life that took place after a mega-extinction event like the Great Dying. First, space (i.e. geographical distance) and not time is a more critical driving force for species to evolve. We see this during the emergence of modern birds and mammals after dinosaurs became extinct, but a more elegant example is that of us humans. We emerged from a stock of early humans who left Africa and spread across all the major land masses. Although humans are not divided into distinct species and we differ very little genetically, the physical differences that we see among the various *races* of humans is mainly due to the separate places that we spread to and lived in. It is space, not time alone, that has given us these different attributes. And these minor differences amount only to differences of race, not species. Had we been isolated and evolved separately, it is possible that we would have evolved into different species.

The second important lesson is that communities of animals remain stable until there is a drastic change in the composition of plant species. Plants are an important evolutionary force which shape how, and which, animals will survive. Ever since the tetrapods (the four-limbed vertebrates higher than the fish) evolved they have been remarkably stable for long periods of time and have changed only when there has been a revolution in the flora. For instance, the appearance of flowering plants and different types of fruits around 90 million years ago (although the ancestors of the flowering plants had emerged 50 to 30 million years before this) led to an explosion of insects that in turn led to the emergence of diverse mammals that would feed on fruits, insects and eventually the flesh of other animals. The destruction of the coal forests 251 million years ago, the Great Dying, and the subsequent revival are all examples of such a floral turnover. The Maleri beds, where some of the best evidence of the advent of the ancestors of modern vertebrates can be seen, show that once new plants emerged, they led to the revival of fauna.

Around 242 million years ago, the Tethys Sea began to encroach inland. Water came to areas where earlier there had been none, and although this gave opportunities for species to colonize new areas, it also posed a hindrance for some. Plants and animals had so far remained largely cosmopolitan—species were widespread and broadly similar across all lands—but when they became isolated by barriers like seas, mountain ranges or vast deserts, they evolved differently in their new settings. The extinction of their competitors, predators or prey created new conditions for these species to evolve. Gradually, new types of reptiles, reptile-like mammals and the early ancestors of mammals emerged from the shadows of the giant herbivores. Among the new reptiles was an early crocodilian that lived on land, was more cunning and better-armed than its contemporaries, and over the next 1 million years it proved to be the nemesis for *Lystrosaurus* and *Thrinaxodon*.

Between 235 and 210 million years ago, land crocodiles entered the water and became its unchallenged overlords. *Proterosuchus* (*protero*: before, *-suchus*: early crocodile) was a crocodilian ancestor with a pronounced overbite which ensured that its prey could never escape. It was nearly 2 metres long and fed on fish, amphibians and small reptiles A *Proterosuchus* fossil was discovered beside the Raniganj coal seam near Dhanbad which, interestingly, also had a foot-long fish alongside it. Crocodiles were perhaps among the first reptiles that could regulate their body temperature and are believed to have been active at all times of the day and night.

There is an interesting anecdote about yet another species of crocodile. In the winter of 1965–66, Sankar Chatterjee, an intrepid and prolific palæontologist, found two near-complete skeletons of an early crocodile called *Parasuchus hislopi* (meaning 'alongside the crocodiles', the specific name *hislopi* honoured a pioneering British palæontologist in India, Reverend Stephen Hislop) in the village of Mutapuram in Adilabad district within the Maleri formation. Inside the ribcage of both crocodiles, Chatterjee noticed the bones of another small reptile that he called *Malerisaurus robinsonae* (Maleri's lizard; the specific name honoured the pioneering work of palæontologist Pamela Lamplugh Robinson). *Parasuchus* had long and slender snouts and were more than 2 metres long, about as long as living gharials. Their hides were tough and covered with sharp triangular ridges (called 'scutes'). They had sharp conical teeth and, like crocodiles, had 25° binocular vision (we know this from studying the placement of their eye sockets in their skulls) which gave them good judgement of depth and distance in water and above it, and would certainly have helped them make a precise strike. These animals were cunning ambush hunters; they would lie still in water or within dense riverine vegetation, and lunge at unsuspecting prey. The two *Parasuchus* fossils of similar size, so close to each other, were called 'the loving couple', and although it is difficult to presume that they hunted in pairs, it is almost certain that they had shared a similar last meal. While removing the soft clay from the fossil, the team discovered that apart from *Malerisaurus*, 'the loving couple' had also feasted on a smaller crocodile. The team concluded that one of the preys was poisonous, but it is not clear which one. These amazing fossils are stored in the repository of the Indian Statistical Institute in Kolkata, but, strangely, have never been shown to the public.

Around 205 million years ago, when dinosaurs began to dominate, the land crocodile retreated and took to water. From these creatures emerged the ancestors of the crocodiles that live today. They developed a secondary palate which enabled them to bite and chew underwater without choking. They also lost their more upright gait which was less useful in water. The early dinosaurs may have outcompeted the crocs on land, but on the water's edge crocodiles terrorized dinosaurs, reptiles and sharks, and their existence today is proof of their enduring success. The deep seas were ruled by the sharks, and although their dominance was challenged briefly by marine reptiles, which emerged 225 million years ago and died out along with the dinosaurs 65 million years ago, they continue to rule rivers, lakes and some open seas.

In the background of these struggles, the reptilian ancestors of dinosaurs and mammals were biding their time. The earliest dinosaurs appeared on Earth around 235 million years ago. They were small, agile and bipedal carnivores that quickly evolved new feeding styles. The other character in the 'loving couple' story, the one that had been eaten, *Malerisaurus*, was a small, long-necked lizard-like reptile with front limbs that were smaller than its hind legs, and a long tail. This suggests that this reptile could take rapid strides while balancing itself on its hind legs (called 'bipedal running') and was capable of climbing trees or taking to water when alarmed. *Malerisaurus* lived along streams and lakes and its teeth suggest that it ate snails, small crabs and insects. *Malerisaurus* was *almost* a dinosaur, but not quite, because it lacked a few critical features in its skull and skeleton.

In 1974, Sankar Chatterjee returned to the P-G Valley and in the red mudstone of the Maleri bed he discovered another reptile with dinosaur-like features. Intrigued, he painstakingly analysed the fossil. The front teeth of this reptile were slender and straight, while the teeth in the sides of its jaw curved backwards like those of the predatory reptilian ancestors of the dinosaurs, although none of these teeth were serrated. This suggested that it was an omnivore with a varied diet, that included insects, small vertebrates and even plants. The skull and backbone appeared to be those of a primitive reptile while the femur (thigh bone) and the ankle bones were clearly those of a saurischian (or lizard-hipped) dinosaur. Chatterjee originally named this dinosaur *Walkeria maleriensis* in 1987, in honour of British palæontologist Alick Walker, a comparative anatomist who developed innovative methods for casting fossils and whose specialization was extinct crocodiles. However, this generic name was allotted to a bryozoan (a type of tiny aquatic invertebrate) which necessitated a new name. In 1994, Chatterjee renamed it with an abbreviated version of Alick's name—*Alwalkeria*.

Although small, Malerisaurus, *which had the ability to walk and run on its rear legs, was a skilful hunter that preyed on smaller reptiles, amphibians and fish.*

Alwalkeria is one of the earliest known dinosaurs. It was well adapted as a fast-running carnivore, with sharp teeth and grasping claws on its forelimbs. *Alwalkeria* was a little less than 1.5 metres long (the tail alone was three-fourths of a metre) and weighed just 3 kilograms. The diet of this reptile consisted of insects and small amphibians but it could occasionally scavenge on scraps that washed ashore on lakes. *Alwalkeria* appeared on the scene around 238 million years ago and became extinct 208 million years ago when other dinosaurs and reptiles began evolving.

Other candidates have been proposed from across the world as ancestors of the dinosaurs. *Eoraptor* (meaning 'dawn hunter') was a carnivore from Argentina that appeared on Earth around 228 million years ago and is regarded among the first 'true' dinosaurs, similar in age to *Alwalkeria*. Could these have originated from a common ancestor or did they evolve separately, in isolation?

These early dinosaurs were built for speed and agility. Their bones were light yet strong, and they were upright with a tail that aided them in balance and defence. Unlike the more sluggish reptiles, they

were highly energetic. The skull of a dinosaur (like *Alwalkeria*) had an extra opening that was absent in reptiles (like *Malerisaurus*) and their hands had specialized bones in their wrists and ankles which helped them grasp better. But the critical feature that distinguished dinosaurs

from reptiles was their hip. Dinosaurs had five fused sacral vertebrae (the sacrum is a triangular bone at the base of the spine, therefore they had five fused bones in their lower back) which formed a strong hip. This accommodated a deep socket for the rounded head of the thigh bone, which gave dinosaurs their upright posture, like a French dandy, with the ability to crouch low when hunting or to stay erect when on the lookout. More importantly, an erect posture gave them (and *Alwalkeria*) a more efficient means to breathe, and also freed their hands, which in the future would be critical in the evolution of diverse dinosaurs and birds.

The earliest dinosaurs, mammal-like reptiles and true reptiles coexisted for about 10 million years. With better 'design', the first dinosaurs adapted well to these still uncertain times. Oxygen levels fell from 15 per cent to below 12 per cent and perhaps even lower, the lowest ever in the past 500 million years. The earliest dinosaurs outcompeted reptiles and mammal-like reptiles on land, first as carnivores and later as herbivores. Most reptiles shrank in size. Some, like the land crocodiles, took to water and others scrambled up trees or burrowed into the ground. Mammal-like reptiles had begun to shed their reptilian traits and transform into true mammals, and take refuge under the ground and in the undergrowth. The earliest mammalian fossils from India (*Gondwanadon* and *Tikitherium*) were discovered around Tiki (Lat: 23°55'N, Long: 81°21'E) in Shahdol district of eastern Madhya Pradesh, and two neighbouring villages, Simaur (Lat: 23°54'N, Long: 81°20'E), and Jora (Lat: 23°54'N, Long: 81°21'E). Tiki lies just north of the P-G Valley in the Son-Mahanadi basin. The Tiki formation is dated between 225 and 190 million years ago (similar to Maleri) and holds within it fossils of mammal-like reptiles and reptiles, but is

An artist's impression of Alwalkeria, *the earliest known dinosaur from India. This lightly built, fast-moving creature could track and hunt down small prey with precision.*

especially rich in fossils of early mammals. Several fossils that have been found here suggest that early mammals were small, no more than about a foot long, and the fierce and menacing dinosaurs ensured that they stayed small. These mammals were ground-dwelling or burrowing (and therefore short-haired); they hunted insects and small lizards, and ate succulent roots and shoots.

It is clear that changing amounts of oxygen in the atmosphere have played an important role in the evolution of life forms on Earth. In the past 600 million years, oxygen levels have risen and dropped several times, and each time it has led to the emergence of life forms which are better able to cope with changing conditions—and the demise of those that aren't. How do we know this? It is not as difficult as it may sound.

Scientists are able to measure oxygen in rock and amber. Rocks that are rich in ores, plant tissue and teeth are good places to begin. Palæontologists have studied gas bubbles trapped in amber secreted by trees and have found that the period when the dinosaurs reigned supreme was also a time when the planet had a very high concentration of oxygen (nearly 33 per cent compared to the current oxygen level of 20.9 per cent measured at sea level). This partly explains why dinosaurs grew so large. Carbon dioxide and oxygen share an inverse relationship: more carbon dioxide in the air or water means a drop in oxygen concentration. In such low-oxygen, high-carbon dioxide scenarios, more heat gets trapped, thus raising the temperature which impacts higher life forms adversely. Animals must adapt (by reducing in size, for instance) or die.

Recently, scientists have begun piecing data from charcoal to calibrate oxygen levels of the past. Charcoal is formed under natural conditions when wood burns intensely in the absence of oxygen. Charcoal is almost pure carbon and provides information about past life on land. Studies using charcoal data indicate that about 400 to 390

Gondwanadon, *one of the earliest mammals discovered in India, perhaps resembled its distant Welsh cousin,* Morganucodon *(above), and lived in burrows or tree holes. Mammals like* Morganucodon *and* Gondwanadon *lived about 210 million years ago and were probably among the earliest common ancestors of all mammals. These small mammals (12 centimetres or so long) passed their DNA to billions of their descendants including you and me.*

million years ago, atmospheric oxygen levels were very low, well below modern levels. Oxygen levels rose rapidly and peaked at about 29 per cent between 300 and 270 million years ago creating conditions for the massive Permian forests to flourish, and remained relatively high until 255 million years ago. A pronounced collapse in atmospheric oxygen concentration occurred following the Great Dying (254 to 251 million years ago) but this decline appears to have followed, rather than driven, the process of extinction. Around 201 million years ago, large cyclical fluctuations in oxygen concentration occurred which saw the rise and diversification of the dinosaurs. Oxygen levels spiked about 190, and then around 100 million years ago, making some dinosaurs truly enormous. From this point onwards, oxygen in the air declined until it reached relative stability at about 21-22 per cent over the next 40 million years. Relatively high concentrations of atmospheric oxygen on Earth strongly impacted the evolution of life and triggered the arrival of vertebrates on land, gigantism among insects at first and then in sauropods, the evolution of placental mammals, and powered flight in both birds and insects.

However rocks do not record spikes in carbon dioxide levels and therefore measuring them is more complex. Plants are highly sensitive to CO_2 levels, and scientists use them to estimate the levels of CO_2 present at various times in Earth's history. Plants tend to favour an environment that is higher in carbon dioxide and lower in oxygen, although in the long-run this can be fatal for plants. In theory, if plants were to utilize all the CO_2 in the atmosphere and none was left to process, photosynthesis would shut down. But in reality this does not happen, simply because when plants die they decompose and release CO_2, which can be used by surviving plants to resume photosynthesis. Scientists have found that plants quickly adapt to fluctuating levels of CO_2 by increasing or decreasing the number of stomata. Stomata (Greek for 'little mouth') are two bean-shaped cells on leaves that act like sphincters to control the movement of gases and vapours into and out of a leaf. They open to exhale water vapour and gases (called transpiration) and close to prevent loss of water during times of water stress. When carbon dioxide levels are high, stomata are fewer and when the levels are low, stomata tend to increase in number and are stimulated to open wider in order to photosynthesize efficiently. The stomata of most species open in daylight and close in the dark, but the opposite happens for plants that grow in arid regions. Here, stomata remain closed during the day to reduce loss of moisture (from evapotranspiration) and open at night to absorb carbon dioxide for use in photosynthesis. Stomatal density varies between plant species and between the underside and top side of the leaves on a plant. You can monitor the increase in carbon dioxide in your locality over the last 50, 100 or 200 years if you have access to an old enough collection of leaves in a herbarium. There are over eighty herbaria in India that have collected and preserved a variety of plant leaves. Once you have identified a plant species from your garden and you have been able to convince the manager of the herbarium (this may be difficult but there is no harm in trying!) to loan you a tiny section of leaf (make sure they are the same species), all you then need is a simple microscope. Prepare an impression of the leaves (both the old and freshly plucked) by coating the leaf surface with a thin layer of nail varnish. Gently peel off the dried layer of nail varnish by using Sellotape and stick this on to a slide. When you view this slide under the microscope you will spot tiles of cells interspersed with pores that are lined by two bean-shaped cells. These are the stomata. Make five counts for the number of stomata on the surface of both the old leaf and the freshly plucked one and average it out for both the leaves. This estimate will allow you to compare the number of stomata that existed in two different time periods and will tell you about the difference in the levels of carbon dioxide in the two periods. Even if you have access to leaf specimens that are twenty or thirty years old, you will still be able do this analysis, but you will need to be very rigorous while counting the number of stomata as the difference in numbers may be small yet significant. A recent study has found that the density of stomatal pores has reduced by 34 per cent in some tropical species as carbon dioxide levels have increased in the last 150 years. One impact of this, among others, is that plants are releasing less water vapour into the atmosphere than before—a sign that the planet is warming.

For nearly 3 million years after the Great Dying, the air kept an asphyxiating hold on the few survivors of the cataclysmic event. What was remarkable about these survivors was that they gave us every single phylum of the animal kingdom which founded all modern families alive today. Our Earth Woman was a little over 44 years old, and over the next two years or so of her life she would witness the evolution of every remarkable animal species alive today.

The P-G Valley contains fossils of the ancestors of almost all vertebrates that exist today, and each layer of rock formation tells the story of how life recovered on land and water after its near obliteration, and the fortuitous events that led to the recovery and revival and created new species—a process that has occurred countless times in the life of Earth. However, much of the land in this hot and arid region, which is often inundated by floods, is farmed, quarried and mined and home to large cement factories, stone quarries, brick kilns and railway lines which cut through the rocks. It is possible that within this area, there may still be fossils of animals and plants that are completely new to science and that could shed new light on the evolution of life. For now, the fossils found here suggest that the fag end of the 50 million years after the Great Dying (250 to 205 million years ago) favoured the take-over of land by a family of highly adaptive and more agile creatures who became the most powerful dynasty ever to rule land—the dinosaurs.

5

THE MAKING OF A DYNASTY

The surface of Earth is an extremely volatile place, constantly moving and shifting, although this is not perceptible to us except during an earthquake. If you were to compress the past 3 billion years or so of our planet's life into a minute-long video, land and water will be seen moving around as hazy blurs of blue and brown and a few seconds of white in between when ice covered parts of the Earth. If you were to make this film five minutes long, it is possible that you may not even spot India until about twenty seconds before the end credits roll. This is because the oceans and lands have changed their shapes several times over these 3 billion years. So much so that even the poles have not always been where they are today. In physical terms, as of December 2015, Earth's magnetic North Pole is in the Canadian Arctic (Lat: 86°29'N, Long: 160°06'W),

Not many places can match the drama of witnessing the colliding or parting of two tectonic plates. One place where you can actually see two plates come face-to-face is in the Thingvellir National Park in southwestern Iceland. Here the Eurasian and the North American tectonic plates pull Iceland apart, by about 2 centimetres a year on average, creating this ravine that cuts across land for over 200 kilometres, dividing even the tops of mountains and extending as a chasm under the clear, icy waters before it disappears into a deep trench in the North Atlantic Ocean. The North American Plate lies to the left and the Eurasian Plate is to the right, and in between is the place where new land is being born, making this a virtual geological no man's land! Thingvellir is literally the progenitor of democracy: it is the place where the first representative Parliament for any country in the world was established in 930 CE. For enthusiasts of more contemporary culture, this rugged and desolate scenery forms the backdrop of the TV series, Game of Thrones*!*

yet the magnetic South is not diametrically opposite to it. Instead, it is in Wilkes Land (Lat: 64°28'S, Long: 136°59'E) in the Antarctic Circle. The magnetic poles shift several kilometres every year and the maritime and aviation industries have to tediously calibrate their navigation instruments each time this happens. Scientists call this changing of the positions of the poles 'polar wandering'. The equator too was not always where it is today. The size, position and shape of continents and seas have been transient. In fact, as we have seen for a majority of the life of our 46-year-old Earth Woman—from the time she was 10 years old till she was a little over 44—all of the continental land masses on Earth were lumped together, separating intermittently for short periods but then joining up again to form supercontinents. At different times between 3.6 billion years and 250 million years ago, the continents joined together to form a single land mass, and Earth scientists have labelled them as Ur (3.6 to 3.1 billion years ago), Rodinia (1.1 billion years ago to 750 million years ago) and Pangæa (300 to 250 million years ago). It was as if nature was playing an accordion with the land masses, drawing them together and then separating them again. The last of these supercontinents, Pangæa, began to split more or less into two halves around 250 million years ago to form east and west Gondwana, and 167 million years ago, the two parts had completely separated, with Africa and South America in the west and the rest of Gondwana (India, Madagascar, Australia and Antarctica) in the east. To the north was Laurasia which comprised North America, Europe and Asia. So it is only in the last two and a half years of our Earth Woman's life that the land masses began to resemble the separate continents in the shapes that we recognize today, although some of these were still attached to each other like pieces of a jigsaw.

So why do continents move? The answer to this came as recently as 1965, in a paper published in the journal *Nature* by Canadian geologist J. Tuzo Wilson, who proposed that the surface of Earth (including oceans) is made of rigid plates that are fitted *together* to form the planet's crust. There are twelve large plates and twenty or so minor ones that make up Earth's surface. Beneath the crust lies the weak and ductile layer called the 'asthenosphere' (Greek for 'weak' or 'fluid sphere') which floats over a viscous blob of magma made of molten iron and nickel. The plates move laterally over these less rigid rocks and are driven by the wobbly, viscous layer below them. The continents and the ocean floor move together, not separately. Continents move and seas expand because the land under the sea (the sea floor) is separating. So the process that takes place is this—hot magma from below the mantle tries to force itself upwards and searches for a weak point in Earth's crust. Deep sea trenches on the sea floor are the weakest points on the crust, made up as they are of a thin layer of rock and water above it, and therefore offer minimal resistance to magma which surges upwards as pressure builds from inside Earth. The heat of the magma melts the thin crust from below and lava begins to pour out on to the ocean floor, melting the area around it. Depending on the intensity of the outpouring of the lava, it can create new sea floor and in doing so push apart the land under the sea. This sea-floor spreading creates wider space for oceans to expand, with new sea floor forming in the middle. The separating sea floor behaves like a giant escalator or a baggage carousel with new land being added between two moving plates. When new sea floor forms at the lagging end of a plate, old sea floor which is at the front end of the plate dives into the depth of the oceans' trenches. These tracts of land (mostly rock) are transported to about 600 kilometres deep into Earth where under pressure and heat they melt into a grey-white, violently bubbling molten fluid. Once this hot mass builds up and the space below is no longer able to contain it, it ascends upwards and oozes out as lava. Other geologists after Wilson presented new data which showed that it is at the boundaries of the plates that Earth's inner energy is released. The plate boundaries are like stitches along the seam of a fat mattress, and when people sit or lie upon it, the mattress accommodates the pressure by bulging and expanding at one place, and flattening elsewhere. Plates also move by a pull exerted by a cold, dense crust and upper mantle which buckles deep into the mantle at subduction zones, and not just by the push from the hot, upwelling magma of a plume from below. As a consequence of landmasses moving away from one another, volcanic eruptions and magma are likely to occur.

The bigger plates are named after the major countries that straddle them—India, Australia, Africa and the like. The smaller ones have a more exotic air to them—Caroline, Cocos and Nazca to name a few. The largest tectonic plate is the Pacific Plate, which is the only large plate that does

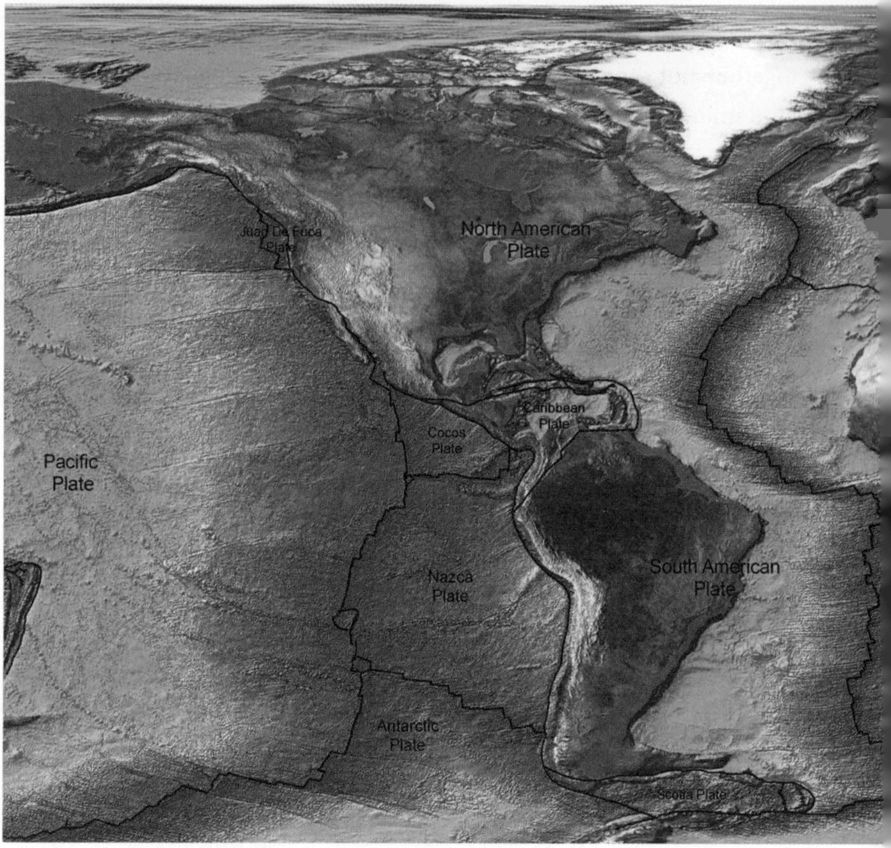

not carry a continent. No major plate straddles an entire continent or covers an entire ocean. These demarcations of countries and continents are made by history and politics rather than geology, geography or even biology. Most volcanoes that exist in the world today lie along ocean trenches, and some jut out to make volcanic islands or archipelagos. Still fewer active volcanoes lie in the heart of a continent; these occur most commonly in South and North America. The Yellowstone National Park, for example, sits over a massive subterranean volcano and is an imminent threat to all of North America.

Earthquakes occur as a result of the rubbing of two boundary plates, but also when massive tracts of land either sink (the effect may be felt on land if the trench is close to land as in Chile, Indonesia, Japan and the Philippines or as a tsunami across seas) or are raised up due to the action of erupting molten rock from below. When a trench runs along the

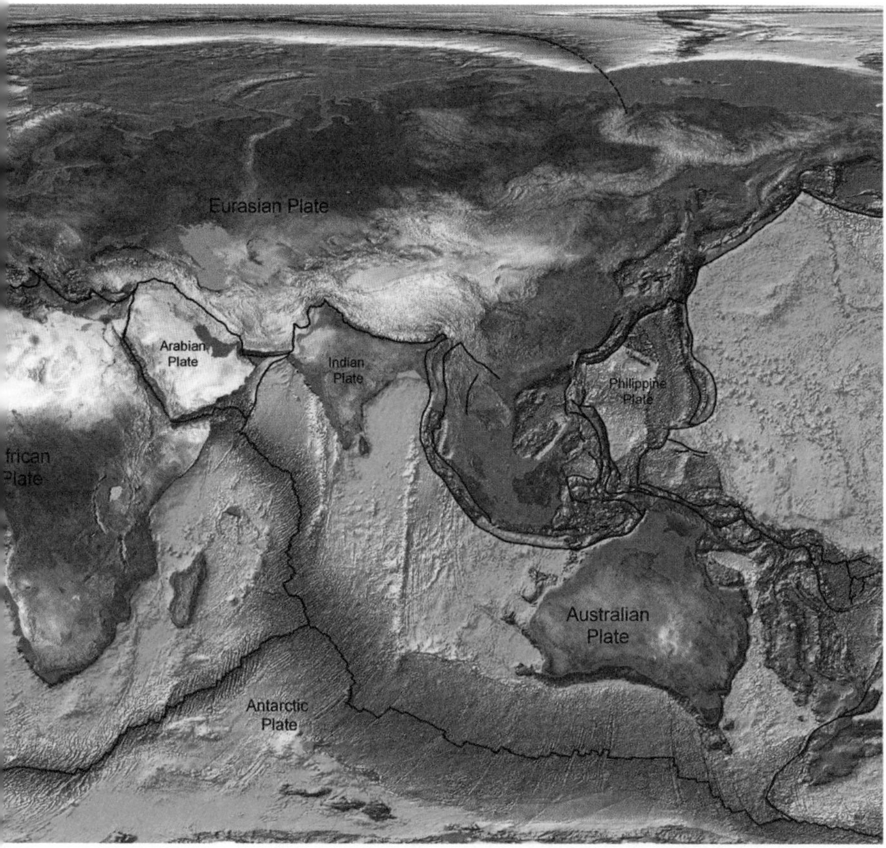

Plates move when they are pulled back into the mantle by a subducting slab or pushed laterally by a ridge system. In this figure, the black lines mark the plate boundaries and the red areas are where sea-floor spreading is actively taking place. The red lines which mark the plate boundaries are places where earthquakes occur most frequently, perhaps they experience small tremors as often as several times a day.

edge of a continent, the sea-floor buckles under the land (called 'subduction'). The combination of sea-floor spreading and subduction means that there are no gaps between plates. To accommodate the mass that has gone under, the continental margin rises up, pushing up mountains. Often these mountain ranges are interspersed with volcanoes. The trench that runs along the coast of Chile is a good example of this. The Nazca Plate which moves eastwards towards the east coast of South America consumes the margin of the continental shelf. The result is the 6000-kilometre-long Andes range, with snow-covered peaks and volcanoes and an average elevation of 4000 metres. Plates therefore grow, shrink, merge, and small plates may even get obliterated over time. The Cocos and Nazca plates are relatively young. Both these plates are remnants of the former Farallon Plate, which broke up about 23 million years ago and at its juncture, a hotspot from below rose up to the surface creating the Galapagos Islands.

A land mass is simply that part of the plate that is raised out of water. Some plates are entirely overland, like the Arabian and Iranian plates. The outer margin (or boundary) of a plate can also be on land where two plates collide and get raised up (like the Himalayan range) which marks the northern edge of the Indian Plate. Where plates meet is a place of geological curiosity and the greatest seismic concern. If one plate invades another, it deforms the boundary of the next, which has cataclysmic repercussions. Plates settle and move in a fixed direction in relation to their neighbours. If you want to know how plates behave, observe passengers who travel on a Mumbai local train during rush hour. Mumbaikars are particularly accommodating and more people sit per seat here than anywhere I know. Daily commuters have an unspoken understanding among themselves and through a gentle gesture of the palm, they request the person occupying a seat to shift slightly and give the seeker a little space to rest his rear. As the journey begins in fits and starts, the seat seeker swivels a little and nudges his neighbour in the hope of finding more room or coming in the path of a waft of breeze from a rickety overhead fan. If his neighbour is compliant, he obliges and shifts; if not, he reacts with firm pressure, defending his claim. Plates behave much like these passengers who have settled in and move in a fixed direction. The Indian Plate, for example, moves north (the Nepal earthquake of April 2015 is evidence of this) and occasionally gets pushed west (like during the 26 December 2004 tsunami which had its epicentre off the coast of Sumatra). To our west, the African Plate is pushing northwards into Europe, and has been responsible for raising the Alps. South of the Indian Plate is the Antarctic Plate which spins slowly from east to west and slightly to the north, like a gently flighted leg break. This constant, fidgety nature of plates prevents them from fusing into one another and becoming immobile. Should this fidgeting not happen, Earth would perhaps become like one of the other planets and die a slow death. The energy that moves plates and creates new land is a manifestation of Earth's internal energy. This energy has set a cycle of destruction and creation of land and oceans over millions of years, which has resulted in climates and environments, and enabled the evolution of life as we know it.

Over billions of years, there has been so much churning of land that many of the continents today are actually mosaics of different land masses. North America rests almost entirely on bedrock which is of Laurentian origin, but for one tiny part of southern Florida and a sliver of eastern Boston which overlie rocks that are Gondwanan in origin. How do we know this? The zircon and basalt that make the bedrock deep beneath Florida match that found in West Africa. About 200 million years ago when Laurasia migrated north, a portion of eastern Gondwana tagged along with it. Over the next 120 million years, as the Atlantic Ocean began to spread, it forced this Gondwanan piece to merge with the land mass which made Florida. The bedrock under Nova Scotia and Newfoundland is the same as that in Norway, parts of Ireland and the Scottish Highlands, and these are all Laurentian in origin. Iceland is split right down the middle—half North American and half Eurasian plates. The political outline of India also encloses within it elements of other plates and land masses. There is a very small portion of the Burmese Plate in north-eastern India and also in the Andaman Islands. The Eurasian Plate crops up in Arunachal Pradesh, east of Leh, north Sikkim and northern Nepal. The Iranian Plate underlies the north and western parts of Pakistan.

Throughout Earth's history, land masses have congealed together to make a supercontinent which subsequently broke away, each episode roughly about 500 million years apart. Most often, continents break along pre-existing lines of weakness which were created during earlier continental collisions. Scientists call the clumping of all continents into one big land mass the 'supercontinent cycle'. We are witnessing a super slow-motion version of this cycle of collision even now. Africa, Arabia and Australia are slowly pressing into Eurasia. Scientists predict that in a couple of hundred million years, barring the Americas and Antarctica, most of these land masses will fuse together again.

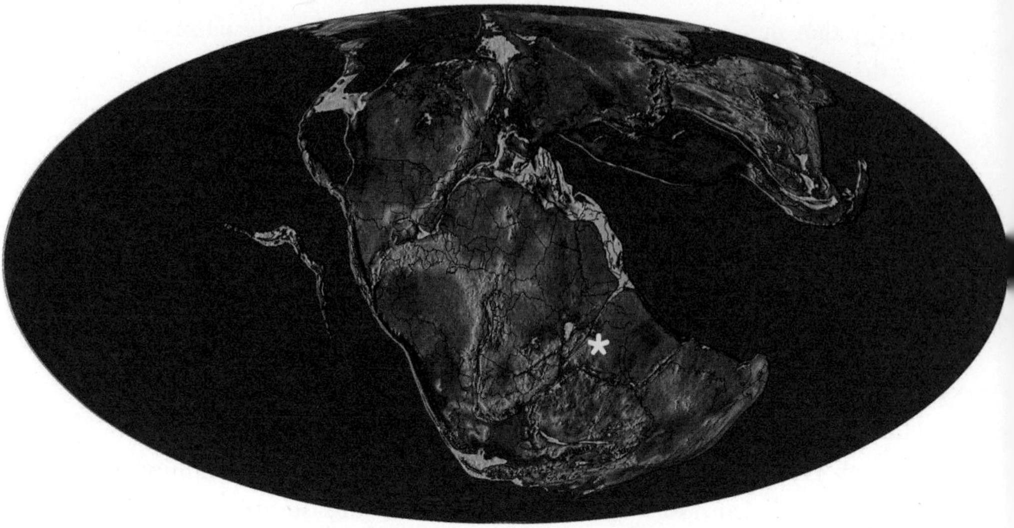

Earth, 200 million years ago. Laurasia had already separated from Gondwana. India (the yellow asterisk) lay nestled alongside Australia, Africa and Antarctica. Between 170 and 167 million years ago, Gondwana broke up into two parts forming East Gondwana (made up of Antarctica, Australia, New Zealand, Madagascar, Seychelles, India and Sri Lanka) and West Gondwana (consisting of Africa, Arabia and South America). An ocean began to open up between these two land masses which later became the Indian Ocean.

We know that around 220 million years ago the vast unbroken supercontinent of Pangæa straddled the equator and was shaped like a thick C or like Pac-Man with the Tethys Sea entering its gaping mouth. During this time, the polar region had no ice, and the ocean currents in the massive Panthalassa Ocean created monsoon forests along the margins where land met sea. But the centre of Pangæa was largely arid, and for another 15 million years the supercontinent had forests only in its coastal regions. A uniform world led to uniform animals, and this is one reason why the early dinosaurs lacked variety. As long as the supercontinent existed, there was very little change. But when change happened, it was quick (in geological terms) and dramatic!

Pangæa began to rotate anti-clockwise and gradually different parts of this land mass rotated in different directions and at different rates. This twisted and contorted Pangæa, and rift valleys (narrow valleys with steep slopes and a flat floor that appear in a straight line) ruptured the land masses. By 201 million years ago the land masses had split wide open. And by 190 million years ago, great rifts separated Antarctica from the southern ends of South America and Africa, developing an arm that extended eastwards from South Africa. This restless period when the continents were on the move is called the 'JURASSIC period' (named after the mountain range of Jura, outside Geneva in Switzerland).

When land masses drift apart and new lands are created, it increases the rate of the weathering of rocks, and this in turn increases the absorption of carbon dioxide on land. Dinosaurs first emerged during the time when oxygen was at its lowest in the last 500 million years or so. For nearly 20 million years they stayed small, adaptive and agile. But times began to change, led by a rapid recovery of plants which began regaining the glory not seen since the Permian coal forests and within 7 million years much of the land became covered by forests with tall and diverse species of conifers, cycads and ferns and the amount of oxygen in the air increased manifold.

During the oxygen crisis, some reptiles had returned to the seas and they became top predators there. These marine reptiles were spindle shaped and many of them became fish-shaped, the most efficient and popular design to succeed in water. Dominant among these were fast-swimming giants called ichthyosaurs (fish-lizards). Ichthyosaurs had the ability to dive deep and chase small fish in groups, much like dolphins do today. Their fossil remains have been found extensively in the Americas, Europe, Africa and China which were partly under deep seas at the time. Until recently only a single vertebra of an ichthyosaurs was discovered in Tiruchirappalli near Chennai. Palæontologists had concluded that because India had remained landlocked or covered by shallow seas between 550 and 170 million years ago such habitats would not be favoured by large, deep-diving reptiles. One group of palæontologists concluded that the marine reptile perhaps reached here when Greater India was beginning to separate from Australia and Antarctica. In 2015, a near complete fossil of an ichthyosaur was discovered by paleontologists at Rajasthan University, Japiur. This marine reptile shows affinities with ichthyosaurs that lived along the western fringe of the Tethys, where Germany, England and France are today.

On land, dinosaurs adapted faster than their competitors, the land crocodiles and other reptiles. Mammal-like reptiles *Thrinaxodon* were extinct by this time and true mammals had emerged but were still too small and insignificant to claim dominance. Around 237 million years ago, dinosaurs began their domination on land. The increase in the amount of vegetation and the number of insects and small animals along with the abundance of oxygen led to smaller dinosaurs growing very large.

The story of how dinosaurs transformed from small, agile creatures to animals of monstrous proportions can be pieced together from fossils found in the rocks of the Pranhita-Godavari (P-G) Valley between the villages of Dharmaram (Lat: 18°12'N, Long: 78°49'E) in Medak district of Telangana, and Maleri and its surroundings. The lower layers of the Maleri formation, dated to be about 237 to 208 million years old, yielded fossils of large amphibians and one of the first known dinosaurs (*Alwalkeria*). Close to Dharmaram and Krishnapur (which is about 20 kilometres south-east of Dharmaram), scientists found evidence of more dinosaurs as they explored younger rock sequences. In 2007, two new, mid-sized dinosaurs were discovered in sediments in the Dharmaram formation, from layers estimated to be 206 to 196 million years old. The bones belonged to a type of early herbivorous dinosaur called 'sauropod'. Early sauropods were car-sized with a small head, a long neck and tail and five-toed limbs (its five toes resembled those of a lizard, hence its name, *sauro*: lizard, *-pod*: feet). In the next 10 million years, some sauropods evolved to become the largest known land animals ever. One medium-sized dinosaur called *Lamplughsaura* (pronounced lam-plo-saura, named in honour of the eminent British palæontologist, Pamela Lamplugh Robinson) measured about 10 metres in length. It lived alongside another sauropod called *Pradhania* (from Sanskrit, *pradhan*: lord or leader). *Pradhania* was a small and slender sauropod (about 3.5 metres in length) and is less completely known.

Not all animals grew large. Some reptiles became smaller and, in order to escape their predators, they became airborne. No vertebrate animal so far had taken to the air and it was the reptiles who took this leap of faith. The first flight in air was accomplished with the evolution of flying reptiles called pterosaur (meaning 'winged reptile'). *Campylognathus* (meaning 'upward hooked' or 'curved jaw'), a pterosaur found in India, patrolled the beaches and coasts along the shallow

The fossil of the only flying reptile (or pterosaur) discovered in India called Campylognathus *suggests that it would have looked a bit like this. Pterosaurs evolved between 245 and 205 million years ago and were the first vertebrates to undertake powered flight instead of gliding. Over the next 180 million years or so, at least 130 species of these magnificent fliers evolved, some as large as giraffes and a few as small as sparrows. Almost each of these acquired one other talent apart from flying—some were skilled waders, others could walk over long distances, some were runners, some could dive-bomb and a few were adept swimmers. This enhanced their ability to catch prey, thus fuelling their active lifestyles and allowing these flying reptiles to thrive in diverse habitats across the globe.*

seas that extended from Rajasthan through Jharkhand till Meghalaya and along the beaches of the Godavari and Krishna seaway, hunting for fish. A fossil of this remarkable flying reptile that lived from 191 to 182 million years ago was found on a stone embedded in a house in the village of Sirolkhal in Kota district of Rajasthan, and another in Chandrapur district of Maharashtra.

Similar fossils have been found in Germany (much of Europe was a shallow sea at this time). These fliers had a small spindle-shaped body and their outstretched wings, supported by thin bones, extended to about 2 metres. They had long tails (about 1 foot long) that ended in a diamond-shaped fin that was used for steering in flight. When they flew they probably looked like large paper kites in the sky! Their skulls had a pointed snout and large eye sockets, suggesting that they had large eyes. Like present-day birds of prey they probably had keen eyesight and possibly even the ability to hunt at dawn and on full moon nights. Their beaks were lined with sharp conical teeth that were used to hold on to fish when they scooped them up from the surface of shallow seas. They did not dive like kingfishers because they would not have been able to lift off again with wet wings.

Pterosaurs like *Campylognathus* began to die out around 174 million years ago and in their place more efficient reptilian fliers and the earliest birds began to take to the skies. Some pterosaurs became enormous, like *Quetzalcoatlus* of South America, which would have stood as tall as a giraffe when resting on its feet and wings, and were capable of hunting small dinosaurs and even the young of large predators.

Around 180 million years ago, the Tethys Sea began to divide the giant supercontinent of Pangæa into smaller land masses, enabling the seas to fill up places that had seen very little water before. This created new forests that were filled with hitherto unknown varieties of trees and plants in places that were previously parched and arid. The availability of new varieties of ferns, gingkoes and cycads (palms and flowering plants were yet to arrive on the scene) transformed the diminutive sauropods into giants.

Between 1960 and 1961, an expedition of three scientists from the Indian Statistical Institute (ISI) Kolkata, unearthed a graveyard of dinosaur bones in a village in Adilabad district. The fossils were discovered in a derelict limestone quarry that had once been a lake or lagoon. After the excavation, nearly 10 tonnes of sauropod bones belonging to six individual dinosaurs of different sizes were transported to Kolkata. The painstaking process of cleaning and sorting began, and in 1977

As its name suggests, Barapasaurus *was a very large dinosaur. It was among the earliest sauropods to have emerged on the scene. The next biggest herbivore (*Kotasaurus*) was about a third of its size. If you (a 6-foot adult) stood next to an adult male* Barapasaurus *you would only reach its thigh.* Barapasaurus *is officially India's largest dinosaur and it was perhaps the largest dinosaur of its time across the world.*

the skeleton of the largest of these individuals was arranged and displayed in ISI. This was the biggest plant-eater in India labelled *Barapasaurus tagorei* (*bara*: big; *-pa*: foot; *-saurus*: lizard or reptile) and it was the labour of love of one of the country's foremost dinosaur fossil hunters, Professor Sohan Lal Jain. Its specific name (*tagorei*) felicitated one of India's greatest modern poets, Rabindranath Tagore, whose birth centenary was being celebrated the same year that a 2-metre-long thigh bone was excavated.

Barapasaurus lived between 176 and 159 million years ago and is the most complete dinosaur fossil to be found so far in India and the first complete, mounted dinosaur skeleton ever to be displayed in the country. The restored skeleton suggests that an adult *Barapasaurus* would have measured around 18 metres and weighed about 7 tonnes. This dinosaur was unique in many ways and was therefore placed in a family of its own (*Barapasauridae*). It had small, spoon-shaped teeth that were useful for chewing the tough but succulent plants that grew around lakes and lagoons. As is characteristic among sauropods, its skull was small in relation to its body, while its legs were massive and pillar-like. *Barapasaurus* was the first of the giant sauropods to appear on the scene, and the biggest dinosaur of its time, before the giant dinosaurs in the Americas and perhaps China were likely to have emerged.

Another large sauropod that existed alongside *Barapasaurus* was *Kotasaurus yamanpalliensis* (meaning 'Kota's lizard'). Kota is a village in Karimnagar district of Telangana after which the formation is named, and Yamanpalli (also spelt Yamanpalle, Lat: 19°1'N, Long: 79°87'E) is the village where the fossil remains were found. *Kotasaurus* was an 11-metre dinosaur and appears to have been a low browser. In 1988, 12 more individual fossils of various sizes were found among river sandstone in Kota, possibly the remains of a herd of *Kotasaurus* that perished together while crossing a river. Because the individuals are of various sizes—probably a mix of large adults to juveniles of about 2.5 metres in length—it is assumed that a flash flood swept this herd into its watery grave. *Kotasaurus* was closely related to *Vulcanodon* from Zimbabwe, but was similar in shape to *Barapasaurus*, although much smaller. It displays features of both prosauropods (the early plant-eating dinosaurs) and sauropods (the more evolved herbivores), and there are some claims that it may be among the very few dinosaurs that displayed such transitional features. A near-complete mounted *Kotasaurus* fossil (except for the skull which was never found) is on display at the Science Centre in Hyderabad. This is only the second and most recently assembled and mounted dinosaur that has been shown publicly in India.

Browsing dinosaurs like *Barapasaurus* and *Kotasaurus* had teeth that were good for cutting vegetation, but not for chewing. Fossil evidence suggests that large herbivores probably used an interesting mechanical assistance of another kind to help them digest their food. It is thought that dinosaurs swallowed small, fist-sized rocks (known as 'gastroliths' or 'stomach stone') and stored them in their gizzards, a muscular chamber sitting just ahead of the glandular part of the stomach. Once the cud entered the gizzard, the organ would contract, moving the gastroliths around to help grind down the food and push it further down the gut.

Sauropods had particularly long, light necks, heads and forelegs and strong muscular tails and hindlimbs. This made them perfect for browsing tall trees. They could squat or stand for a short time on their two rear legs by forming a tripod with their strong tail for support. Sauropods were the largest land animals that ever evolved (though still smaller than the largest animal alive today, the blue whale). They had long tails and massive bodies with powerful limbs and a pelvis that could hold their weight. Early sauropods were bipedal or supported on two legs, and perhaps juveniles could also stand on their hind legs, but their increasing girth meant that later sauropods became fully quadrupedal.

A typical sauropod day was spent entirely in browsing. Herds were formed with the oldest female leading the pack and such herds dominated a large territory, quite like the elephants of today that are led by a matriarch. Fossil evidence suggests that these herds included other herbivores that browsed on lower storeys of trees and plants, who were perhaps encouraged to walk alongside the dominant herd. This was an effective way of developing a mixed herd and sauropods probably used warning signs from other herbivores to detect potential predators. Fossils of herbivores found in the US and Argentina have shown mixed species, with the larger members of the sauropods along the flanks and the oldest females in the front forming a vanguard.

Fossil remains of Indian ankylosaurs are fragmentary but palæontologists believe that it perhaps would have looked like this Kunbarrasaurus *from Australia. Ankylosaurs were large, lumbering, four-legged plant eaters with exaggerated bony armour around their heads and backs, and a heavy club at the end of their tail. Browsing under a hot sun or in a humid tropical forest with these heavy encumbrances could have potentially overheated this herbivore. However, ankylosaurs had an interesting adaption in their nasal passage. They had convoluted tubes in their upper palate which cooled the animal, quite like a radiator of a car or the coils at the back of the refrigerator.*

The sauropods' use of their very small heads that sat on a very long neck to gather food, which was swallowed and macerated later, is similar to the strategy used by present-day giraffes and ostriches. In addition to their powerful gizzards, where food was ground up with the help of stones, it is likely that sauropods had an exceptionally large fermentation chamber (much like cows do, with their four-

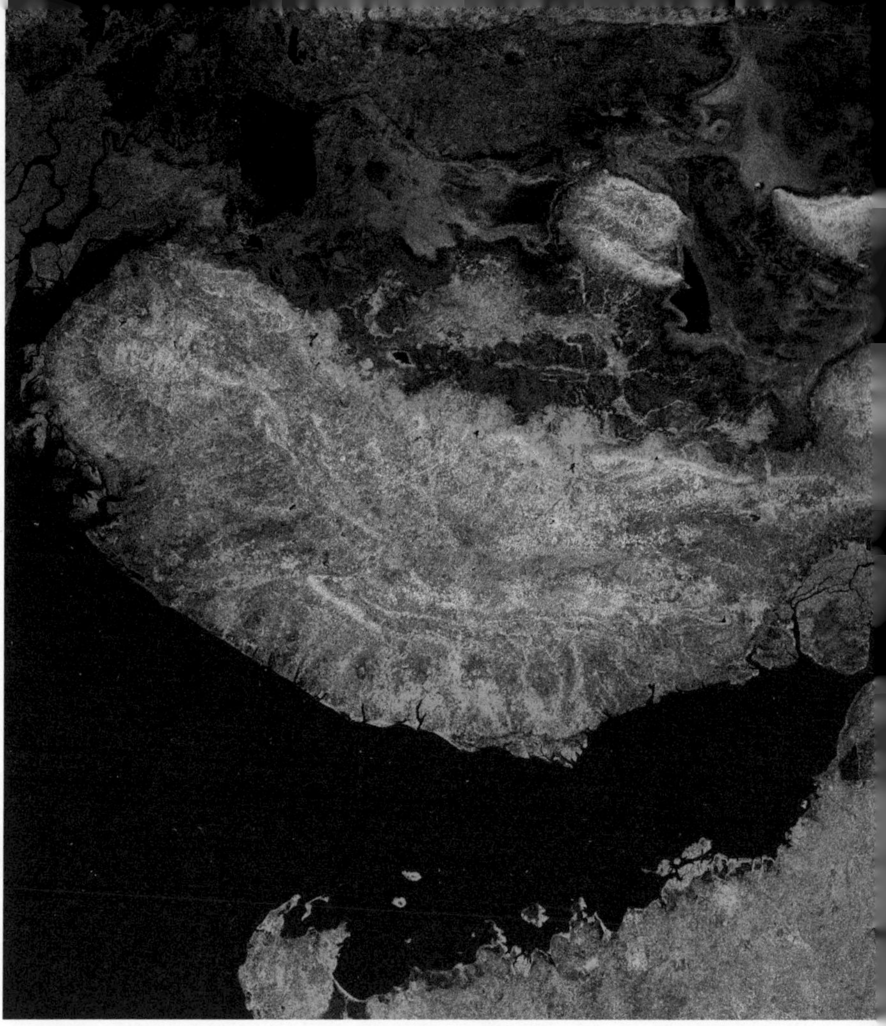

chambered stomachs) with a rich culture of diverse bacteria. These bacteria would have used an array of enzymes to break down plant cellulose and proteins to release the nutrients for their hosts and for themselves. They would also have helped retain food longer in the stomach so that they could extract as much nutrition as possible from such low-quality food.

Further north of Adilabad, in the Bagra Beds near the hill station of Pachmarhi in the Satpura Hills, and on the salt-encrusted flatlands of Kutch, a new species of *Stegosaurus* (*stego*: roof) has been discovered that is unique to the Indian subcontinent. Stegosaurs were quirky herbivores who lived life in the slow lane. They had elongated and triangular-shaped skulls that were small compared to the rest of their bodies. Their backs rose as a crest in the centre and descended into a long tail. On their backs they had large triangular bony plates arranged

The four small hillocks on top, from left to right, are Pachcham, Khadir, Bela and Chorar, which rise above the flat, salt-encrusted sands of the Rann of Kutch. These hills were once the refuge of large herbivorous dinosaurs, and fossils of many other unidentified dinosaurs and reptiles have been found scattered along the seasonal nullahs that cut the south face of these islands.

in a row—small plates behind the neck gradually increased in size till the top of the crest on the back, and the tail ended in fused plates which formed a sort of weapon. Scientists have deduced that these plates did not just serve as armour but also as a means of regulating their temperature. Stegosaurs flushed the plates with blood to warn predators and potential challengers during the season and males perhaps had larger plates than females. Stegosaurs are believed to have survived till the very end of the dinosaurian dynasty.

Another dinosaur that lived life in the slow lane was *Ankylosaurus* (meaning fused lizard). It had an armoured, spindle-shaped body with bony plates and spines that formed a shield across its neck, throat, back and mace-like tail. Ankylosaurs were rotund herbivores and although they appeared slow-moving and lethargic, these dinosaurs were masters of the art of defence, and waited for their enemies to come close before striking a heavy blow with their tails. The ankylosaurs found in India measured about 3 metres, although other ankylosaurs in North America exceeded 8 metres in length. They had a beak-shaped mouth that they used for browsing and for foraging shrubs and low trees. Stegosaurs and ankylosaurs have been discovered only recently in India and more research and fossil finds are needed to know more about them.

Between 174 and 163 million years ago, as Gondwana broke into eastern and western sections, more fantastic dinosaurs appeared on the scene. Further west from the Satpuras in Gujarat, in the Rann of Kutch, lie four hills called Pachcham, Khadir, Bela and Chorar that were once islands. These hills rise abruptly over the flatland of the Great Rann, breaking the topographic monotony of a harsh and bleak terrain with salt pans and lifeless alluvium. After about two hours of travelling north of Bhuj, the capital of Kutch, you spot the highest peak in the province, Kala-dungar (meaning 'black mountain', 465 metres), that lies on Pachcham 'island' and appears like a blunt tooth on the horizon. To its right lies Khadir Hill. On these hills, scientists have found fragments of a claw, hand and limb bones that belonged to the ancestor of a group of sauropods which diverged over 20 to 30 million years, to evolve into giant plant-eating dinosaurs like *Brachiosaurus* from the Americas, and *Camarasaurus* and *Titanosaurus* which roamed in Gondwana including India. The other hills of this region are yet to be explored and could throw up many more bounties of dinosaur bones. We have not yet found the carnivore that preyed on the massive herbivores from these times. Although some teeth and bone fragments of meat-eating dinosaurs have been found, they are too few to be conclusively assigned to a specific species. The fossils also suggest that these carnivores were too small to have preyed on the giant herbivores or even on the smaller but more belligerent herbivores. Could the absence of a top predator be the reason for the gigantism of the herbivores? Or did herbivores increase in size in response to predators that we are yet to find?

Although large dinosaurs dominated the landscape during this period, footprints left behind suggest that small dinosaurs thrived too. If you are close to the city of Bhuj, visit the village of Pakhera (Lat: 23°9'N, Long: 69°52'E), about 60 kilometres north of the city, and the villagers will direct you to the site where two decades ago some geologists had come to study rocks. Here, what at first look like depressions on smooth rocks, on closer inspection are revealed to be the smudged footprints of dinosaurs! These footprints were made on iron-rich sandstone and are of varying sizes. Close to the footprints you can also spot trackways of invertebrates which suggests that this sandstone once lay along a large river or the shoreline of a sea.

A more recent discovery of footprints of dinosaurs was made in Thaiat (Lat: 26°55'50.1"N, Long: 71°03'54.2"E), along the Jodhpur–Jaisalmer highway (NH 168 from Jodhpur to Pokhran and then NH 11 towards Jaisalmer), about 6 kilometres short of Jaisalmer. On this road, just before you reach Jaisalmer, you will come upon a small hillock on your left. There are a few roadside tea shops across the road from this hillock where, if you ask, the villagers will direct you to the area that palæontologists had first explored in January 2014, and where more people

This trident-shaped footprint of a small and light dinosaur (called Grallator tenuis*) discovered on a flat rock in Thaiat village near Jaisalmer in January 2014 suggests that this area was a beach around 80 million years ago, where this opportunistic hunter scanned for stranded animals or dead creatures that the sea washed ashore. A footprint of another slightly larger dinosaur (called* Eubrontes*) was also found nearby by the same team of palæontologists.*

This near-intact fossil of a 5.5-metre-long mollusc munching Jurassic marine reptile called ichthyosaur was discovered near Lodai village (Lat: 23° 22.391′ N, Long: 69° 54.690′ E), about 30 kilometres northeast of Bhuj in the Kutch district. Ichthyosaurs or 'fish lizard' ruled the seas alongside sharks between 250 and 90 million years ago, when the planet was warm and humid. This reptile ate ammonites and other crunchy prey, and perhaps smaller reptiles and sharks. Vertebrate fossils are rare from the Kutch region. This ichthyosaur fossil has been assigned to the family Ophthalmosauridae, *which is widely known from Europe. This new fossil discovery points to the possibility that ichthyosaur may have lived and travelled across the shallow seas that extended from the Americas (covering parts of Canada, the US, Mexico and Argentina) into the western Tethys reaching into western India and beyond to Australia. In this image, the long curved bone is the spine and to the top left of the image is where a partial jaw was found embedded in the soil. The extension you see in the lower left margin, overhanging on a rock, is a complete flipper or the forefin. In the centre you notice delicate radiating bones, arranged like kebabs on a skewer. These are the ribs. On the extreme right of the image are the bones which made up the tail. The fossil is on display at the Geological Museum of the Krantiguru Shyamji Krishna Verma Kachchh University, Bhuj.*

have since come to look at the rocks. If you walk along the crest of the hill, in the ochre and red sandstone you will find layers of shells suggesting that this too was once the shoreline of a shallow sea. On a ledge here you will see different types of footprints, one distinct and a few others that are smudged. These are tracks created by two small dinosaurs about 170 to 160 million years ago. The size and depth of the impressions made on the sandstone have led palæontologists to believe that these tracks belonged to two light, upright dinosaurs walking jauntily along the shore of a Jurassic lagoon, leaving their impressions behind on the sand!

The Jurassic period lasted from 205 to 146 million years ago and saw dinosaurs take over the land completely. Around 155 million years ago, towards the end of the Jurassic period, as the land masses began to part ways, new ocean currents streamed in and began to cool the climate. The South Pole was again covered with ice, although only seasonally, and many kinds of dinosaurs evolved to live in colder climates. The restructuring of land masses gave rise to new plants and forests, which together led to a further increase of oxygen in the atmosphere. This was the time when dinosaurs consolidated and diversified into a bewildering variety of forms, shapes and sizes, and became the most fantastic beasts ever to exist.

6

BEASTS AND BEHEMOTHS

Even though it was still part of the giant land mass of Gondwana, the outlines of what came to be Greater India finally began to be discernible about 145 million years ago. The mountain range that would later split to become the Western Ghats already marked the boundary with Madagascar to the west. In the north lay the craggy Aravallis that extended into a shallow sea to the north and west. In the east, a rift valley was beginning to form between India and Antarctica which would eventually become the eastern coastline of India. This rift valley between Greater India and Antarctica and Australia continued to widen until total separation of the two land masses happened around 122 million years ago. This was roughly when all the continents began to move towards their present positions.

Viewed from space (in fast forward), the continental land masses would have appeared like giant ships sailing away from Antarctica towards the north. Of course, they would actually have been moving at a pace indiscernible to the eye. During the initial part of its journey northwards, Greater India had Madagascar and Sri Lanka in tow. Madagascar had torn away from Africa 128 million years ago, but remained attached to Greater India for another 40 million years. When Madagascar finally went its separate way, it travelled west and rested east of Mozambique, separated by a narrow channel of water from East Africa. It took Greater India 110 million years to traverse 9000 kilometres before it barged into the Eurasian Plate 50 million years ago. That's only a little more than 8 centimetres a year but compared to all the other continental plates, this was a scorching pace!

As all these land masses began to pull apart—not all at once, but each in their own time—seas splashed in to fill in spaces where once there had been dry land. In some cases, new lands arose between the oceans and seas, disrupting sea flows and flooding large portions of the land. Dense tropical forests developed along the margins of the seas but as tectonic forces pushed the land masses away—often violently—it submerged large parts of these forests and through a long process of decaying and 'cooking' under the immense pressure of deep submergence, these forests transformed into coal beds along shallow continental margins. Coal deposits dating back to this period can still be seen in the low hills of Assam and Meghalaya which were part of the northern margin of Greater India. This coal, however, is poor in quality

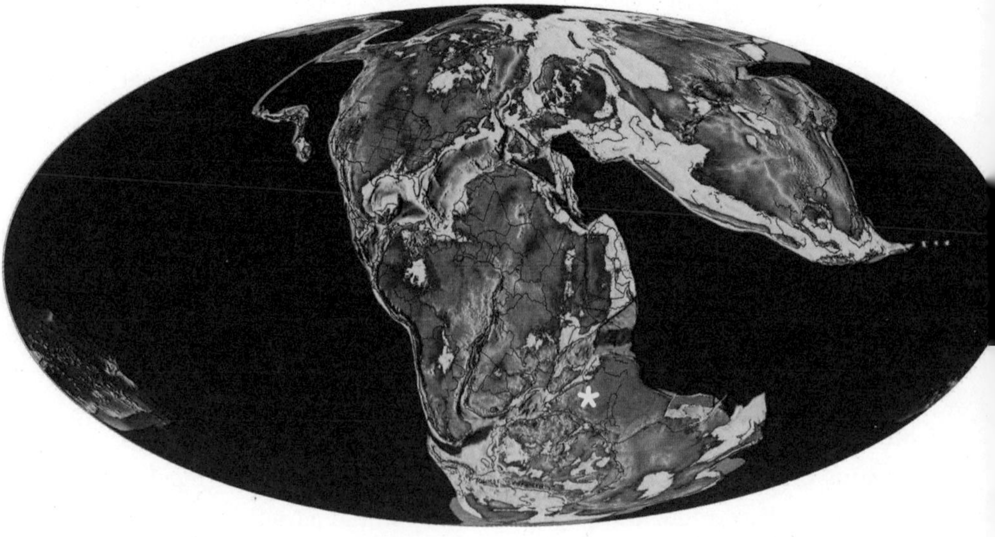

Around 145 million years ago, as land masses parted, seas began to invade the empty spaces. The yellow star marks the position of India. Notice that Kashmir pointed to the east and Kerala marked the western tip of India. The south and east were surrounded by a mass of islands which now make up Antarctica. Madagascar lay alongside Gujarat to Kerala and a shallow sea had begun to separate it from India.

and not as abundant as the Permian deposits that are found over a much larger area in eastern India.

Around 120 million years ago, Greater India would have looked like a very large island with just Madagascar beside it. In place of the towering mountain range that marks its northern boundary today, the northern limit of island India at this time would have been a long, crescent shaped beach where dinosaurs roamed. This beach extended all the way from Rajasthan in the west to Meghalaya and Manipur. Fossils of shells and sea creatures from this period can still be found in large numbers in Jaisalmer and Bhuj, and also in Ukhrul in Manipur. The beaches would have had few trees but further inland there were tropical forests with tall conifers, tree ferns and cycads. The fossilized remains of tree trunks from these forests can be found in Akal in Rajasthan, Sarguja in Chhattisgarh, Sagar and Chhindwara in Madhya Pradesh and in the Cauvery Basin of Tamil Nadu. Dinosaur fossils found in Quetta, Pakistan, suggest that this could have been the northern most point of Greater India or it might have been an island separated from the mainland by sea. Present-day Jaisalmer and Barmer in Rajasthan were under shallow water at this point, but the table-top mountains of Jodhpur were likely to have raised their heads a little above the waters.

If you visit Jaisalmer in western Rajasthan, you will be shown polished vessels of a curious orange-ochre rock that the locals call 'Habur stone' (named after the village of Habur, Lat: 27°19'N, Long: 70°33'E) and earth scientists know as 'Abur limestone'. On closer inspection, you will notice that the polished surface of the cups and bowls is densely decorated with tiny calligraphic squiggles and crescents. These are all the fossilized sections and remains of coiled and shelled creatures that were once abundant in the shallow seas of the region. Vendors will try and sell you these bowls and cups that magically curdle milk without the addition of any culture, lime juice or vinegar, but the reason Abur limestone is able to do so is because it has tiny, microscopic cavities which harbour bacteria and yeasts which do an efficient job of curdling. Abur limestone and some variants of this lovely golden rock have been used to make forts, palaces and temples. If you have visited Jaisalmer Fort, you will have seen that large parts of the fort have been built using this unpolished yellow stone, which gives the city its epithet—the Golden City.

Apart from having a long curving coastline of beaches up north, a map of Greater India would have looked different in other ways too. A large, knife-shaped seaway scythed across the middle of the land mass where the Narmada and Tapti rivers flow today. Evidence of this comes from the shells and fossils of small sea-creatures that are relatively easy to spot in and around the city of Indore and in the neighbouring district of Dhar in Madhya Pradesh. In the villages of Pisdura (Lat: 20°11′; Long: 79°60′) and Dongargaon (Lat: 19°70′; Long: 78°64′) in Chandrapur district of Maharashtra, remains of a variety of fish and large turtles that lived in fresh water and marine environments have been discovered. A little to the east, in Chhindwara district, you can find fossils of planktonic foraminifera—tiny, shelled creatures that drifted in shallow seas. Chhindwara would possibly have formed the southern boundary of this narrow seaway. Fossils of deep-sea fish (like the coelacanths whom you will meet in the next chapter) that have been found in Andhra Pradesh and west Madhya Pradesh suggest that deep trenches and shallow seas existed there too, which is likely to have supported a remarkably diverse marine life. Two other slightly smaller seaways were also formed on the eastern coast when the land mass separated from Antarctica. Today, these constitute the valleys of the Godavari and Mahanadi rivers. These three seaways cut into the Greater Indian land mass and met at a thin land mass in the centre, near Jabalpur. When the seaway began to close in around 105 million years ago, it formed the Narmada, Tapti and Godavari rivers.

This period in the history of Earth is known as the Cretaceous (German, *creta*: chalk), so named because this was a time when gigantic chalk cliffs and beaches were being formed in some places. The Cretaceous period lasted from 146 to 65 million years ago. In the seas, microscopic shelled creatures were absorbing a lot of carbon dioxide to transform calcium into chalk. When chalk is 'cooked' by volcanic heat from below, it becomes marble. The giant seaway that stretched from western Rajasthan to eastern Madhya Pradesh, from where the Narmada flows today, contained a large variety of these microscopic shelled creatures, which explains why we find marble along this juncture today. More than 80 per cent of all building-grade marble

The fossiliferous limestone of Habur is locally known as supari (areca nut, because it resembles a cross section of it). It contains broken shells of ammonites and other marine creatures and is popular for making tiles and handicrafts. The indiscriminate exploitation of the stone has nearly exhausted all the natural deposits and the rock formation is in need of protection.

produced in India comes from Rajasthan. There are at least ten different varieties of commercially extracted marble in Rajasthan. The various kinds are differentiated by their colour—pure white, pink, yellow, grey, rosette, striped grey, patterned greens and browns—because of the way the marble was cooked due to volcanic action and the traces of metals and solutions which leached into it.

As Greater India migrated from the southern hemisphere and crossed the tropic of Capricorn to reach the equator, it alternately encountered cool-and-wet followed by a hot-and-dry climate. Oxygen levels in the air were high but fluctuated during this time—28 per cent (130 million years ago), 29 per cent (115 million years ago), 35 per cent (95 million years ago), 33 per cent (88 million years ago), 35 per cent (75 million years ago), 35 per cent (70 to 68 million years ago), 31 per cent (65.2 million years ago), and 29 per cent (65 million years ago). With increasing levels of oxygen in the atmosphere and changing temperatures and latitudes as land masses parted ways, the

Like the tree fern, cycads are among the oldest surviving trees. They first appeared around 290 million years ago, and seized canopy space during the time of the dinosaurs, eventually receding to the understorey with their demise. Cycads re-emerged from obscurity around 11 to 5 million years ago but remained in the shadows of flowering plants. In India there are fourteen species of cycads but there is only one species in Sri Lanka (incidentally, its only native gymnosperm). The most notable species of cycads are those that produce sago, or sabudana*—edible starchy white balls—that is eaten as porridge, sweets or savoury and is integral to the cuisine, culture and tradition of South East Asia. Sago is vital for several forest-based communities of the region. In India, cycads (like* Cycas circinalis*) and palms (like* Caryota urens*, the fishtail palm) were the original sources of sago, but after the introduction of cassava in the late 1800s, their utility waned. The sabudana khichdi and vada you eat during the Navratras is not the 'real' thing but a substitute that is perhaps manufactured from cassava in Salem, Tamil Nadu.*

Cretaceous became a period when soils underwent a major chemical transformation. The persistent action of humidity and temperature endowed India and other Gondwanan land masses with orange-red soils that are rich in iron and aluminium. Although ores of iron and aluminium originate from the same parent rock, the action of humidity and heat makes them different. Over time, these ores became concentrated and were deposited in thick layers on land from where they are mined today. The iron ore in Goa and Karnataka and the aluminium in Odisha originated during this period.

But the Cretaceous period was also a time when dinosaurs, in a bewildering variety of shapes and sizes, ruled Earth. More than the

Jurassic period (201 to 145 million years ago), which saw the emergence and consolidation of dinosaurs on Earth, this, truly, was the age of dinosaurs, for it was the Cretaceous when most of them attained their massive proportions, ferocity and diversity which allowed them to completely dominate the land. High concentration of oxygen in the air probably partly explains why some dinosaurs grew so large. The ferocious *Tyrannosaurus rex* (*T rex* in popular parlance), in fact, was *not* a creature of the Jurassic. It arrived quite late on the scene (no earlier than 75 million years ago) in the Cretaceous and was witness to the events that led to the extinction of its dynasty. Technically, much of the cast of *Jurassic Park* should not have been in it at all, but in another film called 'Cretaceous Park'!

Dinosaurs of the Cretaceous period developed a number of specializations. To support their enormous size, the sauropods (and the last prosauropods) developed long necks and jaws that opened wide and were lined with rake-like teeth, which enabled them to sway over tree-tops and graze on the foliage without spending much energy. They had a broad frame and pillar-like legs to support them, and strong yet hollow vertebrae which helped them stay on their feet for long periods. Sauropods grew rapidly, adding several quintals to their body mass in the first few months after birth and later adding tonnes as adults. Their bones kept pace with their bulk.

Greater India had its own spectacular array of Cretaceous dinosaurs. The best and richest source of dinosaur fossils from this period is the fossil-rich sedimentary layer along the Narmada river, known as the Lameta formation, named after a bathing ghat which lies on the outskirts of Jabalpur, en route to the famous marble cliffs of Bhedaghat where the river drops as the Dhuandhaar Falls.

The Narmada originates in Amarkantak Hills in Anuppur district of Madhya Pradesh, and travels through Maharashtra and Gujarat, covering more than 1300 kilometres during its journey. For about 200 kilometres on the banks of the Narmada that flows through Jabalpur are marble and dolomitic cliffs that are overlain with sedimentary rocks, and these preserve fantastic fossils from this period. If you want to go looking for fossils in this area, a starting point would be the hilly region around Jabalpur. This area was once flanked by the

Predators like Rajasaurus *and* T rex *hunted by stealth and ambush and not just by ferocity. For many decades palæontologists have wondered what purpose could small arms like those in* Rajasaurus *or* T rex *serve? Like many bipedal dinosaurs, these carnivores supported themselves by a dragging tail, quite like a tripod. When an incomplete skeleton of* T rex *was first unearthed and later mounted, the museum found that its arms were very small relative to its body size, and some believed that these played no role in hunting or eating. Some speculated that the forelimbs were used for mating while others suggested that they provided stability and helped the animal get up if it ever fell down. Some others suggested that the forelimbs were used to pin down the prey. It was also suggested that predators like* Rajasaurus *probably ripped off pieces of meat*

from carcasses with lateral shakes of the head just like komodo dragons and crocodiles do today. But recent discoveries of bones and re-assembling of skeletons have shown that the forelimb bones of T rex *could raise heavy loads, perhaps as much as 200 kilograms and were nearly four times as powerful as those of an adult man. The strength of these arms therefore was useful for holding down prey and also for tearing up flesh when the prey was on the ground. More recently,* T rex *arms are back in fashion. An Instagram celebrity said that selfie pictures make hands appear bent but make them look strangely elegant, quite like that of a* T rex. *This trend is perhaps 'a sign of reverse evolution', as* Marie Claire, *a major society magazine, put it. I am not sure if either knew how* T rex *would have used its hands but I am certain that it would not oblige them for a selfie!*

Narmada seaway to the west, with other rivers originating in the Vindhyas flowing around it. The ebb and flow of river water deposited copious quantities of silt and sediments which settled layer upon layer, preserving the fossils within them. It is therefore not surprising that the first dinosaur to be discovered in India, a sauropod called *Titanosaurus indicus*, was found in a massive sediment horizon in a place called Bara (meaning big) Simla Hill near the army cantonment in Jabalpur. Titanosaurs (or giant reptile) were the giant herbivores of the Cretaceous period. Fossils of bones and eggs of titanosaurs and some other dinosaurs have been found extensively along the Narmada.

Jabalpur cantonment has a second hill close to the Bara Simla Hill called the Chhota (meaning small) Simla Hill where broken bone fragments can be found as you ascend. At the base of the hill, within the boundary walls of the Gun Carriage Factory in Jabalpur, one of the

The Narmada tumbles down and cuts through the marble cliffs near Jabalpur. Strictly speaking this is dolomitic marble, not pure marble. It lies exposed mostly around Bhedaghat as layers of sandstone cover it east and west of here. These white rocks resurface intermittently till Dhar, 600 kilometres west of Jabalpur, after which they disappear. Towards the east, these rocks lie under orange-brown sandstone till Durg in Chhattisgarh, where the dolomite is extracted for processing iron ore in the steel belt around Bhilai.

largest artillery and armaments factories in India, is a temple complex called the Pat Baba Mandir dedicated to Hanuman and other Hindu deities. The temples' precinct offered protection to the bones, eggs and nests that were discovered here because generations of priests and devotees believed that the eggs were signs of Shiva that appeared after he slayed the *asuras* (demons) who terrorized sages in this forest. Tragically, during a renovation in 2011, many nests and eggs were damaged and lost, and today very few fossils remain in the possession of the temples' priests.

The seaways that cut through the middle of the Indian land mass were shallow and dotted with islands, and it was probably easy for large migratory dinosaurs like *Titanosaurus* to wade across these waterbodies. At least seven different species of these gentle, plant-eating giants from the Cretaceous period have been identified in India alone. *Titanosaurus* varied greatly in size and external appearance; there was even a *Titanosaurus* that was armoured, with plates as bony extensions emerging from its skin. The bones of *Titanosaurus* suggest that they were perhaps related to a South American dinosaur called *Saltasaurus*. When *Barapasaurus* and *Kotasaurus* became extinct, *Titanosaurus* dominated as the top browser and is the largest known dinosaur

Sauropods pushed the extremes of terrestrial body size—to the tune of excess of tonnes. In doing so, they challenged the very concepts of anatomy, biomechanics and physiologic design known to science. How these animals coordinated their weight distribution with their exceptional bone structure, nervous system, breathing and digestion, among other functions, has intrigued biologists and compelled them to imagine everything beyond known physical limits. This makes sauropods perhaps the most sophisticated animals

ever to tread the surface of Earth. They lived for nearly 160 million years, from the very beginning to the end of the dinosaur dynasty. They walked on every land mass and evolved into several hundred species. Titanosaurs were the largest herbivores of the Cretaceous from India. Although there is no authentic drawing of these creatures, we can assume that they resembled their Argentinian cousins (shown here) from the same period.

of the Cretaceous period in India. Over 25 metres long and about 12 metres tall, *Titanosaurus* was small in comparison to *Barapasaurus*, but still as tall as a four-storeyed building. For such an enormous creature, its teeth were extremely small and thin and palæontologists believe that they were perhaps used only for stripping leaves and shoots and not for grinding or chewing. That work may have been performed by the gastroliths (stomach stones) in its digestive tract. Like many other

Two Indosuchus *(right) have ganged up to ward off a challenge from the larger* Rajasaurus *(left). In the foreground, smaller* Indosaurus *feed on bits of flesh that they have stolen from the carcass of a juvenile titanosaur.*

sauropods, *Titanosaurus* also had a large thumb-claw that may have helped their young defend themselves against predators. But the biggest weapon these dinosaurs had was their whip-like tail that was capable of stunning any would-be predator. Studies done on trackways and footprints of large sauropod herds in Argentina and the US show that walking in packs with the young in the centre was probably a defensive tactic *Titanosaurus* employed against predators. Despite being widely found, there is no assembled skeleton or even an authentic illustration of any *Titanosaurus* from India.

While the herbivorous *Titanosaurus* lorded over low tropical jungles, small carnivorous dinosaurs like *Indosaurus* (meaning 'Indian lizard') and land crocodiles like *Laevisuchus* (meaning 'light crocodile') lived in dense forests along the Narmada. Another menacing predator from this period was *Indosuchus* which had a skull that measured almost 1 metre, and razor-sharp front teeth that were 10 centimetres long. *Indosuchus* hunted in packs to challenge larger predators. Its fossils have been found at many other sites along the Narmada and a few vertebrae have been found in the limestone beds of Ariyalur district in Tamil Nadu, too.

About 90 kilometres east of Ahmedabad and 70 kilometres north of Vadodara, close to the end of the Narmada's journey, in the village of Raiholi in Kheda district, lies an extraordinary fossil graveyard. Raiholi features prominently on the world's palæontological map because it is one of the best places to see dinosaur nests and eggs. Interestingly, its discovery was almost accidental. In fact, dinosaur eggs appeared rather late—a little over three decades ago—on India's palæontological scene. In October 1982, Professor Ashok Sahni, an incurably curious and widely respected palæontologist, was attending a seminar at the Physical Research Laboratory at Ahmedabad when a young officer of the Geological Survey of India (GSI), Dhananjay Mohabey, asked him about a round rock, about the size of a coconut. Mohabey worked with the GSI's Nagpur office and, while on a survey of the Gujarat region, had heard of the frequent discovery of 'cannonballs' during blasting operations at the ACC Cement factory at Balasinor, not far from Raiholi. The mine managers often decorated their shelves with these so-called cannonballs and used them to line garden paths leading to their site office. Professor Sahni analysed the shell cover of the 'cannonball' Mohabey presented to him and found that it was the egg of a dinosaur!

After this, reports of the discovery of dinosaur eggs began pouring in from other sites around Raiholi and new locations

Spectacular discoveries of dinosaur nest sites and eggs have been made across India, but principally in Gujarat, Madhya Pradesh, Maharashtra and Tamil Nadu. Dinosaur eggs clearly came in a variety of shapes and sizes. Eggs of small predatory dinosaurs varied from long and elongated to teardrop-shaped, while sauropod eggs (the most common ones found in India) were near spherical. The smallest dinosaur eggs were just a few centimetres across (not in India but in Utah, USA, and China) and the largest reported ones are as large as 40 centimetres across, those of sauropods. Recently an enormous, 70-centimetre long, sausage-shaped egg of a giant Oviraptor *from China was discovered. Dinosaur eggshells were made of calcite with microscopic pores which allowed the developing embryo within to 'breathe', much like bird eggs today. But despite these enormous sizes, even the largest dinosaur eggs were small compared to the mother that laid them. The size of the egg is determined by the physiology of calcium and diffusion of gases. Had the eggs been any larger, the shells would need to be thicker and harder, making it difficult for air to diffuse into and out of the shell. A thicker shell would also make it difficult for the feeble hatchling to break through it. The image on the top is from Raiholi and the eggs that have been dislodged from the nest perhaps belonged to a titanosaur; and at the bottom, a dinosaur egg nest is shaped like a* Shiva linga *in a temple near Dhar, Madhya Pradesh.*

in Gujarat, western Rajasthan, Madhya Pradesh and Maharashtra by GSI officers and other researchers. However, Raiholi remains the largest nesting ground of dinosaurs discovered in India, perhaps even in the world. Many nests are clustered together here in close proximity, suggesting that these were communal breeding grounds like the nesting colonies of penguins. The nests were made like hollows in the mud or sand and were lined with vegetation. In each nest the eggs were laid or arranged in a neat pattern so they would not roll around or bump into each other. Tragically, as soon as news of this discovery spread, these sites were looted. Even today, if you stop at a tea stall near Raiholi you might be approached by locals offering to sell you dinosaur eggs. The Gujarat government has set up a conservation site in Raiholi and has made it a recreational park with two massive dinosaur replicas at the entrance to welcome visitors. But a lot of damage has been done to the site by vandals and today only the outlines of eggs within nests can be seen here.

Soon after the Raiholi discovery, a second site rich in dinosaur bones was discovered close by, across the state highway, that came to be known as 'Temple Hill'. Fossil bones are so common here they can be gouged out from rocks with a pen knife. In one particular part of Temple Hill, in a patch of ground only 7 square metres in size, several bones were found which gained attention out of proportion to their size. Suresh Srivastava, a geologist based out of the GSI in Jaipur, worked diligently at the Temple Hill site between 1982 and 1984, painstakingly unearthing bones and carefully noting the position of each one. He found a single braincase located about 3.5 metres away from the backbones. Because the relative sizes of the bones matched, the bones were thought to belong to a single individual. Close to this grave was another set of long bones many of which were broken and which, on closer analysis, were identified as those belonging to several individual sauropods.

Srivastava worked on the bones for several years, cleaning them of extraneous mud and accretions and placing them carefully in cartons. In 1994, Paul Sereno from Chicago's Fields Museum and Jeff Wilson from the University of Michigan were visiting India and met Suresh Srivastava in the GSI office. When Srivastava opened the Temple Hill cartons and showed them the bones of the skull, the two American palæontologists quickly realized that this was no ordinary dinosaur. Wilson, a PhD student then, carefully pieced together the bones under the expert supervision of Sereno and Ashok Sahni. It took them eight months to assemble the skull bones and reveal the apex carnivore of the Cretaceous period in India. They named it *Rajasaurus narmadensis* (meaning 'king lizard of the Narmada'). From the size of its skull, they estimated that this predator was about 10 metres in length. Further analysis suggested that it had a robust build and a very strong skull and neck. Although it was smaller than *Tyrannosaurus rex*, *Rajasaurus* was perhaps more ferocious because it had the framework for greater agility and a stronger bite. It was perhaps a bit like the compactly built but superbly effective boxer Mike Tyson!

Rajasaurus is thought to be closely related to *Majungatholus*, a dinosaur from Madagascar, because their skulls and teeth were similarly shaped and their general appearance probably matched. This, of course, is not surprising, because you will remember that Madagascar at that time was still joined at the hip with western India!

Like all forms of eggs that we know today, dinosaur eggs too were a rich source of protein and were preyed upon by smaller mammals, large lizards, other dinosaurs, land crocodiles and even snakes, which had evolved relatively recently. In 1984 near the village of Dholi Dungri

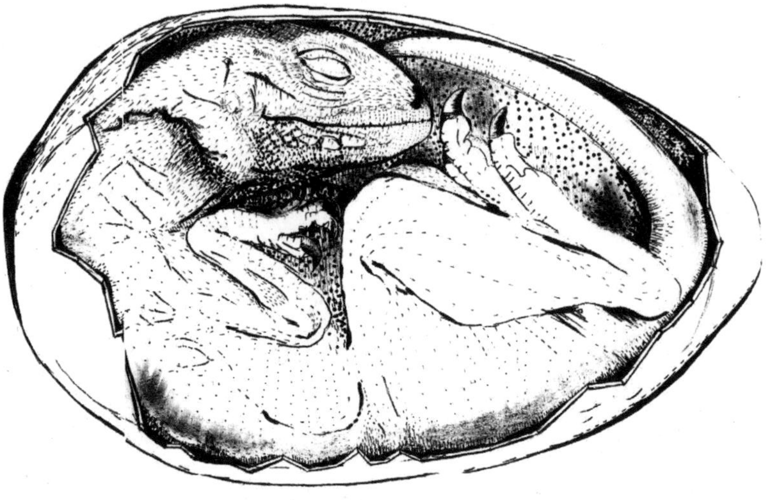

A dinosaur egg would have been like the modern-day reptilian egg which produced a single hatchling. This is an artist's impression of a Rajasaurus *within its egg.*

in Kheda district of Gujarat, an unusual fossil with eggs was discovered again by Dhananjay Mohabey of the GSI. What at first appeared to be a lump of red clay with the calcified bones of a creature wrapped around it, turned out to be three eggs of a sauropod, possibly a titanosaur, but the creature that was curled around the eggs was not as easy to identify and remained a mystery. In 1989, Professor S.L. Jain identified one of the creature's vertebrae as that of a snake but no one paid much attention and the matter was largely forgotten. In 2001, Jeff Wilson reopened the specimen and concurred with Professor Jain's conclusion. It was indeed the fossil of a snake, and a giant one at that! A second block of the fossil that had been collected during the excavation was shown to Wilson and, with a little manoeuvring, it turned out to fit with the other half of the lump perfectly. The snake's twisted, fossilized body became clear as it lay coiled over the crushed fossil eggs. In 2004, the specimen travelled to the University of Michigan's Museum of Paleontology in Ann Arbor, USA, where a further series of tests were performed on it. However, the US-India team wanted to confirm a few other details before they published their astonishing results. So in 2007, a three-member team visited the site from where the specimen was collected and studied the layers and sediments in which it had been found. After a battery of tests, the team's report was published in March 2010. It described the fossil find as an amazing piece of dramatized action frozen in time. The fossil belonged to a snake, little over 4 metres long, (its size was deduced from the dimensions of its skull) which was in the act of eating some sauropod eggs, and had already consumed a hatchling which was more than 0.5 metre long. The sauropod egg was larger than the snake's gape (the maximum extent to which its mouth could open) and it is presumed that it cracked open the eggshell to devour its occupant. This behaviour is rare and has been noted in a small shy snake called the Mexican burrowing snake (*Loxocemus bicolor*) which is known to crack open the eggs of iguanas and turtles before swallowing the embryo whole.

From their study of the site, the team deduced that the snake might have been smothered by an immense slurry of mud. Perhaps a sudden mudslide during a storm had caught the snake in the act of feasting upon the hatchlings, killing them all under the weight of the mud.

This dramatic yet complex fossil also shows a completely articulate sauropod hatchling which is extremely rare to find. Fossil remains of three other large snakes were also found in the vicinity of this site which suggests that they feasted on sauropod hatchlings and eggs as well. Perhaps, for this reason, some snakes preferred to live close to the nesting sites of dinosaurs—easy pickings! The team named the snake *Sanajeh indicus* (Sanskrit, *sana*: ancient, *-jeh*: gape; named for the feeding habit of this early snake). The giant snake was believed to be an extinct ancestor of the macrostomatan snakes (those with mobile jaw joints in their skull, which facilitates a wide gape) that are also presumed to be extinct like *Yurlunggur* and *Wonambi* found in Australia and a few others that were found in Madagascar and South America. This suggests that a common ancestor must have diverged and created new species of snakes in the Gondwanan land masses. Snakes first appear in the fossil record close to the end of the dinosaur era, approximately 98 million

Snakes like Sanajeh *possibly used stealth to feast on vulnerable dinosaur hatchlings and eggs. This life-size model, based on the fossil that was found of* Sanajeh, *recreates the drama of the snake attacking the sauropod nest; the colour of the characters is imagined by the artist.*

years ago, and were the last of the reptiles to emerge. These enemies of the dinosaurs were not as large as the dinosaurs but they depended on cunning and stealth to prey on them.

Along the banks of the Narmada, from Jabalpur, Dhar and Indore, Wardha and Nagpur through Vadodara, until the river reaches the sea, a series of limestone and sandstone exposures hold within them dinosaur bones and eggs, tiny mammal bones and plant impressions. In Tamil Nadu, from Puducherry to Tiruchirappalli, we find fossils from the mid and late Jurassic to the late Cretaceous period.

In 1989, two palæontologists from the GSI found fragmentary remains of the pelvis, vertebrae and the 2-metre-long shin bone of a giant carnivore in Tiruchirapalli district and called this new dinosaur *Bruhathkayosaurus* (Sanskrit for 'huge-bodied lizard') but, later, another palæontologist reclassified this as a sauropod (a herbivore), perhaps a titanosaur. The enormous bones of this creature indicated that it was

the largest dinosaur ever to have lived. It was touted to be a few metres taller and longer than its closest rival, *Argentinosaurus* from South America. Based on the size of its bones, palæontologists and anatomists estimated *Bruhathkayosaurus* to have been 34 metres long and to have weighed 85 to 139 tonnes. But for a creature of such massive size, palæontologists found too few bones, and subsequent excavations have not yielded any more. Tragically (for science), the scientists who discovered the fossils reported that they were unable to transport the fossils back to a safe repository and these fossils were washed away in the following monsoon. The only evidence of *Bruhathkayosaurus* fossils now available are the black-and-white photographs that were taken by the discoverers and published in the journal which announced this discovery. Unfortunately, these images are blurred and do not reveal much.

Some sceptics now claim that these were probably pieces of petrified wood and not fossil bones at all. The credibility of the discoverers has also been in question because the same scientists had claimed to have discovered a dinosaur they called *Dravidosaurus*, which turned out to be a hoax. Based on a few worn-out bones from a skull and a single tooth from a marine bed in Ariyalur district of Tamil Nadu, the scientists had claimed that these belonged to the last surviving stegosaur (stegosaurs had died out in the rest of the world about 15 million years earlier). However, when the fossils were examined closely, they were found to

Bruhathkayosaurus, *a dinosaur who missed his place on the podium! At about 34 metres,* Bruhathkayosaurus *would have been the largest dinosaur to have lived in India, perhaps even the world. Although it has been given a name, scientific protocols were not followed when its bones were allegedly discovered, and hence there is no evidence that this creature actually existed.*

belong to another reptile, not a dinosaur at all. With the credibility of the scientists in doubt and its specimen missing, *Bruhathkayosaurus* was consigned to the rubbish bin of palæontology, and *Barapasaurus* retains the title of India's biggest (confirmed) dinosaur.

Over the 150 million years that the dinosaurs ruled, they demonstrated an amazing flexibility of body plan and evolved into diverse shapes and sizes, many more than any other group of animals. We are attuned to think of dinosaurs as being enormous, but this was by no means universally true. Palæontologists who study the big picture have found a strange pattern in their evolution: during the reign of the dinosaurs, every 25 million years or so, large carnivorous dinosaurs shrank and evolved into smaller versions.

Some dinosaurs stayed small from the very beginning and evolved strategies to survive even as their cousins became outsized and fearsome. Staying on trees was a good option as it offered a variety of insects and smaller reptiles to feed upon. A few gradually took to gliding from tree to tree in order to seek food and escape. Then, around 150 million years ago, the first true fliers (like *Archaeopteryx*, or ancient flier) emerged on the scene. All animals that have achieved flight started out small, and dinosaurs were no exception. According to one researcher, the rate of evolutionary change in the lineage from which birds evolved was 150 times the rate at which other animals were evolving. It was perhaps this adaptability and rapid evolutionary change that would later help the dino-birds survive the mass extinction event that would wipe out the dinosaurs. The early birds acquired a set of features that probably helped them survive better. These included their ability to regulate their body temperature and stay warm in a possible post-impact nuclear winter-like scenario, use flight to escape from wildfires and noxious emissions and search for food more efficiently than terrestrial survivors; and their reduced size also meant they could sustain themselves on much less food than their ancestors. Another important feature was their ability to live in trees or underground. Tree-living requires small bodies, and reduced body size enables other characteristics that enhance success in the trees. Dino-birds developed larger eyes with three-dimensional vision and an enlarged brain to cope with diverse habitats and to feed and fly more efficiently. Their feathers helped them stay warm, which allowed some of them to hunt and feed at night.

Over a long period of time, isolated islands such as Greater India and Madagascar became fascinating natural laboratories of evolution. Oxygen (or its deprivation) is not the only factor that determines the size and extinction of animals. The area of a land mass and its location in terms of latitude, its relief and proximity to other lands, also exert a powerful influence on biodiversity and the size of animals. For many years, evolutionists have debated the 'island rule' which states that large animals tend to become smaller and small animals become larger when they are isolated on islands. At a meeting in Vienna in November 1912, Baron Franz Nopcsa (1877–1933), a dashing nobleman and an avid fossil collector, suggested that this theory was applicable to dinosaurs as well. Nopcsa noted that on the palæo-island of Haţeg turtles and crocodilians were noticeably larger than their normal size, while Transylvanian dinosaurs in the late Cretaceous nearly always remained smaller than 'normal'. Haţeg was a small island which later joined up to form Romania. Nopcsa also observed that the largest dinosaur from Europe (*Magyarosaurus dacus*) was a puny 6 metres long compared to other, more representative sauropods that measured 15–20 metres. Actually, Nopcsa's idea was not completely original. British zoologist J.B. Foster had first propounded this theory in 1902, applying it to modern mammals and calling it the 'island rule'. When Nopcsa presented his idea, other palæontologists agreed that this seemed to hold true for extinct mammal fossils from the Mediterranean region as well. A review of such studies in past and existing vertebrates confirms that the island rule applies to many, if not all, modern cases. For instance, the Asian elephant in Sri Lanka is 12 per cent smaller than elephants from Kerala and Tamil Nadu. Dwarfism probably occurs because adults cease to grow beyond their juvenile state, perhaps due to competition and limitations of food, mates and space. Many rodents and omnivores such as bears tend to grow in size, while large carnivores and herbivores shrink in size in island conditions. The reasons for these are diverse and debatable, and biologists are yet to arrive at a common ground. Could the island rule apply equally to the dinosaurs of India and Madagascar and explain why the titanosaurs or *Rajasaurus* were (relatively speaking) smaller in size than similar dinosaurs found in China or the Americas?

Cretaceous fossils of dinosaurs that have been found preserved in soft marine and lake sediments, like those found along the Narmada, and in other places, like the marine beds and low hills around Ariyalur in Tamil Nadu which remained unscathed by the Deccan lava flows, have, apart from fantastic bones and teeth and eggs, thrown up another interesting type of fossil called 'coprolites'. The high-sounding coprolite is just a polite term for fossilized dinosaur dung! In the village of Pisdura and in an area around a radius of 80 kilometres in Chandrapur district of Maharashtra, just south of Nagpur, there is a bounty of coprolites within the Lameta layers. Coprolites are important because they allow scientists to discover what dinosaurs ate, and their abundance has helped palæobotanists identify and reconstruct the plants that existed in these Cretaceous forests.

Until recently, scientists believed that dinosaurs survived by feeding on the needles of conifers, on ferns and mosses, and that grasses were completely absent from their diet. The conventional view was that grasses like rice originated about 30 million years ago in China, long after dinosaurs had died out. But the coprolites revealed another story: a team of Indian scientists found that 71 to 65 million years ago dinosaurs were consuming the progenitors of bamboo and rice, among other grasses. You might well wonder how scientists are able to divine the nature of plants that have been chewed, digested and excreted many millions of years ago. The answer lies in traces of the most abundant element found on the surface of the soil—glistening silica. Each blade of every single species of grass is lined with a razor-sharp edge that is embedded with a neat arrangement of microscopic 'tiles' of silica known as 'phytoliths'. What's more, the precise arrangement of tiles in phytoliths is unique to each species. This research conclusively debunked the old notion that grasses originated in Asia and later spread into Gondwana by way of India in the late Cretaceous. We now know—courtesy dinosaur dung—that grasses like rice came up something like 35 million years earlier and quite possibly in India. The picture that is emerging is of Greater India as a Noah's Ark of plant wealth, probably due to its isolation during its long, solo journey towards the Eurasian Plate. Among the plants that evolved here were probably rice and bamboo—or certainly their ancestors—which dispersed into Eurasia only when India bumped into the plate about 50 million years ago.

Every few years, new discoveries are made and gaps filled, which improves our understanding of the life and times of creatures from the Cretaceous. In 2013, scientists discovered the teeth and bones of a small land crocodile (related to the *Simosuchus*, pug-nosed crocodile) in Kallamedu village in Ariyalur district of Tamil Nadu. This crocodile was about the size of a spitz (the white, furry dog commonly seen in your neighbourhood, often mistakenly called a Pomeranian) and ate shoots, leaves and the occasional insect. The discovery of similar crocodilian fossils in western Madagascar and some Indian Ocean islands suggests that animals and plants were able to travel between land masses even when lands were separated by spreading seas. Or perhaps the species from Kallamedu got isolated from a common ancestor and evolved into distinct species.

This lettuce-munching land crocodile called Simosuchus *from Tamil Nadu was not menacing like others of its kind and, had it not had a bony armour, would probably not belong to the family of crocs.* Simosuchus *lived on land and its crouched posture and wide body suggest that it was not very agile or fast, and perhaps hid from its potential predators. A study has found that if crocodiles run out of prey in and around water, they will readily eat fruits, fleshy vegetation and even grass. Feeding on fruits in rivers and lakes aids in the dispersal of seeds and therefore crocs also play an important ecological role in maintaining the population and diversity of trees. This new discovery partly explains why crocodiles are such successful animals. It should therefore not surprise you to know that the Nyanyana crocodile farm (a farm that is used to produce hides for fashionable leather bags rather than to save these extraordinary animals) that is situated along the shores of the world's largest artificial lake (Lake Kariba in Zimbabwe) is raising them on a vegetarian diet.*

Like beetles and ray-finned fish, dinosaurs are an elegant example of adaptation and diversification. In the first 20 million years of their arrival on the scene, dinosaurs were small (between 1-3 metres in length) and not particularly diverse. They quickly took to walking on two legs, and evolved designs that were more efficient at breathing and running and this helped them survive better than other animals and also enabled them to evolve faster when the opportunity arose. For 50 million years from 201 to 150 million years ago, the carnivorous bipeds, the long-necked herbivores and diminutive omnivores continued to diversify. The growing challenge of carnivorous dinosaurs, and the bounty of limitless food in the form of foliage in the forests (which also maintained high oxygen levels) allowed or perhaps even propelled dinosaurs to increase in size. Around 180 million years ago, the prosauropods, forerunners of the sauropods, were beginning to grow large. These giant herbivores lived in large herds and may have had social orders, much like elephants do, and the imminent threat of being preyed upon was overcome. Alongside, a few herbivores like the stegosaurs added armour and plates to their appearance, which made them seem slow-moving and bizarre but these were abiding designs that helped them outlast many others who fell by the wayside of evolution.

The stock of dinosaurs that would eventually become birds emerged at this time. Around 160 million years ago, gigantic, earth-shaking sauropods emerged and were followed by meat eaters who needed to bulk up to match them. Somewhere between 140 and 125 million years ago, the giant grazers began to die and smaller grazers like titanosaurs, stegosaurs and ankylosaurs took their place. The mid to late Cretaceous period (125 to 65 million years ago) saw the emergence of diverse types of dinosaurs, including most large meat eaters like *Rajasaurus* and smaller carnivores. Dinosaurs ruled the land for over 150 million years and in the last 80 million years (from 146 to 65 million years ago) their domination remained unchallenged.

Sadly, just as the dinosaurs were beginning to hit a high note, the skies literally fell on them! In a span of 3 million years, between 67 and 65 million years ago, a combination of extraterrestrial, tectonic and climatic factors conspired to bring about the downfall of these mighty creatures. Almost as infamous as the dinosaurs themselves is the story of their fall. Dinosaurs are remembered as much for their sudden death as they are for their splendour, variety and longevity. We often use dinosaurs to depict things that do not work (Google Chrome displays a dinosaur icon when it is unable to connect to the Internet) but this analogy seems a little unfair. Dinosaurs were among the largest animals ever to roam on land, the most specialized in terms of occupying every corner and, most importantly, they ruled more successfully and for longer than any other terrestrial vertebrate did before or has since.

Extinction does not mean failure. Over 99 per cent of all creatures that have ever lived, including our ancestors, are extinct. So the next time someone compares you or your ideas to a dinosaur, smile. It may not be meant in quite that way, but you are being likened to the most successful large beings Earth has seen! The demise of the dinosaurs left a void, and the battle for succession to these giants set off a new evolutionary drama, with some very unlikely contenders staking their claim to dominance.

7

ISOLATION

It does seem highly improbable but India, Australia and South America—not to mention Antarctica—were once not just neighbours but joined at the hip and it is only in the last three or four decades that we have begun to piece together a better understanding of the supercontinent 'Gondwana' and how it might have broken up. The name Gondwana is, of course, an allusion to the Gond tribe of central India, and its coiner was a professor of geology at the University of Vienna, Dr Eduard Suess—no relation to *the* Dr Suess of *The Cat in the Hat* fame but, in his own right, a prolific inventor of labels such as the 'Tethys', 'Panthalassa', and a host of geological neologisms like 'sima', 'syntaxis', 'eustasy' and 'biosphere'. Professor Suess, in 1885, was among the first people to hypothesize that all the major land masses of the world were once fused together in a supercontinent or, at the very least, connected by land bridges and he borrowed the name 'Gond-wana' from British explorers' accounts of this remote jungle-dwelling Indian tribe!

Our understanding of the break-up of Gondwana hinges on evidence left behind by three major volcanic events that occurred in a span of

These ancient charnockites formed deep within Earth's crust lie exposed at the southernmost tip off the Indian mainland, at Thiruvalluvar Rock in Kanyakumari. Geologists call this point the 'Gondwana junction', because this marks the place where India, Madagascar, Sri Lanka, East Antarctica and Australia were once joined together, forming the supercontinent, Gondwana.

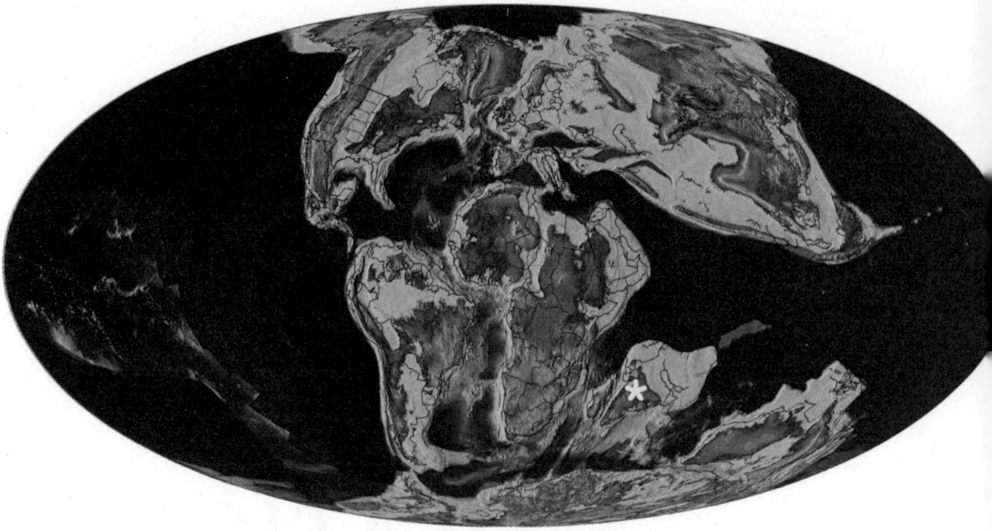

Around 88 million years ago the land mass (Madagascar-Seychelles-India marked here with a star) was ruptured by a large undersea volcanic event that took place where the Marion Island lies today in the Indian Ocean (called the Marion hotspot). By 83 million years ago Madagascar had completely separated from Seychelles-India and got welded to the African Plate, while Seychelles, India and Sri Lanka together continued their northward drift. Seychelles would continue its journey with India for another 20 million years. These newly liberated land masses crossed latitudes and shaped new oceans and current systems, and these constantly changed global climatic patterns.

roughly 100 million years, from 180 to 80 million years ago. The first of these eruptions pared off Africa and South America from the rest of the Gondwanan land mass 180 million years ago. The second eruption 118 million years ago prised Australia and Antarctica away from Greater India (comprising modern-day India, Sri Lanka and Madagascar) and this is when the Marion and Crozet hotspots (a 'hotspot' is an area of intense volcanic activity, usually under the ocean floor) emerged from beneath south-east India and Madagascar between 107 and 105 million years ago. Around 88 million years ago a third major volcanic event sundered India from Madagascar. In each of these volcanic events, immense lava flows continued for millions of years and left behind unmistakable vestiges that have helped us fit together pieces of the jigsaw puzzle.

There is one special place in India which was ideally situated to witness the formation and eventual break-up of Gondwana. This is a hemispherical rock that lies about half a kilometre off the southernmost tip of India in Kanyakumari, Tamil Nadu. On this rock stands the Vivekananda Memorial which is a popular tourist spot. People also flock here to see the merging of three waterbodies—the Bay of Bengal, the Arabian Sea and the Laccadive Sea. The rock pops out of the sea like the humped back of a whale and is made of a single massive block of charnockite, a rock used to build churches (such as St Thomas's Mount in Chennai) and memorials (like that of Job Charnock, the founder of Kolkata, after whom this rock is named) in parts of southern and eastern India. This oval islet-rock (about 534 metres long and 400 metres wide, roughly the size of the Chhatrapati Shivaji Terminus in Mumbai) rises about 60 metres above the sea. Lying close to it is a second smaller rock upon which there is a statue of the sage Thiruvalluvar. From 870 to 680 million years ago, massive outpourings of magma began here and at different points between Balasore in Odisha and Arcot near Chennai, where India was fused with Antarctica and Australia. Between 180 and 118 million years ago, when India separated from Antarctica, it left behind this islet, a desolate witness to the merging and breaking of Gondwanan land masses. Geologists call the Vivekananda Rock Memorial 'the Gondwana junction', because it marks a place where India, Madagascar, Sri Lanka, East Antarctica and Australia were once joined together.

The volcanic eruption 126 million years ago that separated Africa from Antarctica opened up the sea floor and created a cluster of new volcanic islands. The largest and best known of these islands is Bouvet Island (the underwater magmatic event from which it was created is known as the 'Bouvet Plume') which lies north of Antarctica and southwest of South Africa today. A plume is essentially a rapidly rising neck of lava that is shaped like a matchstick with a clear 'head' and it generates so much heat that it can melt Earth's crust from below. Bouvet Island is a small glacier-covered island, inhabited almost exclusively by fur seals, and with an active volcano still at its centre. It lies so far away from any other land mass that it is a prime candidate for being the most isolated place on Earth! The opening of the sea floor initiated by the Bouvet Plume also set in motion the train of events leading to the formation of the Indian Ocean 167 million years ago.

At the time the Bouvet eruption hived off Africa, the rest of the supercontinent (Antarctica, Australia, Madagascar, India and the

Indian Ocean islands) was still fused together, separated, at best, only by shallow seas. India's present-day eastern coast—from Bengal to Tamil Nadu—snuggled close to Antarctica. The extreme eastern corner of Greater India—roughly where Shillong and Guwahati are today—was a mere stone's throw away from where Perth is now, in Western Australia. Throughout this time dinosaurs reigned supreme across all the continents and could have easily walked from Chennai to East Antarctica.

The second massive volcanic event occurred around 118 million years ago which formed a gigantic plume that rose to the surface, creating the Kerguelen Islands (and a large underwater Kerguelen plateau) in the southern Indian Ocean. The force from this massive plume further ruptured the supercontinent and Gondwana gradually began to unzip. Starting from the north-eastern end, Greater India began to separate from Antarctica and Australia, thus creating the eastern coasts of India and Sri Lanka. The pressure from below was so much that it raised the plains by 300 metres and formed the Shillong plateau. The relics of this volcanic eruption can still be seen as small magma chambers in Sylhet in Bangladesh, Meghalaya and in the Rajmahal Hills of north-western Bengal and eastern Jharkhand.

The Rajmahal Hills are made up of dark volcanic basalt rocks (called 'trap') and form a group of rolling hills in eastern Jharkhand, occasionally popping up in parts of Bihar and West Bengal, too. The volcanic rocks near Sahebganj are very similar to the basalt flows and columns that can be seen along the coastline of the port city of Bunbury in Western Australia. Bunbury lies more than 7000 kilometres away today but evidence suggests that sometime between 119 and 113 million years ago, it was probably contiguous with Sahebganj. About 116 million years ago, after 2.5 million years of intense eruptions, the Kerguelen event began to wane and as the lava cooled, it formed enormous volcanic formations of interlocking basalt columns. Only a few of these magma chambers remain in Sahebganj today because most have been quarried and destroyed.

This volcanic event created a geological feature with a quirk. The village of Katthghar (or Khutt-ghar, Lat: 25°22'N, Long: 87°74'E), near the historic town of Rajmahal about 70 kilometres south-east of Sahebganj district's headquarters on National Highway 80, is a sleepy

The underwater volcanic event—the Kerguelen event (120 to 117 million years ago)—raised the eastern margin of the Indian subcontinent and formed a series of hills and plateau that extends for more than 500 kilometres from north to south and east to west. Around Shillong it forms flat-topped mountains which continue as the rolling tea gardens of Sylhet in Bangladesh (the Sylhet traps). The western extremity is marked by the Rajmahal Hills in Sahebganj in Jharkhand. Until the 1950s, these hills were regarded as a continuation of the Vindhyas but studies have proved that they are not only physically detached but also geologically different. Geologists believe that massive alluvium deposits brought down by the Ganga and the Brahmaputra have submerged many magma chambers. Only a few, like in this illustration of a chamber somewhere between Panchet and Simra in Jharkhand, made in the 1870s by a British geologist, still remain exposed.

little settlement, speckled with ponds lined by toddy palms and rice fields. Not much stirs in Katthghar except for cows slowly wending their way home or slim young men shinning up toddy palms to get at their sweet sap. Things start to happen, as they usually do elsewhere in India, when there's a wedding, and especially when the bride comes from somewhere outside the village. Young children are sent off to perform a standard errand, to collect a special kind of gravel that is found only here, in the ponds and streams in Katthghar, and nowhere else. The children know what to look for: thin, white shards of quartz that look exactly like grains of uncooked rice. After the wedding ceremonies, the bride is asked to cook this 'rice' for her in-laws and is made to understand that her culinary skills are on trial. Imagine her frustration when she finds that the rice remains uncooked no matter how long she boils it! The rice, of course, is the quartz, rice-shaped gravel that the children had collected, and her frustration plays out to the delight of the bridegroom's entire family. The prank always ends in laughter and much bonhomie, and has become a ritual with which the bride is welcomed into her new family. It is important, of course, that the bride be from another village and know nothing of Katthghar's uncookable rice.

The origin of the faux rice goes back some 118 to 116 million years ago when the Rajmahal volcanic event occurred at the tri-junction of modern-day India, Australia and Antarctica. Glassy droplets of silica

Rice-like volcanic gravel is found in stream beds and the bottom of wells in some places near the Rajmahal Hills. Residents of Katthghar believe that several centuries ago a ship laden with rice sank in the seas and the rice that got washed ashore was carried inland by the streams. These quartzy grains were formed during the first episode of the Kerguelen event, which was short, explosive and violent and the magma poured out to cover a small area (about 4000 square kilometres). A volcano's explosiveness depends on how quickly the gas it holds within its magma escapes. Magmas with high silica content tend to erupt violently because they are viscous and trap more gas. This perhaps occurred in Katthghar as well. Toba, Krakatoa and Tambora in Indonesia experienced similar and violent volcanic eruptions recently. In contrast, basaltic magma (like the one in the Deccan 65 million years ago) contained less silica and was more fluid, and therefore did not erupt explosively but rather flowed out of vents and fissures.

were expelled hundreds, perhaps even thousands, of feet in the air in immense, violent eruptions of lava. As the droplets started to fall back to Earth, they were elongated by the effect of gravity into flattened spindle shapes. Over time, these pieces of gravel were worn smooth and rendered uniform in size, the better to fool unsuspecting brides (who haven't learnt their ancient volcanic history). So next time you find grit in your rice, remember Katthghar's rice that is made up entirely of grit!

Even after Greater India parted ways with Australia and Antarctica, Madagascar still remained attached to it as part of the Dharwar Craton, the oldest rocks in the Indian subcontinent. Due to the violent magmatic force of the Kerguelen plume in the southern Indian Ocean, Madagascar reached its current position with respect to Africa and, along with Greater India, remained close enough for the next 20 million years for animals and plants to hop from India into Madagascar and other small islands to reach Africa (and vice versa). Despite its proximity to Africa today, Malagasy life forms show a closer affinity with India (and even more emphatically with Sri Lanka), Malaysia and even Australia, than Africa because while Madagascar separated from Africa around 135 million years ago it stayed close to India for the next 47 million years. This period saw an explosion of flowering plants, along with grasses that tried to keep pace with them, and benefiting from this floral abundance were small mammals which were steadily growing in size and ambition. The life forms that are still found both in India and Madagascar today include the fly-trapping pitcher plant (*Nepenthes*, found in the Khasi Hills of Meghalaya), sundew (*Drosera*, found in the moist hills and ponds of Tamil Nadu and Karnataka), and many kinds of palms and other trees. There are creatures like chameleons and killifish which migrated from Madagascar into India and further afield into South East Asia. They also exist in Africa but they are an exception. Madagascar's creatures preferred to hop eastwards, across the Indian Ocean islands, rather than cross the narrow but deep Mozambique Channel.

Around 90 million years ago, a triangular piece of land broke off from northern Madagascar to form the Seychelles plateau. Over the next 20 million years or so, this disintegrated into several smaller islands that are scattered through the western Indian Ocean between Madagascar and India. But the really 'big one', the singular event that levered the

separation of India and Madagascar, was a massive underwater volcanic eruption called the Marion hotspot that began 88 million years ago and continued for another 2 million years. Open any good atlas and you will notice that the eastern coastline of Madagascar fits snugly with India's western coast like pieces of a jigsaw puzzle. If you peer more closely, you should be able to spot a 30-kilometre break in the Western Ghats in Kerala called the Palakkad Gap, which corresponds *exactly* to a gap in the central highlands of Madagascar. Both these features fit nearly perfectly into each other.

Another distinctive feature you'll notice on the maps is that while all of Madagascar's major rivers flow west from its central highlands, those originating in India's Western Ghats travel eastwards. One reason for this is that the separation of Madagascar from India began from a point in northern Madagascar and it unfastened like a zipper along the range of mountains which tore down the middle—one part becoming the Western Ghats on the Indian side and the other the central highlands of Madagascar. The powerful event that split the thick, rocky spine into the Western Ghats and the central highlands of Madagascar was the Marion volcanic event. The Marion event did not cause outpourings at

The pitcher plant of the Khasi Hills bears long narrow pitchers, while those found in other parts of tropics like Borneo or Sumatra can be larger and appear ferocious, few even bearing red teeth-like protrusions that give them a more terrifying appearance. The one found in the Khasi Hills is diminutive, yellowish-green in colour when young, changing to red-green as it matures. It only flourishes on marshy ground on the margin of small pools of water in damp forests. The seeds germinate in shallow water and the young plant springs from the boggy ground. The bright pitchers are visible from a distance and perhaps attract insects the same way flowers do. The pitchers secrete copious honey-like exudate through epidermal cells under the surface of the lid and around the mouth of the pitcher. Insects wander on to the interior surface of the orifice only too readily. If you gently open the lid of a pitcher, you will notice that the delicately fluted rim drips and glistens with this sugary juice and a few struggling and floating insects lie within. The inner face is smooth and precipitous, and rendered so slippery by a coating of wax that the straying visitor slips down to the bottom of the pitcher and falls into the thick syrup where it drowns in no time. Only a few manage to pull themselves out and crawl back on to the waxy inner walls, but they too invariably slip back and repeat the process, until, eventually, they die of exhaustion. This liquid is a weak acid and begins to digest the prey. Once an insect dies it descends to the bottom where more acidic fluid is secreted by special cells. If this liquid from a pitcher is collected in a glass vessel, it can dissolve a piece of flesh. The digested visitors are absorbed by cells that line the bottom of the pitcher and the nutrients extracted from them are sent to the roots and stems. The quirky pitcher plant from the Khasi Hills of Meghalaya is fast vanishing because forests are being cleared, and there is rampant illicit trade of this protected plant. It is possibly easier to find the plant in the bylanes of Galiff Street in Kolkata, where they are sold, than in the forests of Meghalaya.

The Palakkad Gap is a relic of India's attachment to Madagascar. The gap is a deep pass that cuts into the Western Ghats between Coimbatore in Tamil Nadu and Palakkad in Kerala. It is just 140 metres (with the Nilgiris and the Anamalai Hills rising considerably higher on either side) but is 30 kilometres wide. This broad gap exerts a profound effect on weather patterns in the region. During the day a tropical sun heats the area in and around Bengaluru, located on the Mysore plateau at an average elevation of 900 metres, creating a pocket of low pressure. Moisture-laden winds from the Arabian Sea whoosh in through the Palakkad gap and bestow a cool mist in the late afternoon to the hills of Ooty and then descend towards Bengaluru where they shed their remaining moisture. The sea breeze travels about 400 kilometres to reach Bengaluru and gives the city a shower almost every day. Despite this daily shower Bengaluru still gets only about a third of the rain that Mumbai receives in just the five months of the monsoons.

first but simply ripped the mountain apart in two halves. The lava flow occurred later and started in north Madagascar. Eventually, when the magma poured under the Indian Ocean, it cooled quickly and created new land on the bottom of the sea, thus causing sea-floor spreading which pushed the land masses away from one another. New sea floor (or oceanic crust) is created when large slabs of Earth's crust split apart from each other and magma, viscous like treacle, wells up to fill the gap. This magma is rapidly cooled by seawater, and this new rock anneals together to form new sea floor.

So it is not that land masses float away like buoyant rafts, but that they are pushed away by new land that is being created under the sea. This was what occurred under the eastern Indian Ocean when the hot Kerguelen Plume pushed Greater India, Australia and Antarctica away from each other and this happened again during the Marion event. As the sea floor spread, Greater India began moving away from Madagascar at a tremendous speed (18–20 centimetres per year, breakneck pace by geological standards), a whole lot faster than any of the other Gondwanan land masses. This was because two massive volcanic events—the Marion and the Réunion-Deccan events 67 to 64 million years ago—had melted away the founding rocks on which Greater India rested, making the continental crust lighter and thinner. All the Gondwanan land masses today have a thick crust, somewhere between 160 and 300 kilometres deep, but India (mostly western and central India) has a thin layer of basement rock (the lithosphere) that is only about 100 kilometres deep. A lighter Greater India moved faster simply because, with its lighter crust, it yielded much more readily to the force exerted by lava flows at the bottom of the sea. Of all the constituents of Gondwana, only Antarctica remained more or less stationary while all the other Gondwanan land masses went off in separate directions, at their own speeds. After Greater India collided with Eurasia, it slowed down considerably, to about 5 centimetres per year, but even then it has been moving faster than Australia and Africa which are clocking a mere 2–4 centimetres a year.

Though much of the evidence of the Marion volcanic event lies inland in Madagascar or deep below the southern Indian Ocean, an eloquent remnant persists as an island off the western coast of India that is home to a unique formation made by lava flows 87 million years ago. About 60 kilometres north of Mangalore in Karnataka, off the coast of the fishing village of Malpe, lie six small islands called St Mary's Islands. Of these, the northernmost is called Coconut Island which rises a mere 10 metres above sea level and from a distance appears to be deserted for the most part with only isolated coconut palms and a few noisy tourists who flock to its northern beach. But it is in the south and east of this island where the real action lies! Here there are clumps of jagged rocks and an outcrop that rises straight out of the sea. From far away the outcrop looks unremarkable but as you draw closer these rocks begin to look like carved pillars or a stack of broken colonnades from an abandoned palace. The pillars emerge upright from the sea, and the tallest ones are about the height of a coconut palm. Over the years, waves have eroded the columns so that they seem to gradually descend into the sea, forming what looks like an irregular stairway.

This grey-brown rock with a reddish tinge is characteristic of volcanic basalt, which is basically solidified lava. As the lava slowly cooled upon being exposed to the air and seawater, it cracked and contracted, creating the polygonal design of these columns—a fairly common phenomenon with volcanic rock—known as 'columnar jointing'. The width and height of the columns has to do with how quickly the lava cools—the slower the cooling, the larger the columns. The entire island rests on top of peninsular gneiss, similar to the basement rock of western Karnataka, and this suggests that the lava flowed over land and did not rise from under the water. St Mary's Islands are the only remnants of the Marion event in India that are above the sea, but the north-western shoreline of Madagascar and the Marion Islands have identical basaltic rocks, conclusive proof that the lava from Marion not only pushed away the two lands but also spilled over to create (part of) the floor of the Indian Ocean. Greater India pirouetted 30 degrees anticlockwise after it broke off from Madagascar near the Marion hotspot and then headed determinedly northwards like it had an appointment to keep!

Plumes and hotspots are associated with places where land masses were separated from one another, because they represent fundamental zones of weakness in Earth's outer shell (the lithosphere). Pick up a good atlas and compare a map that shows the faults and sutures on the sea floor with a physical map of the Indian Ocean. You will notice that strings of little islands are situated precisely along the fault lines. These islands mark the place where the upwelling of lava (the plume) took place and created them. Bouvet Island actually marks the point where Africa separated from East Antarctica, Kerguelen Island indicates the point where India, Australia and Antarctica parted company from one another, and the Marion Islands are present where India said goodbye to Madagascar. The last major volcanic activity

The chain of St Mary's six islands (four large and two small ones) often gets submerged during high tide, although the tallest columns of basalt that rise to about 10–12 metres above sea level remain exposed. These islands are remnants of the fag end of the lava flow (Marion event, which formed the islands of the same name, Lat: 47°S, Long: 38°E) that occurred in Madagascar 88 million years ago.

(the Réunion-Deccan event, 64 to 67 million years ago), marked by Réunion Island, deserves a separate chapter, because this was the event that shaped much of contemporary peninsular India, but, more importantly, in terms of the great evolutionary drama, caused the extinction of the dinosaurs.

If you want to get a sense of how far India has travelled since it broke off from the other Gondwanan land masses, you can measure the distance by using the scale on your atlas. Take care when you measure distance, because a mile means different things on land and sea. A mile on land is a standard 1760 yards or 1609 metres, whereas a sea or nautical mile is calculated in degrees first and is always one-sixtieth of a degree of latitude. This means that a nautical mile varies from 6046 feet (1843 metres) at the equator to 6092 (1857 metres) at 60° latitude. The difference may be small, a mere 50 feet or so, but it adds up when you measure across several degrees of latitude. So a standard nautical mile is taken as 6080 feet or 1.51 statute miles. To make matters even more complicated, measurements also differ between Imperial and American systems. It might be simpler to abandon the use of miles and use kilometres (as measured on land) instead. So on your map, take a piece of thread to the legend of the map and do your measuring in kilometres. You will find that the Marion hotspot, marked by the Mascarene Island in the southern Indian Ocean, lies 3900 kilometres south of Thiruvananthapuram, and this marks the distance that India travelled in 38 million years before it came to its current position. Now that you know the distance and how long it took, you can calculate the speed at which Greater India moved after each event.

Amidst all the violent churning that separated India and Madagascar, a gentle survivor has lived to tell its tale. In August 2000, an unusual frog was discovered in the plantations of the Nilgiris that scientists called the Indian purple frog (*Nasikabatrachus sahyadrensis, nasika*: nose; *-batrachus*: frog; *-sahyadrensis*: of the Sahyadris or the Western Ghats). It looks like a fat eggplant; with its smooth, thick skin, small head and plump body that resembles an inflated balloon, it is unlike any other frog found in India. Its skull is small but thick and ends in a blunt nose with a tiny protrusion. The Indian purple frog has short limbs which it uses to burrow into termite mounds. This is where this frog lives, as deep as 12 feet, near termite mounds because termite tunnels are light and airy, easy for the frog to dig through, but also because termites are an easy meal for a frog that almost never leaves its underground home. The male comes to the surface for only about two weeks in a year during the pre-monsoon rains to mate with females.

The closest relative of the Indian purple frog is a frog found in the Seychelles—Sooglossidae (from Greek, *soos*: safe, sound, unscathed, unwounded; *-glossa*: the entire smooth margin of its tongue). Genetic tests have shown that these two types of frogs, separated by a large expanse of ocean, are distant cousins, far removed. Nearly 130 million years ago, the Indian Plate and the Indian Ocean islands were home to a large number of amphibians. As the islands began to separate from a drifting Greater India around 88 million years ago, some families of frogs remained in India and Madagascar, while others were marooned on the Seychelles. Over time, and in complete isolation, these frogs (and other similar, marooned life forms) evolved into distinct species, but retained a few ancestral features. The family of the Indian purple frog disappeared from the Indian Ocean islands, perhaps because of volcanic activity there, but its relatives survived and continue to live in the Seychelles. The discovery of the purple frog in the Nilgiris confirms that the islands acted as land bridges between India and Madagascar enabling some life forms to disperse and evolve uniquely. Like the tuatara, the only reptile native to New Zealand, or the kiwi, the Indian purple frog is a unique product of a long and splendid isolation.

Evidence of India's shared geological history with its Gondwanan cousins is also manifest in, of all places, jewellery shops in busy downtown Kochi in Kerala. Here, a jeweller sifts expertly through a small heap of pink rubies from across the world, and is able to discern almost exactly where they are from. These glittering stones hold in them an important lesson on how the ancient land masses were arranged millions of years ago. Nearly 600 million years ago, a continuous layer of rock ran from the western margin of present-day Australia, through India, Sri Lanka, Madagascar to southern Ethiopia. At this time, the western margin of modern-day Kerala was joined with Madagascar, with a teardrop-shaped Sri Lanka attached to it. Within this immense ribbon of near-contiguous rock were gems belonging

The purple frog is adapted for life under ground. It digs as deep as 4 metres below the surface of the soft litter-rich forest soil and comes out only for a few nights in the monsoons to mate. The males call out to attract females as most frogs do but what is peculiar is that after the first heavy downpours they begin to call from underground, even as they are digging out of their earthy lairs. The female's response is quite inaudible to most researchers. The purple frog faces threats of habitat loss and frequent slashing of forest undergrowth, to make way for cultivation and tilling, which are reasons for conservationists to worry about this fascinating creature.

to a family of stones called 'corundum' which comprises jewels like rubies, sapphires and chrysoberyls. Corundum is a rare mineral made up of densely packed aluminium and oxygen atoms, which are normally colourless. When other mineral atoms replace a few of the aluminium ones, brilliant hues emerge. Minute amounts of chromium impart a deep red colour to rubies (termed 'pigeon blood'), traces of titanium and iron produce the stunning blue of sapphires, and chromium and ferric iron create the delicate orange tints of the extremely rare and prized *padparadscha* (from Sinhalese, *Padma*: lotus; *-radschen*: blossom).

Rubies are red, but they are also extremely rare—even more so than corundum—and stand apart from all these other precious stones by virtue that they contain hardly any traces of the most abundant elements found deep inside Earth—silica and iron. In fact, the magical effulgence of a good ruby under ultraviolet light has a lot to do with the absence or near-absence of iron. Instead, they bear hints of exceedingly rare metals like chromium. Thus if corundum is rare, rubies are rarer still. That the ribbon of near-contiguous rocks of Mozambique, Madagascar, India and Sri Lanka contains deposits of rare rubies shows that all three regions underwent common, intense geological activities. They may be slightly different from each other—Sri Lankan rubies are clear and ruddy because of an extra dash of chromium, Mozambican and Madagascan rubies are made lustrous by a hint of gallium, and Kerala's smaller rubies are distinctly unremarkable by comparison—yet all these glistening stones are clearly cousins from the same parent rock.

Kerala provides another elegant example of ancient kinship from its placid backwaters. One enduring species of fish from the southern seas of Gondwana are the cichlids (pronounced 'sick-lids'). Scientists believe that a common ancestor of cichlids lived in lake waters of all Western Gondwanan land masses from around 180 million years ago. As the land masses went their separate ways, the fish travelled with them and evolved into many distinct species. Cichlids did not reach Australia because it had drifted away by the time they began to migrate eastwards, and therefore only a few island-hopping reptiles like the monitor lizards managed to cross over into Australia after this. The cichlids continued migrating into Indo-Madagascar and from here into Africa, South America and the Caribbean Islands. Among the places where they found permanent sanctuary were the rivers and lakes of Africa about 110 million years ago. About 20 million years ago the Rift Valley of Africa created deep lakes like Lake Malawi and in each of these lakes a unique set of distinct cichlids evolved, all parented by just one or two founder species. Over several thousand years and most recently between 400,000 and 17,000 years ago, the alternating mixing and periods of isolation created hundreds of different species of cichlids. The Red Sea, the Gulf of Aden and the East African Rift Valley, with its rift-induced volcanoes (like Kilimanjaro) and rift-depressed lakes (Victoria, Malawi, Tanganyika and several smaller ones), meet at the triple junction of plates off the southern tip of Arabia, in the shallow inlet of the Red Sea. This rifting process isolated the cichlids and created two new freshwater species which occur in Ethiopia and Iran. Besides these, there is a telltale trail of fossils in the Arabian Peninsula suggesting that cichlids migrated northwards as well. The lakes of Africa have nearly 1500 species of cichlids, of which Lake Malawi alone is currently home to more than 500. Many cichlids are still to be named. An ancestor of modern cichlids journeyed westward from western Africa into South America (these two continents separated about 150 million years ago) and reached their western limit in the Caribbean Islands, parts of South America and southern Texas about 70 million years ago. Their isolation and adaptability have made cichlids the most diverse family of vertebrate animals on record, with more than 2500 species. From a single ancestral species in the ancient Indian Ocean, several thousand species of cichlids have evolved, ranging from fish with heavy jaws that can crush giant African snails to slender and swift cichlids that feed on minute plankton; from smaller ones that feed on the dead exfoliating skin of wading hippos to others that graze on algae growing on rocks. More than 500 species of cichlids evolved in the lakes of Africa alone in just over a few hundred thousand years. So rapid and recent is the evolution of these species that there is hardly any genetic variation between them, but they are still considered different species because they cannot mate with one another, and are also completely different in their diet and behaviour. How do we know that these existed when Gondwana was still in one piece? Because cichlids live entirely in fresh water with only a few varieties able to tolerate the salt in mildly brackish water and therefore they could not have undertaken the journey across the seas after the Gondwanan break-up.

India is home to four species of cichlids, the most common of which are those found in the backwaters of Kerala. Less abundant species are found in streams around Mangalore (*Etroplus canarensis*), and Ratnagiri in Maharashtra (*E. maculatus*), and a few inhabit the backwaters of Andhra Pradesh and Chilika Lake in Odisha. The three species of cichlids (*Etroplus spp.*) which are common to India and Sri Lanka prefer brackish water, and differ from their six closest cousins from Madagascar (*Paretroplus spp.*) which live in freshwater lakes. Although they bear a striking physical resemblance to each other, they differ in their feeding habits—the Madagascan are omnivores while the ones in Kerala are strict vegetarians.

One more remarkable creature was silent witness to the changing geology of Earth—to the tumult of sea-floor spreading, the widening of the Indian Ocean and the tearing apart of the Gondwanan land masses—from where it was perhaps best placed to see all of these momentous happenings: the very depths of the ocean. The coelacanth is a primitive armoured fish that has remained virtually unchanged for 310 million years. Once a ferocious hunter of the deep, it was considered extinct until it was accidentally caught by a South African trawler in 1938, creating a terrific stir in the scientific community! The coelacanth lives a reclusive and near-retired life deep down in the Indian Ocean where sunlight barely penetrates, a zone which divers call the 'twilight zone', swimming lazily off steep rocky vestiges. In the daytime these fish gather together in volcanic outcrops, leaving these rocky retreats only at night to feed on small, soft-bodied prey like squids and octopuses. Since the first time it was hauled in, the coelacanth has been found in the Comoros, Mozambique, Madagascar and, most recently, off the northern coast of Indonesia. As the sea floor spread, the coelacanth was gradually carried with the separating land masses and developed into isolated populations of coelacanths in the southwest Indian Ocean and Indonesia which are now nearly 2600 kilometres apart. The Indonesian population reached its current refuge around 6 million years ago. A near-complete fossil of a 200-million-year-old species of the coelacanth

*The orange chromide (*Etroplus maculatus*) and green chromide (*E. suratensis*) or the pearlspot are prized in the fish market of Alleppey (or Alappuzha) in Kerala. An average adult is small, about the size of a grown man's hand (about 24 centimetres). They are served up as a scrumptious dish known as 'karimeen' in Malayalam (kari: dark mud which is found at the bottom of lakes, canals and paddy fields; -meen: fish). Karimeen is Kerala's signature dish and in 2010, the pearlspot became the official fish of the state, and the year was declared 'The Year of the Karimeen'.*

(called *Indocoelacanthus*) was discovered in the Pranhita-Godavari Valley along the Godavari river of Andhra Pradesh and the fossil suggests that it would have been similar to the living coelacanth.

While the Greater Indian land mass was rushing northward in a tearing hurry (relatively speaking), it was dominated by a variety of dinosaurs between 88 million years ago to about 68 million years ago. The two large islands (India, a temporary island from 88 to 50 million years ago; and Madagascar, a permanent one) and several small ones that were created in the Indian Ocean, became laboratories for the emergence of new forms of life. Evolutionary biologists term the isolated Greater India an 'ark' or 'biotic ferry' that carried with it unique life forms which evolved independently to disperse into new lands.

The coelacanth lives in deep trenches of the Indian Ocean and is among the oldest surviving vertebrates in the world. It has not changed its appearance in over 300 million years. It is widely believed that an animal that descended from this fish gave rise to all back-boned animals that walked on land.

But something portentous and destructive was waiting to happen as Greater India drifted across the Tethys Sea. Around 68 million years ago, the world was rocked—literally—by one of the most massive volcanic events ever to take place. It happened in the Indian Ocean and continued in waves for the next 4 million years. These lava flows would play a huge role in shaping the landscape of Greater India and, more dramatically, in bringing all life on Earth perilously close to being obliterated in its entirety.

8

DECCAN'S INFERNO

Ellora's superlative rock-cut caves lie some 35 kilometres north-west of Aurangabad city in Maharashtra, close to the village of Verul. Its name is derived from 'Elapura', a Kannada appellation given to these caves by Rashtrakuta rulers who added the last Jain temples to the complex of Buddhist caves and Hindu temples in Ellora between 800 and 1000 AD.

Ellora represents the epitome of India's rock-cut architecture. For over five centuries, generations of Buddhist, Hindu and Jain monks designed their sanctuaries and put tens of thousands of craftsmen to work, carving monasteries and temples along a 2-kilometre-long escarpment of dark rock. It was an extraordinarily sustained act of venerating a natural site that was considered eminently suitable for monumental carving. There are thirty-four caves on the vertical face of the horseshoe-shaped Charanandri Hills: twelve are Buddhist (5th–7th century AD), seventeen Hindu (carved out between the middle of the 6th century to the end of the 8th century) and five Jain (9th and 10th centuries), running in linear sequence. The proximity and preservation of these temples are testimony to the mutual tolerance and respect that existed between rival faiths during this period.

Between 68 and 65 million years ago, copious amounts of lava flowed out of fissures, which spread and deposited over a large area in the western part of Greater India creating majestic, 1400-metre-high mesas like this named Konkankada (Marathi for the Konkan cliff) in the western Ahmednagar district, Maharashtra.

Cave 16 or the Kailash temple in Ellora is among the greatest architectural feats in India and it owes its grandeur to the softness and consistency of the volcanic rock from which it is hewn.

The most exquisite craftsmanship in the Ellora complex is seen in the Hindu temple called Kailash (Cave 16, as archæologists label it). It depicts Mount Kailash, the abode of Lord Shiva, carved from a single massive rock that sprawls across twenty-two tennis courts! Sculptors working on this cave would have had to remove 200,000 tonnes of rock and it is hardly surprising that it took nearly a century to complete. Kailash should not be called a 'cave' at all because it is open to the sky. While similar in design to other temples of the period, it was hewn from the top down instead of being built up from the bottom.

When the temple was commissioned on this site, this was a bare basalt hill, towering 68 metres above its surroundings. The temple was not built but rather excavated, carved, hewn and chiselled, with not a single block of rock brought from outside to the site. The 'construction' began with two parallel trenches dug through the shoulder of the bare hill, cutting deep within. A third trench, about 46 metres, connected these two. Together the three trenches were 200 feet long, 100 feet deep and 15 metres wide. These trenches created space for the sculptors to work the rock. No scaffolds were used to carve the temple, and as the contours of the rock were hewn to carve out steps, from top to bottom, the sculptors laboured alongside while the trenches were still being dug. The plan of the temple, its layout and artwork were modelled on two smaller temples, the Virupaksha temple in Pattadakal, Karnataka, and the Kailasnatha temple in Kanchipuram, Tamil Nadu.

Ellora's spectacular temples and caves were carved from the dark-grey rocks that were piled up into a series of high plateaux. These rocks were made by massive lava flows known as the 'Deccan Trap' and their story takes us back some 68 million years to a critical juncture in Earth's history. Greater India, you will remember, was still in stately motion, riding on its plate at the rate of a few centimetres every year. Something momentous happened when it reached the Réunion hotspot, about 700 kilometres east of the coast of Madagascar, exactly where Réunion Island stands today in the southern Indian Ocean. As the plate bearing Greater India nudged northwards with its leading edge corresponding to where modern-day Sindh (in Pakistan) and Kutch (in Gujarat) are today, it was as if this part of the subcontinent had tripped on a landmine. An enormous, bulbous lava head lifted the land above the hotspot and ripped open the surface to disgorge copious quantities of lava.

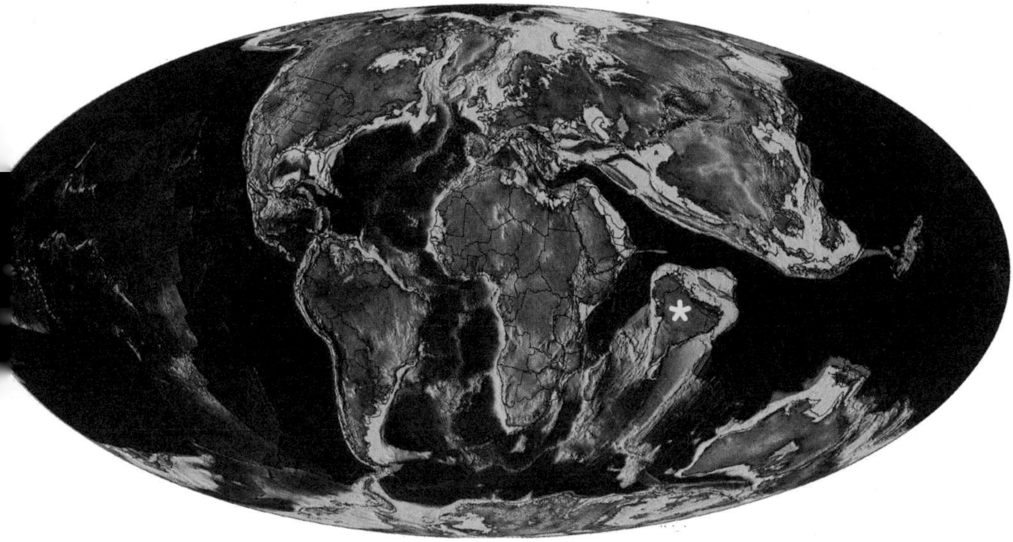

After separating from Madagascar, Greater India embarked on one of the longest journeys undertaken by any land mass. While drifting northward, around 68 million years ago, Greater India stepped over the Réunion hotspot causing massive outpourings of lava over land and sea. For 4 million years a volcanic event below the Laccadive-Chagos Ridge in the Indian Ocean pushed Greater India and Madagascar apart and widened the newly created ocean. As the sea floor spread further, Greater India reached breakneck speed (in geological terms) and hurtled towards Eurasia.

Massive outpourings of lava welling up from the hotspot raised the western margin of Greater India, further tilting it slightly from west to east, much as one might raise the corner of a table by lifting one of its legs. It is because of this tilt that the rivers of peninsular India flow west to east. All of them—with two prominent exceptions—originate in the Western Ghats, just a sliver away from the Arabian Sea, but then flow all the way across the peninsula into the Bay of Bengal. The exceptions are the older rivers, the Narmada and Tapti, which originated in the east and cut through ancient rift valleys to open into a seaway, and these have stayed on course since.

But this was only the beginning of the pyrotechnics. Around half a million years after the first upwelling, the volcano entered its most ferocious phase, erupting in four major waves that ended 64 million years ago. The first eruption, close to the end of 68 million years ago, was relatively minor and poured lava over the north-western part of Greater India as it glided over the Reunion hotspot. Small, dark, dome-shaped mountains scattered across the north-western border of Kutch and

eastern Pakistan mark the first lava flows from the first episode of this Deccan volcanic event. To see the best example of what these eruptions left behind, travel to Anjar (Lat: 23°11'N, Long: 70°03'E), a small trading town in Gujarat, known for its Kutchi food. North-west of Anjar, about an hour's drive away, lies a large, isolated hill that rises abruptly 220 metres from a flat plain. This grey, rocky hill is Dhinodhar (or Dinodhar, as some anglicized travelogues would have it). From its summit, Dhinodhar (Lat: 23°45'N, Long: 69°35'E) offers spectacular views of the Great Rann and Chhari Dhandh (Gujarati for 'salty marshland') which attracts water birds from near and far after the first monsoon showers. What is left of Dhinodhar today is an exposed magma chamber that serves as a dramatic reminder of the first episode of the Deccan outburst.

A magma chamber is essentially a large, bulbous reservoir of molten rock that has cooled slowly under the surface of Earth. When a magma chamber is formed on the surface, it creates a rounded hillock within which lies a core of still-molten rock. As the magma releases heat, it 'cooks' the soil and rocks it has raised from below and forms a thin pie-crust of rock around itself. Over time, the rock erodes to expose the underground magma chamber. It's a little like a pimple that pushes up from under your skin, desiccating the skin surface but continuing to cap and hold the ooze within. When the sides of the hill (or the skin in our pimple analogy) erode over time, they expose column-like formations of cooled lava underneath. Unlike a mere pimple, however, Dhinodhar is an awe-inspiring sight. Dark, pillar-like columns dominate the entire hill like excessive scaffolding and look particularly out of place where they stand tall in the flat, barren landscape. Like something that has been transported straight from the underworld of the Orcs of Middle-earth or, to give an analogy from closer to home, from an *asura*'s backyard!

Not surprisingly, there are legends associated with this singular volcanic hill. One of them tells the story of a hermit called Dharamnath who lived here in the sixteenth century. In the manner of Hindu ascetics who choose to undergo intense penance and privation in order to gain spiritual powers, Dharamnath stood on his head for twelve years until

From a distance, the dark-grey Dhinodhar Hill lords over the flat salt pans and scanty grasslands (opposite) but up-close, it looks ominous with its upright columns of basalt (below).

the gods themselves made him stop to prevent him from becoming too powerful. This angered the hermit so much that when he opened his eyes, he scorched the land with his incandescent gaze, causing the sea to recede and leaving the whole of the surrounding region barren and desolate. The followers of Dharamnathji founded a cult called the Kaanfata (slit ears) and his story is preserved in naïve temple art on top of the mountain.

Before the first eruption of the Deccan lava, Gujarat comprised five separate islands. One large island was made up of the surroundings of Bhuj, the capital of Kutch, a second large island lay to its east, extending along the Narmada from Kheda to Vadodara, about 75 kilometres long and 16 kilometres wide. Three smaller isles lay a few kilometres north of this second island. To these slightly older rocks which date from the Jurassic period (201 to 145 million years ago), the cooling volcanic lava added new rock. Gujarat gradually began to look like it does now when the Tethys Sea drained and, from 38 to 9 million years ago, the Indus river deposited huge amounts of sediment around these volcanic islands. In geological terms, that is just yesterday.

Although the first volcanic event was relatively low-key, its impact on plants, both terrestrial and aquatic, would have been cataclysmic and far-reaching. On land, many species of ferns, cycads and conifers would have been catastrophically affected, making it possible for a completely different kind of plant—the angiosperms (from Greek, *angeion*: vessel, *-sperma*: seed)—to inherit the land.

Angiosperms are the flowering plants that we know so well today, but they were latecomers in the history of life on Earth. The emergence of flowers was a significant development in the history of evolution. The question of why flowers originated and how flowering plants displaced the gymnosperms has long mystified scientists. In July 1869, Charles Darwin wrote a letter to his botanist friend Joseph Dalton Hooker, saying 'that the rapid development as far as we can judge of all the higher plants within recent geological times is an abominable mystery'. Darwin was aware that flowers had arrived relatively recently on the scene but had spread very quickly thereafter, outcompeting their gymnosperm rivals, especially in warm climates. Palæobotanists estimate that flowering plants began gaining a foothold around 125 to 105 million years ago in the northern land mass of Laurasia but were largely absent in Gondwana. Many of these are now extinct, like *Montsechia*, a water plant whose fossil was discovered in the limestone quarries of France and which has pushed back the date of our understanding of when flowers and flowering plants evolved. Others which survived have lived in isolation, like the shrubby *Amborella* tree which produces large numbers of small, white flowers and lives in the understorey of the rainforests of New Caledonia. Flowering plants in and around water like the common water 'weed' (*Ceratophyllum*) and palms evolved in Laurasia but spread across the tropics once Gondwana began to break up.

The discovery of new fossils of plants with flowers has thrown new light on the timing and evolution of plants. Early forests which colonized land roughly 450 million years ago were made up of tree-like plants like lycopods (clubmosses) and horsetails that are today pushed into obscurity. Then, around 385 million years ago, another group called the 'gymnosperms'—plants with unenclosed or bare seeds, such as conifers—emerged and began to take over 250 million years ago, almost immediately (in geological terms) after the Great Dying. Genetic studies on the *Amborella* suggest that the first angiosperms probably appeared because of a genetic error in an ancestral gymnosperm which led to the emergence of tiny, inconspicuous gymnospermic flowers that became the ancestors of the flowering plants around us. This occurred perhaps close to 200 million years ago (the oldest flower fossil discovered so far is about 160 million years old). In the fossil record though—gleaned mostly from the persistence of silica-rich pollen—there is a sudden spurt of flowering plants that appeared during and after the Deccan event 67 to 64 million years ago. Is this a story about destruction and new beginnings? We don't quite know this for sure, but palæobotanists have raised the possibility that the Deccan lava cataclysm made it possible for the angiosperms to emerge from the shadows of gymnosperms, and led to their diversification and eventual supremacy.

How do we know any of this? Mostly by tirelessly sifting through the sequence of layers from the Deccan lava flows and looking at what the pollen tells us. The pollen record in the Deccan lavas is an efficient timekeeper. In case you're wondering how molten rock can possibly preserve delicate pollen, it really is the sedimentary rocks that formed

Fire fountains spewed lava, layer upon layer in the Deccan for nearly 4 million years. How the lava is released to the surface determines the characteristics of the rocks it will make when it cools. The second and third Deccan episodes produced the hottest form of lava (more than 1000 to 1200°C) which had less silica and more iron and magnesium, and was not very viscous. This formed basalt. Slightly cooler lava (880 to 1000°C) has more silica and is more viscous and makes a rock called andesite, which is typically seen in the Rajmahal Hills and the lower Aravallis, especially around Chittorgarh,

but rarely found in the Deccan. The third rock is made from even cooler lava (650 to 800°C) and is high in silica, low in iron, magnesium and salts, and therefore highly viscous, containing a lot of gas. This forms rhyolite which is part of the 770 million year old Malani volcanics and is commonly seen in and around Jodhpur. The Mehrangarh fort stands on a hill made of rhyolite. The dark-grey wrinkly folds of basalt you see in this illustration are string-like pahoehoe that can be seen around Pune and Satara.

between successive Deccan flows ('inter-trappean rocks') that provide us with this information. The study of fossil pollen from several inter-trappean sites such as Pisdura in Maharashtra and Mohgaon in Madhya Pradesh have helped identify the flowering plants which exploded on the scene and went on to make up the jungles that existed between successive volcanic episodes.

The most intense of the four Deccan volcanic episodes was the second one. Many lava vents and magma chambers that formed during this time around the rift valley and seaway flanking the Narmada and Godavari rivers can still be seen in western Madhya Pradesh, northern and eastern Gujarat, and a few smaller ones in Maharashtra. For example, there is an exposed 80-metre-tall magma chamber 30 kilometres west of Indore before the town of Manawar (Lat: 22°24'N, Long: 75°09'E) on

One of the last remaining magma chambers in western Madhya Pradesh, this one close to the town of Manawar near Indore is being detonated and quarried for gravel (bottom right) to lay roads.

the Agra–Mumbai National Highway. From a distance the hill looks like any other. But as you approach it, you'll notice that the exposed half of the hill reveals tall, yellow-grey basalt columns rising from its base and converging towards its narrow top. The hill is currently being mined to make gravel for expanding the highway. We don't know exactly how many such magma chambers exist but many are being mined and obliterated for a variety of reasons. There are no laws or policies to conserve these ancient geological monuments. Perhaps we need babas and sadhus—not babus—to save these wonders, as they had done for Dhinodhar!

The upwelling of magma also had an impact on the geography of several regions. In places where magma could ooze out it formed magma chambers, but where the layers of rock above it were harder and thicker,

An illustration of an exposed magma chamber with basalt columns near Koteda (Lat: 22°76'94"N, Long: 75°59'93"E) in Jhabua district, Madhya Pradesh, drawn by a British geologist in 1888. This has now been levelled to make way for a state highway that connects Indore to Vadodara.

the magma, swelling from below, was unable to break through and ended up raising the height of the existing land. The upwelling along the Satpura range (along the axis from south of Khandwa to north of Mandla) in Madhya Pradesh and accumulation of stresses along the peninsular cratons, raised the land by several hundred and perhaps even a few thousand metres. Pachmarhi, the verdant hill station in the heart of Madhya Pradesh, owes its elevation to this push.

The second eruption lasted for about a million years, between 66 and 65 million years ago, and produced nearly 80 per cent of all the lava released in the Deccan, decimating almost all large animals on land, most large reptiles and many fish in the seas. Try and imagine what a million years of lava floods would entail. After the initial intense heat and acid rains, the eruption would have produced enormous quantities of noxious gases. Dust would have filled the upper atmosphere, creating dark gloom over the whole Earth. The Sun's rays would not have been able to pierce through the pall of dust, and that crucial process we know as photosynthesis would have been severely compromised. Large swathes of forests would have died en masse and with no large trees left, the last remaining sauropods subsisting on the foliage of these forests would have quickly followed them into oblivion.

It is difficult to overstate the extent of death and destruction. The disappearance of plant-eating sauropods would have been followed by the death of the smaller (but more ferocious) carnivorous dinosaurs who preyed upon them. Among the few survivors would have been small animals like lizards, snakes, crocodiles, turtles and most mammals. Some, like the frogs that lived underground and who developed the ability to aestivate (summer sleep, the opposite of hibernate) were especially successful during these difficult times. For once, being small would prove to be, literally, a lifesaver.

To add to the misery of the survivors, a meteor 5–15 kilometres wide crashed near present-day Yucatan in the Gulf of Mexico. This collision happened about 100,000 to 200,000 years after the second episode of Deccan lava flows. The impact of the meteor called Chicxulub (pronounced chik-shoo-loob; Mayan for 'tail of the devil') created mega-tsunamis reaching halfway across the world and its heat would have scorched the land and evaporated much of the oceans' water. The tremendous impact of Chicxulub would have charred the dead and dying trees, killed most of the remaining large animals and submerged low-lying areas. But even this was not the end of the hammering that life on Earth was subjected to. A third and fourth wave of Deccan lava flows followed around 65 and 64.7 million years ago and they would have completely obliterated any large animals that might have survived the first two cataclysmic events. These later eruptions were relatively short, lasting only for a few thousand years each. But coming as they did in rapid succession, they would have prevented any recovery and, in all probability, made the final extinction of the dinosaurs a certainty.

The Deccan eruptions were the greatest lava floods in Earth's history. Each outpouring of lava spewed between 100 to 150 billion tonnes of sulphur dioxide and other noxious gases into the atmosphere. But the signature of the volcanic flow is its basalt rocks which, over 5 million years of volcanic activity, piled up to form a plateau more than 3350 metres high covering much of central and western India. The four episodes of eruptions in the Deccan also overlaid successive layers of thick basalt across the Indian Ocean. The exposed Deccan lava flow in India covers an area of over 500,000 square kilometres (roughly the size of Spain), and an equal or perhaps larger area lies under the Arabian Sea which is roughly equal to the combined size of the states of Alaska and Texas and is about 1 kilometre thick.

What we know today as the Deccan trap region ('trap' is the Swedish word for stairs, coined by W.H. Sykes in 1833 to describe the step-like terraces peculiar to this terrain) is only an intensely worn-down remnant of this gigantic outpouring of lava. This volcanic event created the largest and longest lava flow on Earth that traverses central India, originating from close to Mahabaleshwar, the picturesque hill station in Maharashtra, and flowing eastwards, passing through central Maharashtra and then criss-crossing between the rivers Krishna and Godavari until it reaches Rajahmundry in Andhra Pradesh. However, this lava is not uniform across the region. It has been shaped by elements of nature in different ways over time. In the western parts of Maharashtra, the lava pile in some places is about 1.5 kilometres thick which reduces to a few metres as you travel east and you can see remnants of the Deccan events scattered over a huge compass: it makes for most of the table-top plateaux and dark cotton soil

The second outpouring was massive and spread across millions of square kilometres. It would have resembled this scene from Tolbachik volcano in the Kamchatka Peninsula of eastern Russia in 2012 which spread only across 20 square kilometres. Once the pouring stopped, the magma froze inside the chambers creating columnar basalt, like the ones in Dhinodhar and Manawar. Notice the dark-grey wrinkles of basalt on the bottom right, quite like the ones we see around Pune.

in the plains of Maharashtra; the black and dark-grey magma chambers of Kutch and Saurashtra in Gujarat; the broken funnel-shaped mountains of western Madhya Pradesh; the massive boulders and rocks strewn across gently sloping plains of Andhra Pradesh and northern Karnataka. The dark volcanic rocks at the bottom of rivers like the Chambal near Dhar and the Narmada in Omkareshwar are similar to rocks that can be seen along Worli Sea Face, Nariman Point or Carter Road in Mumbai. There is even a magma chamber in the heart of Mumbai called 'Gilbert Hill' which has a temple on the top but is overrun with shanties and encroaching colonies. In Rajahmundry in Andhra Pradesh, the lava perhaps entered a shallow inland sea around 65 million years ago. We know this because some basalt was colonized by shelled creatures whose fossils are still found embedded in these rocks and it is relatively easy to gouge them

out. But the most common geographic feature of the lava flows is the flat-topped plateaux, created by the layer upon layer deposition of lava that resembles a layered cake or the Goan delicacy Bebinca. These plateau-like hills have alternating strata of brown and grey bands and begin from the Dangs in Gujarat, run along the west coast of Maharashtra into Goa and end in northern Karnataka. The dark basalt rocks that dot many popular beaches like those in Palolem, Benaulim and Colva in Goa are remnants of the lava flows and from a distance they look like viscous blobs (which is what they actually were) that have flowed into the seas.

Amazingly, despite the succession of cataclysmic events, life on Earth showed resilience. It helped that the volcanic events were not

The Deccan plateau was formed by the overlaying of successive lava flows and hence resembles a layered cake, like this mesa (the Salher fort) near Nashik.

Lava from the Deccan that flowed into the sea can be seen as dark basalt on the beaches along the west coast, like in Gokarna, Karnataka. The orange-red soil on the hilly mound is iron-rich laterite formed during the late Cretaceous. This flow marks the southern margin of the lava flow along the west coast of India. The northern limit of the Deccan flow is near Udaipur in Rajasthan and a small portion lies exposed not far from Lalitpur in Uttar Pradesh.

During the intense second episode of eruption of the Deccan volcanoes, copious amounts of lava flowed across the peninsular India. The magma chambers and vents extruded lava in throbs which flowed over long distances, often as much as 200 kilometres away. Depending on the intensity, timing and chemical composition, each pulse created a variety of basalt formations. These soils are rich in nutrients and support lush forests even along steep slopes like these hills near Mahabaleshwar in Maharashtra.

relentless and continuous and there were short periods of quiescence ranging from a few hundred to a million years. During these intervals of reprieve, plants and animals that had somehow clung to survival quickly recolonized the land or water. Our knowledge is based on information preserved *between* the lava flows, in sediments deposited by fresh water. These fossil-rich beds tell us intriguing stories about what happened not only to dinosaurs and early mammals, but also to shells and fish and other small creatures.

Earth scientists call these sediments 'inter-trappean' beds because they represent the gaps—in both time and space—between the trap flows. Most of these beds are 1–3 metres thick but may have been as thin as 15 to 20 centimetres when they were formed in shorter intervals of calm before the lava floods recurred. Geologists and palæontologists have been studying these layers since the 1880s and have found fossils of frogs, turtles, crabs, fish and the shells of snails. In 1998, a team of palæontologists found crocodile eggshells (in Greek, *krokodeilos* means 'lizard') on a spit of land where Mumbai's Malabar Hill and Worli's office district are located today. Crocodiles have proved to be one of the most doughty survivors ever to have lived. When dinosaurs ruled the land, crocodiles ruled the water along with a few other reptiles of the deep. The fossilized remains of crocodilian ancestors dating back over 200 million years have been found to bear an unmistakable resemblance to their modern counterparts. On small islands, which would later become the modern metropolis of Mumbai, the inter-trappean sediments reveal that crocodiles basked on these shores where fresh water mingled with sea water at high tide. The rocks around Mumbai hold more treasures: on Malabar Hill, in Chowpatty and Santa Cruz (where they were found behind a car garage!), layers of inter-trappean shale and sandstone have preserved delicate fossils of varieties of frogs, among other small creatures.

The Deccan lava flows were not absolutely uniform, varying slightly in terms of their minerals, texture and flow to form rocks with noticeable differences. Two main types of basalt rocks are 'a'a' (pronounced 'ah-ah'; Hawaiian for 'lava with stones') and the distinctive pillow-shaped rock called 'pahoehoe' (pronounced 'pah-hoh-ey-hoh-ey' in Hawaiian, and referring to the texture of an animal's intestines which are convoluted

The eastern margin of the second Deccan episode can be seen here as the basalt traps in Gowripatnam near Rajahmundry. These are arranged in conical mounds along the course of the Godavari river and reach the mouth of the Krishna–Godavari Basin. Notice the three layers of basaltic flows (grey rock) each separated by sedimentary rocks (the brown horizons or inter-trappean beds) which are made up of sandstone, greenish-grey and light-purple clays and fossiliferous limestone. The fossil-rich limestone suggests that this was once an immense estuary. This massive flow (truck for scale) suggests that the Deccan event may have been larger than previously estimated, since a lot still lies buried.

but smooth!). A'a is formed when some part of the lava begins to cool faster than the rest to form lumps or boulders within the cooling mass. Typically, a'a flows appear rough and blocky, with small holes or vesicles created by pockets of gases trapped within the lava. Pahoehoe, on the other hand, is smoother, with greater consistency. Ellora's temples and caves could only have been carved out of pahoehoe because, unlike a'a, it is relatively soft underground but hardens when exposed to air. This made it easier to cut through the pahoehoe rocks to carve the temples and stupas.

All across central India, this basalt rock is used to make the foundations of homes and temples, including the massive columns you sometimes see at the entrance to temples in Maharashtra. Blocks of Deccan basalt were used by the British to build the foundations of the Gateway of India and the Chhatrapati Shivaji Terminus (formerly Victoria Terminus) in Mumbai. The same grey rock was used to make railway stations like the one in Kolhapur (built in 1888) and the quaint

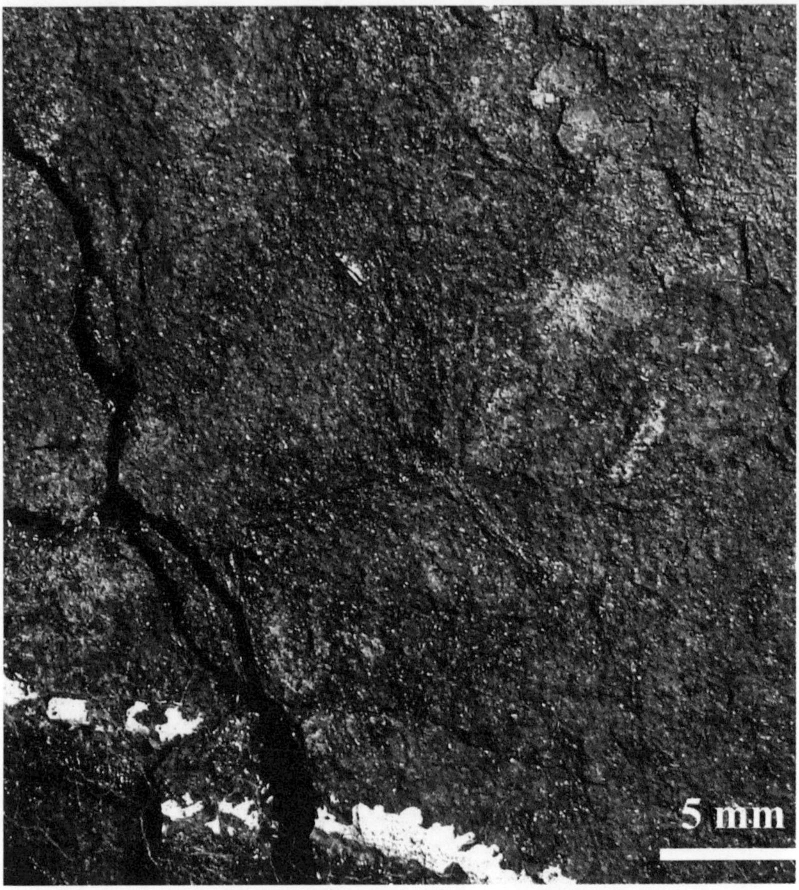

Spot the frog? Fossils of frogs and marine creatures were discovered in Mumbai in 1847 and new ones have been reported from here since. The layers of sandstone which house these fossils are called frog beds. These resemble the ground frogs from Australia and New Guinea and include some quirky members like the cannibal frog, pouched frogs, the brooding frog (which allows tadpoles to hatch in its mouth) and giant burrowing frogs. The ancestors of these frogs are believed to have left India via northern Antarctica and South America, and finally made it to Australia where they continue to survive. This delicate outline and fossil of a frog is from Malabar Hill, Mumbai.

little stations of Dahanu in Maharashtra and Valsad in southern Gujarat. The Mumbadevi and Sitladevi temples, Elephanta Island caves and Kanheri Caves in and around Mumbai, the Ajanta Caves near Jalgaon and several monumental rock-cut temples like the Pataleshvara temple in Pune are among the many examples where pahoehoe basalt was used. So if you were to mark the thousand-odd Buddhist and Hindu caves in Maharashtra that are hewn in rock on a map, you will see that most of these lie in places where pahoehoe rocks exist. The predominantly pahoehoe region lies in the centre of the plateau, extending from Mumbai in the west till Pune in the south, and from Aurangabad in the east to the Tapti river in the north. The spectacular Maratha forts typically use both types of volcanic rock—the heavier dark-grey pahoehoe basalt which makes for a stronger foundation and the lighter brownish-grey a'a rocks to make walls, balustrades and towers.

For nearly 4 million years—from 68 to 65 million years ago—the intense heat of the Deccan lava flows singed the island of Greater India and lava flowing into the seas warmed the water and turned it acidic. Fumes rising from the lava and from the water's edge where the lava met the ocean made the skies rain down acid. Persistent acid rain and cataclysmic changes of climate spelled doom for the big creatures that had previously ruled Earth, but for a few species of animals it proved to be an opportunity and a boon. Sea turtles, for instance, did not just survive the extinction event, they *thrived* under these conditions. Turtles had left their moist sand burrows and taken to water 170 million years ago. Their ability to dive deep into the sea, their cosmopolitan diet and laid-back metabolism turned out to be key ingredients for their survival. After the massive destruction of the Deccan events, the sea would have been littered with carcasses of the predators that had once terrorized the sea turtles, leaving the turtles to feast off the misfortune of their former enemies. Extinction in the sea was a gradual process, with one species dying followed by another and then another, providing the omnivorous turtles with a continuous supply of food. With a warmer climate, the re-arrangement of land and sea, and the creation of tropical ocean currents, the turtles' time had come and they began to colonize new regions.

It was not just the turtles who benefited. Ray-finned fish, the most diverse and ecologically dominant group of vertebrates on the planet, also made the most of this time. The fish survivors of this period were creatures of the open seas (or pelagic as marine biologists call it), away from the ocean deep and the margins of land where lava was pouring in.

The dinosaurs had ruled Earth for 160 million years but their reign ended abruptly and it now was a time of change. The Gondwanan land masses drifted away from one another and opened up the seas. This created new ocean currents which gradually began to transform the global climate. It would take another million years for the world to heal

Ammonites were perhaps the most enigmatic creatures in the seas, along with trilobites. They appeared around 410 million years ago and, like the dinosaurs evolved into amazing varieties during the Jurassic and Cretaceous periods, and died with them 65 million years ago. These molluscs were more closely related to modern squid and cuttlefish, than their modern shelled prototype, the nautilus. They get their name from their coiled rams' horns shape, which reminded Greek naturalist Pliny the Elder of the horns of Ammon or the Egyptian god Amun. They lived in the open seas and the deep; some were filter-feeders, and a few were predatory like squids, and ranged from a cookie to the tyre of a giant truck in size. They were prolific breeders. Female ammonites were up to four times the size of males, presumably to make room for their eggs. Given their diversity and fecundity, ammonites survived several episodes of extinction. Their fossils are found in many types of marine sedimentary rocks and are relatively common and easy to identify. This makes them a preferred choice for an indicator fossil and can be used to distinguish intervals of geological time of less than 200,000 years duration—a rather precise timeframe in geological terms. They are among the most abundant fossils that one can find in India and are commonly encountered within the round rocks along the riverbeds of the Himalaya and between the sandstone of Kutch and Tamil Nadu. They became extinct because of a sudden change in climate which caused the death of the planktons and other small creatures on which they fed.

itself from Deccan's devastation, but this was merely the blink of an eye in terms of Earth's history. With widespread death, both land and water awaited new claimants. In the absence of the mighty dinosaurs, two unlikely candidates began to stake their claim for dominance—birds and land crocodiles, both far-removed cousins of the dinosaurs. And watching this internecine battle from the sidelines were meek mammals that had not at this time the weapons or defences to compete with ravenous birds or menacing land crocodiles.

9

HUMBLE BEGINNINGS

Lava outpouring in the Deccan caused mayhem on a grand scale and left much of the subcontinent singed and charred. Viewed from space, most of western India—what is now Maharashtra, southern and eastern Gujarat, Madhya Pradesh, Goa, north Karnataka and parts of Andhra Pradesh—would have looked dark and desolate. Only a few slivers of forests, which escaped being charred by the lava flows, survived in the deep valleys along peninsular rivers. Greater India at this time had moved up into equatorial latitudes from where noxious emissions could spread more easily, not just to the rest of the world, but to the outer layers of the atmosphere as well. Every animal heavier than 25 kilograms perished. The trapped gases suffocated and smothered life, and the climate for a short while on Earth was similar to Venus's, with dust and carbon dioxide creating a mild greenhouse effect. The days were hot and humid with a bleak Sun that would rise above poisoned skies, and the nights, though cooler, provided little respite. Life became difficult

Magnolias were among the earliest flowering trees to emerge. Their large blooms with copious pollen and nectar attract insects and small mammals. Beetles, for instance, enter the youngest flowers which have only just opened and help themselves to the sweet juices exuding from the stigma as well as the pollen from the anthers. When the magnolia flowers open in the bright midday sun, beetles linger and warm themselves, and when evening comes and the upper petals close, they have no reason to leave their floral quarters because they remain safe and warm inside the flower. Many beetles continue to stay in their floral B&B until the petals fall off!

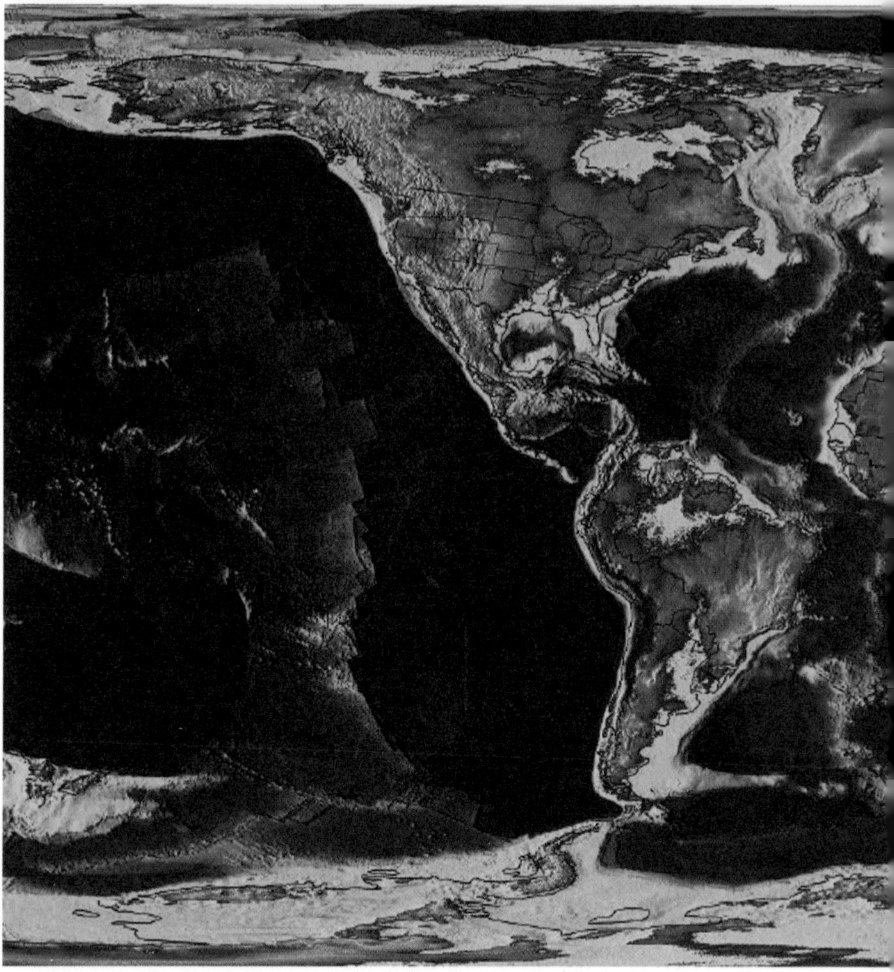

for those who survived, as vast numbers of both animals and plants that depended on one another for survival had died out. Every creature had to adapt quickly in these suffocating times. Or perish.

Over a span of 5 million years (from late 68 to 64 million years ago), each pulse of volcanic activity under the Indian Ocean pushed the Gondwanan land masses further apart. Africa, Madagascar and South America moved further west from India, and Australia had begun its journey eastwards. Each land mass carried with it a depleted set of survivors. The rearrangement of land masses created new ocean currents and as these grew stronger, the mild greenhouse effect began to wane and Earth began to cool again. Falling temperatures returned

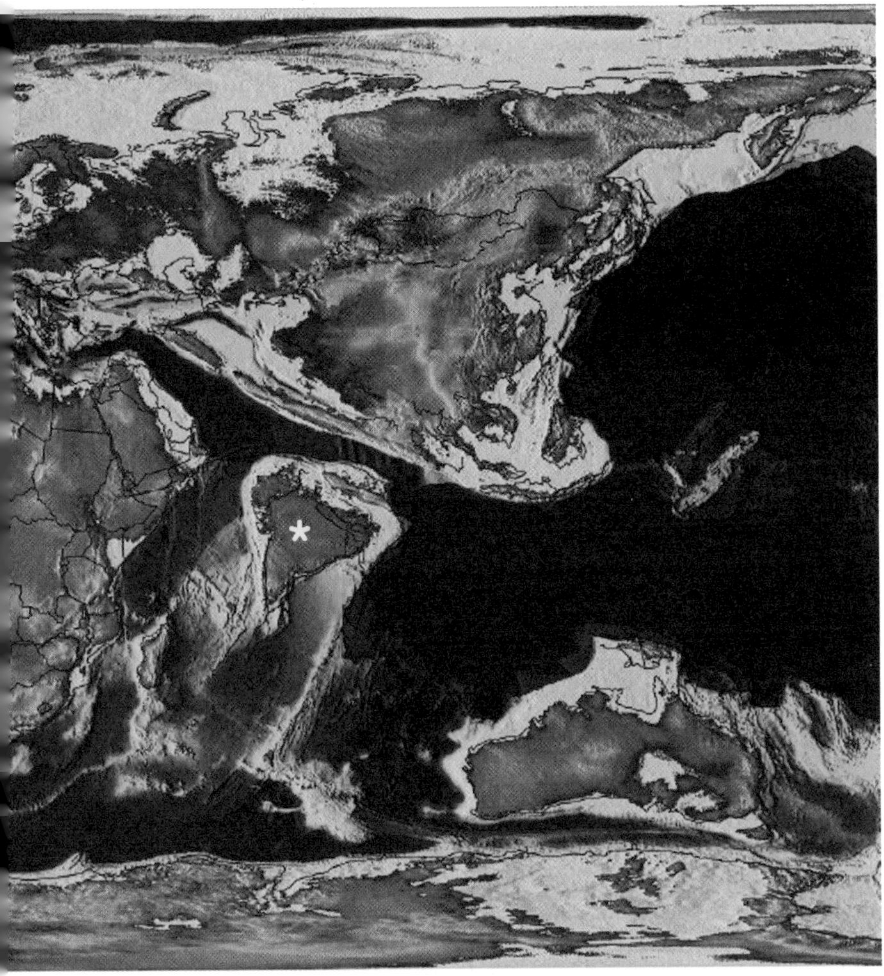

Nearly 56 million years ago, Greater India (marked with the star) was moving at a tearing pace towards Eurasia while constraining the Tethys, which had until now spread halfway around the globe, connecting the Pacific to the Atlantic. Along the Tethys, a long sinuous coastline began to develop, forming several shallow lagoons and islands. These islands and shallows became niches for new life forms to evolve and spread to recently created land masses. New Zealand and Antarctica (and Australia to some degree) had parted ways earlier and remained in splendid isolation.

moisture to the skies causing rain, and the seasons became defined once again. None of this happened quickly. It took nearly 10 million years after the gas holocaust before the world began to heal itself, but when life resumed it did so in full gusto and in the mid latitudes, in particular, tropical trees regained their former glory.

As dinosaurs became extinct, two very unlikely candidates battled to become top predators—birds and long-limbed land crocodiles. Early birds had descended from an ancestor of the bird-like dinosaurs and were covered in feathers and some fur. How they survived the wipe-out that decimated the dinosaurs remains a mystery. Perhaps they took refuge in the hollows of trees or in large burrows. The massive sea crocodiles from the dinosaur era had perished, but a few smaller land crocodiles had survived. Fossil evidence suggests that in Madagascar and northern Europe, flightless birds emerged as the carnivores-in-chief whereas Africa and Australia were dominated by giant land crocodiles that could run on land and hunt mammals and smaller reptiles. And in South America, the terror birds ruled alongside land crocodiles. But what of the Indian subcontinent? Currently, no fossil evidence of terror birds or menacing land crocodiles from this period has been found in India. But the ancestors of ostriches and another large flightless bird called the 'elephant bird' originated at a time when India and Madagascar were joined together. When the two land masses went their separate ways, the elephant bird evolved and dominated the forests of Madagascar, while the ancestors of ostriches were carried aboard India and made their way west into Africa only when India collided with Eurasia.

In the early 1980s, palæontologists worked tirelessly to piece together minute fragments of fossil bones from the skull, forearm and teeth of a mammal found in remote villages of Maharashtra, Madhya Pradesh and Andhra Pradesh. These were the bones of a tree-dwelling insectivore named *Deccanolestes* that had survived the Deccan event and its aftermath. For 160 million years—between 225 and 65 million years ago—dinosaurs ruled the world, but living in their shadows were small furry creatures like *Deccanolestes* (from Greek, *-lestes*: robber, presumably named because of its foraging behaviour) and its cousin *Sahnitherium*. These were ancestors of modern mammals.

Sometime between 280 and 248 million years ago, reptiles began to show mammal-like traits which included distinct mammal-like cusps in their teeth and jawbones, walking with a raised gait (instead of slithering), the ability to regulate and maintain warm-bloodedness, and stronger social instincts. The first true mammals emerged before dinosaurs had begun their domination and these

Early mammals like this insectivorous Deccanolestes *probably hunted for insects and worms on trees or the forest floor at night. It is perhaps from these times that mammals developed a keen sense of smell, sight and hearing.*

were the ancestors of some of the earliest mammals that emerged in Gondwana and other southern land masses. Early mammals from India like *Gondwanadon*, *Tikitherium* and *Nakulodon* lived between 240 and 210 million years ago in the Triassic period. They were small and shrew-like and lived in tree holes or underground, which would have served them well at a time when oxygen levels were low, and predators were menacing and mean. These difficult times forced mammals to diversify in order to survive. Scientists estimate that early marsupials—pouch-rearing mammals like kangaroos and wallabies—and placental mammals—all the rest of the mammals that we know, including ourselves—split into two separate lineages around 175 million years ago. Then around 125 to 80 million years ago, in a kind of revolution in the undergrowth, the ancestors of nearly all the placental mammals began to diversify and evolved into a variety of forms and shapes. And after the demise of dinosaurs 60 million years ago, within 5 million years, the ancestors of most modern families of mammals had established themselves on land.

All mammals of these times continued to be small and lived under the shadow of the new predators—carnivorous birds and land crocodiles. The largest mammals were about the size of a rabbit or a civet cat, and their kind had remained unchanged since the time of the dinosaurs.

Life for *Deccanolestes*, and other mammals like it, could not have been easy. A typical day would have been spent in terror. Every morning at break of day as the Sun's rays dissolved the mist on the forest floor, mammals would venture out timidly to hunt for insects or an occasional small amphibian or lizard and to forage for nuts, seeds and fruit. Although there was abundant food, they needed to collect as much as they could for themselves and their litter before giant reptiles and birds arose from their slumber in the warmth of the morning Sun and made a meal of them. Many early mammals evolved an acute sense of hearing and smell. This helped them detect small movements on the forest floor

and enabled them to catch their prey or locate a fruit lying in leaf litter or to scuttle away in time from an approaching predator. Mammals were agile and some had already developed hands and functional tails, and many could stand on their hind legs, and hop or climb trees.

Although giant birds emerged soon after dinosaurs died, by 52 million years ago they too became extinct in in all land masses except for the Americas. Terror birds like this Titanis *survived in the grasslands and open forests of South America and entered North America when a volcanic eruption created a land bridge (or the Isthmus of Panama) about 5 million years ago after which it colonized North America.* Titanis *fought off sabre-toothed cats, dire wolves and other large carnivores for its share of prey that included small horses, before it became extinct 1.8 million years ago.*

The giant flightless birds could charge through the undergrowth like rugby players outpacing any prey they sighted. They had enormous beaks and powerful talons and they hunted alone. Their fierce cries and heavy stomping would have caused panic among faint-hearted mammals, making them scurry for cover. Any mammal caught unawares would have been extremely vulnerable. The giant birds and crocodiles were fiercely territorial and produced a clutch of a few eggs every year. They also raided the nests of others and ate the young of their competitors.

Some mammals and smaller reptiles like snakes began to develop a taste for eggs, and stealing these and newly hatched chicks from nests and devouring them became an easy and nutritious option for them. Giant birds and land crocodiles produced very few eggs each year, and when a large proportion of these began to be eaten by mammals and reptiles, their numbers diminished. Also as their dominance over forests began to wane and therefore their access to food lessened, the birds and crocodiles began to shrink in size. Mammals, on the other hand, who were not only getting bolder but were also feeding on a nutritious diet of eggs, began to grow in size and started to challenge the territories of those who had once terrorized them. Over the next 20 million years—between 60 and 40 million years ago—fossil records show mammals growing in size, with bodies that would have made them more efficient at hunting, except in South America and parts of Australia where big predatory birds continued to dominate, and flightless birds gradually became severely restricted in their range and had to give up their lifestyle and find new strategies to survive. By now, birds on trees had mastered the art of flight and radiated into spectacular new species; to this day, they remain largely unchallenged in the air. Mammals were able to combat vicious birds and overbearing land crocodiles and other reptiles by stealth and adaptation. The meek had inherited Earth.

Mammals did not do this alone. Help came from an unlikely and unassuming ally. A quiet revolution had taken place in the plant world that aided the mammals' quest for domination. This was the development of tropical flowering trees. Before this, the landscape had been dominated by towering conifers, gingkoes and tree ferns.

Around 65 million years ago, flowers on tropical plants changed from being pale or white and largely inconspicuous to becoming large and fragrant. Until now, flowers had offered little by way of reward to attract pollinators, and relied on wind and water rather than insects or animals to help fertilize them. Gradually, some families of flowering plants began to make extravagant investments in developing flowers and fruits. Insects had already evolved more than 400 million years ago. The first flies had appeared when the first dinosaurs and mammals were evolving 248 to 208 million years ago, the first moths and butterflies around 130 million years ago, and wasps, which later evolved into bees and ants, around 100 million years ago. But with new conditions and a bounty of flowering plants, a riot of diverse insects exploded on the scene after the Deccan event. Beetles were most adept at taking advantage of new conditions and have since captured every niche possible—from feasting on rotting flesh and pollinating tiny flowers to eating inconspicuous fruits and nesting in the dung of herbivores.

With other kinds of food becoming hard to come by, some mammals too began to develop a liking for flowers, fruit and nuts. Insects were now paying regular visits to flowers which had become brightly coloured with interesting patterns and attractive scents. Mammals found a larder of nectar and pollen within flowers and of course, the luscious offerings of fruit and insects. Magnolias with their large and sweet-smelling blooms were among the earliest families of trees with large flowers. Others like the lotus, water lily and hyacinth evolved underwater or on its surface and developed flowers with colours and scents to attract insects that lived and bred in or around water.

By facilitating the exchange of pollen from one flower to another insects helped mix pollen, enabling the creation of new species. Fruits are an interesting and productive feature of plants. As a fruit developed, it remained bitter, hard and dull-coloured until its seeds came of age. Then it ripened, became brightly coloured and soft, and is transformed into a tempting proposition for fruit-eating mammals. Once the fruit was eaten, its seeds were dispersed along with the dung of the mammal or simply discarded on the forest floor where columns of sunlight offered them an opportunity to spring up and mature into trees. In close-canopied forests where little light reached the forest floor, trees developed large, fleshy fruit which bore large seeds. Each seed came equipped with its own larder of nutrition that helped it survive a long gestation as it waited for conditions to turn favourable for germination. The fleshy pulp also helped the seed travel long distances if carried in streams or in the gut of herbivores. This principle holds true even today—trees in dense tropical forests are endowed with brightly coloured pulpy fruit with large seeds and most of these are dispersed by birds and mammals. The tasty pulp serves as a reward for its visitors. It is a rich source of lipids, proteins and sugars and provides nutritional support for mammals during different stages of their life cycle. Mammals need more protein and less sugar while conceiving, but need more sugar when hunting to feed their new litter.

Plants in grasslands and deserts, on the other hand, developed other strategies to conquer new land. Their seeds depended on the wind or hitch-hiked on the fur of animals to venture into new places, and therefore these plants had small fruit with light seeds which could disperse far and wide.

Over 12 million years after the Deccan event—between 65 to 53 million years ago—flowering plants diversified into bewildering varieties that offered mammals a vast choice of pollen, nectar, fruit and foliage. As mammals increased their visits to trees, some began to evolve prehensile tails, opposable hind toes and mobile ankle joints which helped them climb and reach the fruit on higher branches. A few mammals developed a taste for tubers and roots, worms and insects, and they began to live closer to the ground, and gradually the length of their tails and body hair became shorter but their bodies grew larger.

Meanwhile, the humble grasses were biding their time, literally, at the feet of giant flowering trees. Grasses may seem to be an inviting proposition, just waiting to be eaten by a herbivore, but they have an efficient deterrent that prevents them from being indiscriminately devoured by all herbivores. The blades of all varieties of grass have a lining of silica along their margins called 'phytoliths'—tiny siliceous appendages that appear like tiles under the microscope and are neatly arranged all along their edge. This gives grasses a microscopically serrated edge which makes some grass blades razor sharp. Over several million years, as herbivores evolved, so did grasses, each with their own set of phytoliths with their unique shape and structure, enabling *specific* herbivores to eat only certain types of grasses. Very few herbivores are cosmopolitan.

Grasses had challenged the dominance of mosses and conifers but had been pushed back by trees when wet and humid conditions returned. As the climate became increasingly variable with cold and dry periods and less and less rain, conditions began to favour grasses once again. Grasses and their seeds need far less water, can survive heat, fire and freezing and, above all, withstand rough treatment from herbivores, all of which tropical trees cannot do. Some trees that endured and adapted to these changing times lived in clumps and grasses exploited the spaces opened up by dead trees. Roots of grasses hold on to the soil and prevent it from being blown or washed away, thus protecting the topsoil. Their shorter but denser networks of roots enable them to take up water faster and more efficiently than most other plants. Grasses also use diverse strategies to spread. Many varieties survive being uprooted and in some species even a single stalk of grass can strike root and give rise to a new plant. Tufts and clumps of grass grow from the base of the stem, close to the roots which ensures that if a stalk or blade is ripped off by a grazer, it can be replaced. The more grass is grazed upon, the more it gets pruned and the more robust it becomes over successive generations. Some grasses can even survive the chewing action and the assault of digestive acids in ruminants, so that when they are passed through dung, their seeds strike root in the soil. All this gave grasses a head start when forests were struggling to endure the vagaries of a changing climate. With sunlight striking the ground wherever the canopy of trees was punctured and few heavy browsers present, grasses had a field day! All grasses are flowering plants with small but copious flowers which produce enormous numbers of seeds. The pollen of near and distant cousins travelled through the air and by dint of insect pollinators, which spurred greater genetic exchange and over time, hundreds of new species of grasses emerged.

Although they developed an appetite for fruit and flowers, mammals did not take to leaves and grasses until 55 million years ago. Digesting leaves and grass requires larger stomachs where they need to remain for a long time for microbes and specialized acids to break them down, and being small, early mammals probably could not consume grass in large quantities. The small herbivores of today can eat grass and leaves despite the relatively reduced size of their gut because they have a more efficient chewing mechanism and possess a more evolved digestive chemistry.

As their diet of fruit, flowers and grasses and an occasional insect or a lizard became increasingly diverse, mammals began to develop different kinds of teeth. Every grass eater, from dinosaurs to mammals, evolved teeth to suit the type and intensity of its browsing and developed specialized teeth that could chew for hours. Some animals evolved high-crowned, deep-set teeth, and in some cases these continued to grow from the base when the crowns wore away. The back of the jaws of *Deccanolestes* had teeth evolved to cut and grind while its front teeth were designed to shear and slice. This became the template for early insectivores and omnivores, enabling

*A mammal perhaps similar to a linsang was among the first to develop a taste for meat, and it evolved to become the earliest true carnivore. Linsangs are slender, graceful and beautiful but are shy and rare and very little is known about their habits and behaviour. There are two species of linsangs in Asia and one in Africa. The species of the linsang (*Prionodon pardicolor*) shown here is found in eastern Nepal, the sub-Himalayan mountains of north-east India and Indo-China.*

them and their meat-eating descendants (ancestors of cats, wolves, hyenas, civets, toddy cats, pandas and bears) to consume their specialized diets. The herbivores that had seized the opportunity to feed on new types of grasses that evolved also developed a set of new grinding teeth.

In the past decade, scientists have made significant discoveries which have helped them reconstruct the tropical forests of this time and their denizens. Along the coast of Gujarat, in the Gulf of Cambay, near the town of Vastan, some 30 kilometres north of Surat, lies a lignite mine. In 2009, scientists discovered pieces of amber embedded between the layers of lignite which were formed 52 to 50 million years ago. Amber is hardened resin, a viscous gum that flows from the trunks of some trees. Some creatures, especially ants and beetles, are attracted to the sweet-smelling resin and often they stray into the path of this hypnotic golden blob, get trapped inside the viscous fluid and are permanently entombed. Because no microbe can now reach these insects in their amber tombs, they remain beautifully preserved with all their details amazingly clear and intact. Over 100 species of scorpions, spiders, bees and flies have been found in amber excavated from the Vastan lignite mines. The amber here was formed from the resin of sal and other large dipterocarp trees, and not just small insects but also small snakes, seven new species of bats, a possible marsupial, three new primates (a monkey called *Asiadapis* and two recently identified primates) and a tapir (*Cambaytherium*), among other mammals, have been found in Vastan! A single feather recovered from Vastan amber is thought to belong to a small, tree-dwelling bird which scientists have named *Vastanavis*. Most of the insects and other small creatures found in the amber show affinities to those found in northern Europe, South East Asia and even Australia. Mammals (especially simians like *Asiadapis*) were all closely related to their cousins from Eurasia. This cache of amber suggests that Greater India, despite being far adrift in the ocean, was not so isolated after all. A string of islands connected South East Asia and eastern Africa with India, which enabled species to hop across to these continental neighbours, and accounts for the carnival of plants and animals that migrated into Greater India over several million years. After the Deccan plateau cooled, its devastated plains and hills offered virgin ground for colonization and these island arcs served as the

About 50 million years ago, Vastan in Gujarat could well have looked just like this reconstructed scene of Messel, a famous fossil site in Germany. The dominance of land crocodiles and giant birds ended as mammals grew bigger and bolder. Birds had to

reinvent themselves and take on several shapes, sizes and fill newer niches just as mammals had done before them. But for crocodiles, going back to water and ruling it was their safest bet.

highway for many different plants and land animals to cross over into Greater India. Studies suggest that chameleons and lorises came to India from Madagascar, while lizards and frogs skipped across to Madagascar and Africa from India and were perhaps swept away from land to sea or transported on tree 'rafts' and ended up colonizing new lands.

Vastan fossils and amber also suggest that the ancient forests in that part of the country would have looked much like the rainforests of Malaysia or Borneo of today. The dominant trees in these tropical forests would have been dipterocarps (or 'two-winged fruit'), of which the handsome sal tree from the sub-Himalayan region is an emblematic representative that has survived to this day. In the region that formed southern Gondwana, starting from the Seychelles and encompassing Sri Lanka and India, 13 genera and 470 separate species of dipterocarps existed, some of which spread towards the north-east into southern China and the South East Asian islands (3 genera, 40 species). Some dipterocarps migrated west from the Indian Ocean islands into Africa and then reached South America (where only one genus comprising just a few species are found).

Several other tropical trees have also been identified in the Vastan amber from the lignite mines. Some of these perhaps had arrived from Africa via Madagascar and others from South East Asia. Amidst the rumbling of trucks and excavators that scour the lignite mines of Gujarat and Rajasthan to fuel the thermal power plants, scientists are working tirelessly to find new plant and animal fossils which will help reconstruct these glorious ancient tropical forests.

Mammals inherited the world not because of their size or ferocity but due to their mutable varieties of shape, size and feeding habits. This capability was complemented by growing intelligence arising out of relatively larger brains. With larger brains, mammals acquired superior social and survival instincts compared to reptiles and other challengers. The role of mothers became critical and it was their investment of carrying their babies longer in their womb and nursing them for a longer time that paid off as their children developed bigger brains, better social instincts and better programming for collaborative behaviour. As a result, mammals matured later but lived longer. Their strength was their ability to adapt and take care of their young, ensuring the survival of successive generations.

After this period, the landscape would not only be dominated by mammals but also be shaped by them. But in more subtle ways, it was flowers and grasses that were dictating how the mammals would evolve. The battle between flowering trees and shrubs, and the grasses was not only changing the landscape, but also influencing the design and size of mammals. During the floral revolution of the 60s (-million years ago!), mammals had begun exploiting every habitable place on land and water. They even pushed the boundaries of competition by taking to the air in order to reach more flowers and fruit. Bats evolved into the only mammals that truly mastered the art of flying, though a few other mammals took to gliding and parachuting to reach floral bounties. Grasses held a major attraction for the mammals on the ground. Over a period of 40 million years (between 46 and 6 million years ago) as herbivores grew bigger on a diet of grass, many carnivores too became larger, and to escape predation the herbivores grew larger still. This ushered in the age of giant mammals and the battle to rule the land commenced among mammals themselves.

The evolutionary drama of mammals was taking place against the backdrop of Greater India's collision with Eurasia. The sporadic thrusts of Greater India raised the highest mountains and plateaux and etched out the largest rivers in the world, but their most profound impact was to cool the atmosphere and by doing so, they created a new climatic order—one which would give rise to all modern life as we know it.

10

MOVING HEAVEN AND EARTH

If I were to ask you to pick one geographical feature that defines India, chances are most of you would choose the Himalaya. The northern boundary of the Indian subcontinent is demarcated by these mighty mountains and they give rise to the subcontinent's oldest river—the Indus or Sindhu—which lends its name to India. Yet less than 50 million years ago there was no Himalaya; there were only a few scattered volcanic islands in the north-west that had been created some 750 million years ago and whose remains can still be seen as flat-topped mesas in Jodhpur today. Barely half a year ago in the life of our time keeper, the 46-year-old Earth Woman, the *only* imposing mountains in the vast tract that lies from the Aravallis in the south to Siberia and even beyond were a few broken mountain ranges in Turkmenistan (the Altaids) and these, of course, had no influence at all on India in any way. And yet, you would be right in choosing the Himalaya as the defining geological feature of the country because, though it is only a very recent feature of India's geography, this stupendous mountain range has played a huge part in shaping India's landscape and climate and even its forests.

So how did this imposing, spectacular, literally life-changing feature appear out of nowhere? To tell the story of the Himalaya we need to

Batil Koh in north-west Pakistan was possibly the last volcanic eruption to occur in the subcontinent before Greater India collided with Eurasia. This picture of the vestiges of a volcano was taken by British geologists in July 1900. There are about forty small volcanic cones like this one in Baluchistan in northern Pakistan. Viscous lava oozed from these hemispherical cones which did not flow a long distance from them, and was confined close to their base. Over millions of years, the southern face of many of these volcanoes have been eroded, creating innumerable ravines or exposing basalt which have crumbled under the action of heat and cold in this mountainous region.

return to the grand ark of Greater India and follow its progress as it transformed neighbouring land masses and how these, in turn, shaped life as we know it.

Nearly 88 million years ago, after Greater India had broken away from Madagascar, intense volcanic activity under the ocean set it off hurtling northwards towards Eurasia (the unified and undifferentiated continent of Asia and Europe) at a great speed of about 230 kilometres per million years, which amounts to a little more than 31 centimetres per year—not the speed of your average car but the greatest speed that any large land mass ever achieved! In comparison, the subcontinent is crawling today at a speed of less than 5 centimetres a year.

As Greater India pushed onwards, it met an obstruction in the form of an arc of small, volcanic, Japan-sized islands. Between 80 and 65 million years ago, the subcontinental land mass slammed into these islands with such crushing force that its leading edge slipped *under* the islands, making the islands buckle and fold till they were stacked over the north-western edge of Greater India and completely subsumed

A volcanic dome in the foreground with Batil Koh in the backdrop. Notice the three colours of rock. The lower layer comprises a grey-white limestone through which a maroon and brown-grey pillow basalt spilled over about 52 to 47 million years ago. On the right of the maroon basalt are Buddhist rock paintings which are 1500 years old. Many of these have been vandalized, as here by education propaganda in Urdu for free books and education for children by the local government (it reads: मुफ्त किताबें, मुफ्त दाख़िला, मुफ्त तालीम, अपने बच्चों को आज ही बेसिक एजुकेशन कम्युनिटी स्कूल में दाख़िल करवाएं).

in the process. Geologists call this 'obduction' (*ob*: over; *-duct*: to flow). We can see remnants of these islands in the Kohistan-Ladakh and Dras arc, a low, 800-kilometre-long bow-shaped mountain chain starting from the Kohistan Mountains west of Nanga Parbat in Pakistan, passing through Ladakh and terminating near Nainital. Even today, these volcanic mountains stand out distinctly amidst the surrounding mountains and flatlands because of their composition—bare, dark-grey basalt in the midst of rich brown sedimentary rocks made from marine and riverine deposits, rich with vegetation. Some of these volcanic mountains have an overlying layer of marine sediment, a telltale sign that they were once surrounded by, perhaps even submerged under, a warm tropical sea. A small grey hill that lies west of Ladakh was unmistakably once a submarine volcano, where you can find fossils of shelled sea creatures. These volcanoes were once active but a slab projecting from the approaching Indian Plate severed the neck of the magma chambers like a knife, cutting off the supply of magma from below and sending the volcano into a long, cold sleep.

As Greater India continued its stately journey northwards, it experienced a second major episode of volcanic activity around 48 million years ago—the first being the massive Deccan lava flows 65 million years ago. This second round of vulcanism began under the east Indian Ocean where the Burmese and Indian plates started to rub against each another. Intense tremors broke off parts of the mountains of the Burmese plate and oozing lava melted these rocks together with rocks found on the ocean's floor, creating an amalgamated rock called ophiolites (*ophio*: snake, *-lite*: stone). Together, these broken bits of mountains and oozing ophiolites created the Andaman Islands (in Malay the Hindu monkey-god Hanuman is Andoman) that today lie roughly equidistant from Chennai and Kolkata in the Bay of Bengal. The Andaman Islands contain a series of conical volcanoes that are made from re-cooked rocks.

Over the past 70 million years, the sources of magma for all but two of the Andamanese volcanoes have dried up. Only Barren Island and Narcondam (from the Tamil *naraka-kundam*, meaning 'Hell's pit'), which rise 2000 metres from the bottom of the Andaman Sea, show some signs of volcanic activity. The Narcondam volcano has not erupted for at least 2 million years, which makes Barren Island the only active volcano within India's boundaries. Barren Island is covered by lush evergreen jungles on its southern and eastern sides, with freshwater springs at its base. The island is inhabited by feral goats, fruit bats, rats and parakeets. Since 2004, the Barren Island volcano has been spewing out ash with increased frequency. There are also some mud-spewing volcanoes in Baratang in the middle Andamans. Mud volcanoes are essentially mounds that exude water and mud which are forced out by hot gases from the depths of Earth, but they contain no molten lava or ash.

The Nicobar Islands lie south of the Andaman Islands, separated by a 150-kilometre-wide gap called the 'Ten Degree Channel'. The Nicobar are smaller than the Andaman Islands and were created by the violent thrust and rubbing together of the plates that lie below western Indonesia. The Deccan eruptions had propelled Greater India faster along its western edge and rotated it clockwise but the tectonic activity that created the Andaman Islands countered this swerve and pushed Greater India northwards at higher speeds along its eastern edge.

Barren Island is India's only active volcano. It is the northernmost along the explosive volcanoes which make the Andaman-Sumatra-Java arc. This region has witnessed some of the most violent volcanic eruptions of recent times, like Krakatoa in 1883; Tambora, the most powerful eruptions in recorded history which occurred in 1815; and the cataclysmic explosion of Toba 74,000 years ago. Barren Island may well be a sleeping giant. Before the illustration by British geologists in 1887 (top), there were other etchings by European explorers, although none as accurate as this one. Since early 2004, the volcano has simmered and the cinder cone has spewed a'a lava from its 500-metre summit into the sea. The activity has gradually increased since the 26 December 2004 Indian Ocean earthquake and tsunami. This photograph from 2013 (above) shows the lava flow entering the sea.

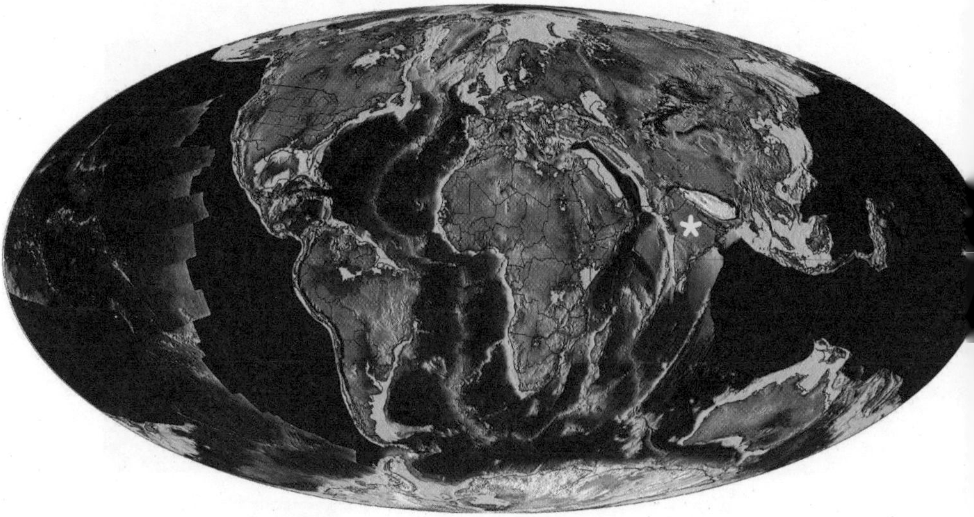

After India (marked with a star) subsumed the volcanic islands of Kohistan, the first contact between India and Eurasia was a 'soft collision' between 52 and 40 million years ago. As sea-floor spreading gathered force beneath the Indian Ocean, the Indian Plate rammed harder into Eurasia at several different points, raising the Tibetan plateau. The second episode of sea-floor spreading (along the Carlsberg Ridge) commenced around 34 million years ago and it led to the first upheaval of the Himalaya. A third phase of sea-floor spreading (between 16 and 11 million years ago) was intense and made the Himalaya a formidable range. As India got pushed intermittently for the next 8 million years, it folded new mountain ranges and raised the Himalaya even further.

As Greater India inched closer to the Eurasian Plate, pressure began to build all along its northern boundary as well as in Eurasia. As the two land masses drew closer, they exerted more and more pressure on each other until the stress became so elevated that one of the land masses had to give in. Greater India had travelled some 6000 kilometres northward by this time and, being lighter in density, was the one that literally buckled under the obduracy of the more dense Eurasian Plate and slipped *under* it, in a process known as 'subduction' (*sub*: under, *-duct*: to flow). This had the effect of hoisting higher a plateau that already stood 300–400 metres above sea level on the southern edge of Eurasia which created the Tibetan plateau, and its elevation continued to increase by another 200–300 metres over the next 10 million years or so. Because a part of Greater India lies directly below it, the crust under Tibet is *twice* as thick as most other continental crusts.

The relentless force of Greater India pushing northwards raised mountains and plateaux, and shaped the margins of Myanmar and South East Asia and Arabia, but also had one other important corollary. It squeezed out a sea.

A map of the world about 50 million years ago would have featured a wide body of ocean stretching halfway across the globe—roughly from where Gibraltar lies today all the way to Indonesia in the east. Of course, the ocean didn't have a name then but earth scientists have posthumously dubbed it the 'Tethys Sea' because we need labels! When the gap between Greater India and Eurasia began to close in the east with the Shillong plateau nudging Eurasia, the Tethys Sea was forced westwards. As the land masses of South East Asia, India, Arabia and Africa moved inexorably northwards, the Tethys started to break up into a string of salty lakes. These lakes were incubators for the evolution of many of the ancestors of a variety of modern mammals, fish, crocodiles, snakes like the python, and turtles. The ancestors of whales also emerged along the banks of the Tethys.

Remnants of one such massive lake, rich in calcareous sediments, can be found in an unexpected place. As you drive from Jaisalmer towards Barmer, about 10 kilometres short of Barmer town, you will come up on the village of Bothia (Lat: 25°9'N, Long: 71°4'E). Some 2 kilometres off the highway lies a local mine from which a pale white calcium-rich clay called fuller's earth is excavated. Fuller's earth is a calcareous sediment that was formed in an erstwhile Tethys lake. Popularly called *multani mitti*, fuller's earth is found at a depth of 30 to 50 metres under the arid land and has been traditionally used as a skin-exfoliating cosmetic by women in the subcontinent. In Hindustani, multani mitti translates as 'Multan's mud' and was so called because it is abundant near Multan in Pakistan from where it was exported to all over the subcontinent. From a palæontologist's point of view, what is interesting about fuller's earth is that it entombs and preserves specimens of living things beautifully. Life forms are preserved best in conditions that have very low oxygen like fuller's earth, for instance, because they retard the process of decay without deforming it. You can find complete fossils of coconuts, flowers and fruits, fish, snakes and even bats, which provide an insight into the flora and fauna that existed as the Tethys began to drain away.

Another fossil you can find here is of the custard apple (*Anona squamosa* or *sitaphal* in Hindi) or perhaps a close relative. Custard apples were believed to have been brought to India by the Portuguese in the sixteenth or seventeenth century from South America and the Caribbean, along with other fruits and vegetables like chillies, potatoes and tomatoes (can you imagine trademark Indian curries like dum aloo without these introductions?). But the presence of 46-million-year-old custard apple fossils in Barmer suggests that some of these tropical plants existed in India a long time ago and probably vanished due to changes in climate that were a direct corollary of the closing of the Tethys.

Even after India (no longer 'Greater India', because by now the land mass had fused with Eurasia) collided with Eurasia, sea-floor spreading in the Indian Ocean did not let up and this meant that the pressure of India pushing against Eurasia was unrelenting. India continued to slip under the Eurasian plate until it reached a depth where it encountered hot, molten magma. The 300 metre-high Tibetan plateau could not absorb the shocks and gradually layers of rock began to crumple and fold all along the Indian margin. It is these rocks which,

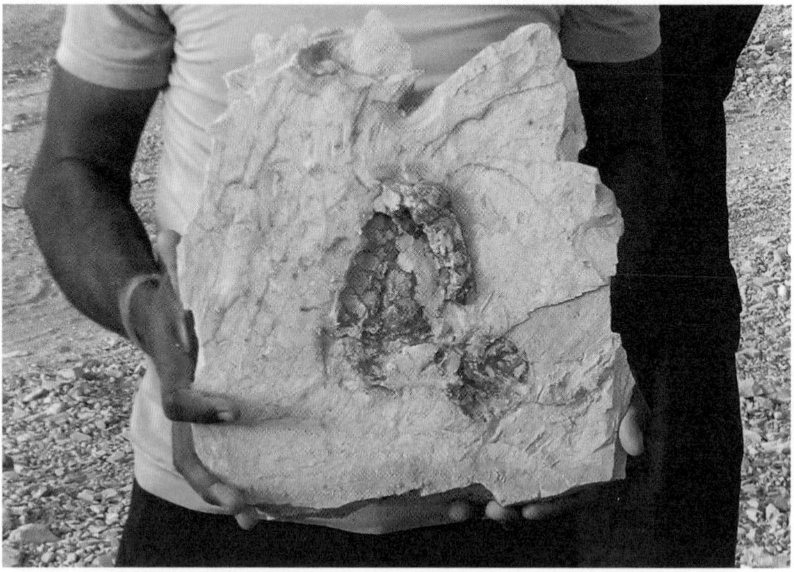

A custard apple fossil from 46 million years ago found in a fuller's earth mine in Barmer, Rajasthan. This fine-grained clay is found in several parts of India like Jharkhand, Gujarat, western Rajasthan, eastern Karnataka and Andhra Pradesh. Each deposit has a curious collection of fossils which are mostly imprints since this clayey mass is soft and hardens only when exposed to air. But do try to find a mine near you. You may soon be holding your first fossil of a plant, fish, crab or prawn. What's more, it is relatively simple to find one!

quite literally, rose to form the world's mightiest mountain range, The Himalaya. In some places, sediments from the Tethys coastline and seabed were also pushed up along with these rocks and were sandwiched between the rising mountains on both sides to form magnificent ranges.

It is important to understand that the entire Himalaya was not all created together or at once. There were several minor shake-ups but it was primarily three major upheaval events that fashioned the Himalaya. The first major mountain-building event occurred from 41 to 32 million years ago, the second from 13 to 9 million years ago and the third and final one from 4 million years ago to about 300,000 years before the present. These three tectonic upheavals gave the Himalaya its present form and played a decisive role in shaping the climate, physical environment and, indeed, the life of all modern mammals on the subcontinent.

Open a good physical map of the Himalaya or view it on Google Earth and you should be able to discern three distinct ranges. The Tibetan plateau is usually rendered as a dark chocolaty-brown landscape 4300 metres high with a frosting of snow. Though this defines the northern fringe, it does not count as part of the Himalaya. Just below lie the near-continuous tumble of grey-blue mountain ranges covered in perpetual snow. The northernmost of these ranges holds the highest peaks, in excess of 5000 metres, and is known as the Great Himalaya, varying in width from a little more than 200 kilometres to a few hundred kilometres. Next lies the 90-kilometre-wide belt of the Middle Himalaya with significantly lower peaks—between 3600 to 4600 metres high on an average—covered with snow only in winter. Lastly, the verdant Lower or Outer Himalaya, also known as the Siwalik Range, is made up of small mountains only 900 to 2800 metres high. The Siwalik vary from 10 to 45 kilometres in width and were the last to be formed, between 7 million and 300,000 years ago.

Geologists have delved deeper to study rocks that have raised the Himalaya. During the collision of Greater India with Eurasia, India moved northwards for a further 2000 kilometres, resulting not only in compression within the young mountain chain to the south of the Indus suture, but also uplifting of the 5-kilometre-high Tibetan plateau. The geography relates well with the rocks that lie beneath. Geologists

have found three major divisions (and several minor types) of ancient rocks south of the suture with Tibet. From north to south, these are the Tethyan Himalaya (rocks here are made from marine sediments), the Greater Himalaya (entirely made up of superheated like granites) and the Lesser Himalaya (largely sedimentary). These are separated by major faults—the South Tibet Fault system (also called South Tibet detachment) which separates Greater Himalaya from Tethys sediments; the

The mud-like rocks you see between the more 'rock-like' rocks were formed when sediments from the Tethys were being pushed up and squeezed between the two plates. After subsuming the Kohistan-Ladakh volcanic arc close to where the Makran-Indus trench lies today, India met with Eurasia roughly where the Shyok-Tsangpo rivers emerge. All along the Shyok and Indus you will find these Tethyan sediments sandwiched between the Eurasian boundary (on the right of the picture) and the invading Indian land mass (left). As subduction of the Indian Plate continues in fits and starts, mountain ranges (comprising the Karakoram, Himalaya, Tien Shan and the Burmese mountain chains) are being endlessly created and destroyed.

Main Central Thrust which delineates lesser Himalaya from the Greater Himalaya; and the Main Boundary Thrust which makes a boundary of lower hills from Lesser Himalaya. South of the Siwalik lies a trough or basin which runs parallel to the mountain chain and this is filled up by sediments that has been carried down by fast flowing rivers and deposited over the past 20 million years or so, and this region separates (called Main Frontal Thrust) the features made by the collision of Eurasia and Greater India, from the Peninsular shield.

The Himalaya rose from below. The rubbing together of the immense plates and the monumental crushing-buckling-folding of land produced a tremendous amount of heat and caused magma from below to ooze out of deep fissures which opened up on the surface. This melted and remelted granite, and pushed it upwards to the surface. As the granite slowly cooled, successive batches of molten granite thrust their way up, forcing the older granite slabs higher. Over time, this process created a pedestal for mountain building. Because the 'cooking' process varied (different types of granite are cooked at varying depths), the densities of rock slabs also differed. This created large cracks or 'faults' along places where the continental crust rasped, ground and pushed slowly onward. The melting and cooling of rocks and the movement of the plates continue till today, though on a smaller register.

You can see the building blocks of a Himalayan mountain in two different ways. The first requires you to climb to a high vantage point from where you can observe an entire mountain from its base to the very top. I recommend the east view of Kangchenjunga (Tibetan for 'five hidden treasures') from Pelling in West Sikkim because it is relatively easy to access. Once you get over the breathtaking majesty of the mountain, on a clear day you should be able to see that it resembles a three-layered cake. From the base rises a broad grey layer, about 6700 metres high, which is its granite foundation. Next is a 1220-metre yellow-tinged layer made up of limestone and marble. The limestone originated at the bottom of the Tethys Sea where it was formed between 251 and 240 million years ago and was scooped up in the process of mountain building. Hot molten granite cooked the limestone, turning it into marble which forms the distinct yellow band 150–200 metres wide that you can see. Right above this lies a thin layer of white-grey

It is seldom that you see the Everest without its cover of snow and ice. But it is only then that the famous yellow band is visible. The light-grey summit of the Everest contains fossils of small creatures that lived on a tropical shore about 350 million years ago, but these now sit 8500 metres above sea level. The rocks that underlie and make the base of Everest are granite which is about 50 to 30 million years old and were buried, melted and extruded up to the surface raising the older layers and that of the Tethys above them. Notice the parallel lines at the bottom of the mountain which reveal the episodes which cranked up the granite pedestal.

limestone followed by a band of dark-grey limestone only about 400 metres thick. This forms the peak of the Kangchenjunga and is made up of marine limestone that is rich in fossils of simple soft-bodied and shelled sea creatures.

There is, however, another, much easier, way of looking at what a Himalayan mountain is made of that does not require you to do any climbing. All you need to do is take a walk along a fast-flowing Himalayan river like the Ramganga near Corbett National Park. The rounded rocks that you will see in the riverbed are broken off bits of the face of a Himalayan mountain. As they roll down and are carried down by the force of rushing mountain streams, the rocks scrape against one another, which makes them smooth and rounded. You will also see a few large boulders and some rock debris scattered in the middle and along the banks of the river. If you manage to get close to one of these sooty grey boulders, you will see that they are embedded with small glass-like gemstones. The bluish and sometimes reddish glassy material is kyanite and the greenish one is garnet which is embedded in gneiss. These gems are formed in rocks that are buried 40 kilometres or more below the surface of Earth where they get pressure-cooked. Along the stream you will also notice large grey rocks with prominent white streaks. This is granite made up of silica and other minerals which was also cooked at great depth by molten magma and ejected out to make the base of the mountains. You will find no gemstones embedded in granite. Amidst all these rocks, there will also be some grey, rough and irregular boulders that do not contain any gems either. This is likely to be grey limestone which is less hard and comes from the top of the face of the mountains. All of these rocks in the stream are clues to what the various layers of a Himalayan mountain are made of. You don't see them in ordered sequence like you would if you could see the whole mountain, but to a geologist's trained eye, these riverine rocks harbour the whole story.

Palæontologists tend to view the Himalaya from a slightly different perspective. This breed of scientists is always on the lookout for preserved signs of life. Starting from their base up to nearly half their height, most peaks in the higher Himalaya are made of granite which holds no fossils within it. On Mt Everest, for example, the world's highest mountain at 8848 metres, the lowest slab of 4270 metres is a granite pedestal that is completely devoid of any signs of ancient life. The granite was formed about 50 to 30 million years ago and was buried, melted and extruded up to the surface, effectively melting and obliterating any fossils that might have been present in the Tethys bed that was lifted by the granite. Above the granite zone—starting at 4270 metres from the base—is the main body of the mountain. This is made up of a vast pile of rock of compressed mud, appropriately known as 'mudstone', and sandstone made from sediments that were deposited in the shallow sea at northern-margin of Greater India and would have held fossils of a variety of sea creatures including trilobites, were it not for the fact that these rocks are now too deformed to preserve any fossils. Just below the base of the summit pyramid is the famous 'Yellow Band', once a stromatolite-bearing reef of Cambrian age (540 to 490 million year old). Ascend onto the summit pyramid itself and you are in well preserved Ordovician rocks (which are 485 to 445 million year old) which contain trilobites and other marine shells, standing in silent testament to how the bottom of the ocean has become the top of the world. So, effectively, the bottom 4270 metres of Mt Everest is made up of more recent granite pedestal, while the top 4500 metres or so comprises layers of rocks which are 550 to 450 million years old.

If you can't go to the top to hunt for fossils, you can do the next best thing and look for them in a stream bed. To find an ammonite, look out for a round rock about the size of your palm. Chances are that when you crack open the rock along its rounded edge with a hammer, you will find an intact fossilized ammonite. It is relatively easy to find these fossils in the river valleys of Spiti and Ladakh, and along the rivers that course through Nepal. They are so common that roadside curio shops in Kathmandu sell ammonites of all shapes and sizes at modest prices. Ammonites survived all previous extinctions including the mother of all extinctions, the Great Dying, 251 million years ago, but they died out with the dinosaurs 65 million years ago. Why they died at this time and how they survived the Great Dying earlier remain a mystery for palæontologists.

The impact of the collision of the Eurasian and Greater Indian plates was not confined to the Himalaya alone. North-west of Kashmir lies a point called the Pamir Knot (from Persian, *pa*: foot, *-i*: of, *-mir*: the eminent one) which forms a terminus from where seven mountain ranges radiate. From space the Pamir Knot must look like the filamentous head of a hydra, spreading in all directions. About 2000 kilometres west of the Pamir Knot, in Iran, the shock of the collision created a wave-like mountain range called the Zagros. In the far east where the Himalaya peter out, the plates invaded deep under the modern-day Mekong region of South East Asia, raising a series of parallel mountains—the Arakan and Pegu Yoma—which run perpendicular to the Himalaya and through which flow major tropical rivers—the Salween, Mekong and Irrawady that empty into the Indian Ocean. On a map, if you follow the course of the Indus with your finger moving west from Mt Kailas, and trace the course of Tsangpo eastwards till it reaches the Red river (somewhere between the Irrawady and Salween rivers), this would roughly make for the margin of Greater India. All the land north of this margin was subducted under Tibet. Some geologists estimate that as much as 40 per cent of the Greater Indian land mass was pushed under Eurasia and was swallowed up by the mantle.

Rivers are an important part of the Himalayan story. The Indus and the Brahmaputra were already formed when the great mountains began to rise. Although gigantic tectonic forces caused the rise of the mountains and calibrated the height of the Himalaya, it was water that etched its boundaries. The water of streams that descend from mountains may seem much too innocuous to have any significant impact on the mighty Himalaya but steady erosion by water is a

Mount Kailas is the oldest mountain in the Himalaya, and from its base emerge the Indus and Brahmaputra rivers, which roughly outline the northern margin of Greater India before it collided with Eurasia. In the foreground is a herd of the docile Tibetan wild ass or the kiang. The rising mountains isolated an ancestor of the kiang around 2 million years ago and as the mountain rose, they were further isolated into different groups. This led to the evolution of three distinct sub-species about 1 million years ago, or perhaps even more recently.

powerful force and has a cumulative effect. The closing of the Tethys had created small lakes and rivers into which water from the Tethys had drained. When the Tibetan plateau was raised up, old rivers in Tibet began to flow towards the south and south-east. Even when the mountains began to rise, these rivers stayed their course. As the mountains folded and rocks began to emerge from below, the rivers cut through these young rocks and continued on their paths, overcoming or subduing new obstacles as they were created. In a sense, mountains developed *around* the Indus and Tsangpo/Brahmaputra rivers rather than the other way round. The Himalaya thus came pre-endowed with the Indus and Brahmaputra and the valleys, gorges and sheer faces—some as deep as 4570 metres—that you see today were etched by the power of these tumbling streams and fast-flowing rivers relatively recently.

The Indus is the oldest Himalayan river and its course was etched out in soft marine sediments between 45 and 36 million years ago after the Tethys closed. The Indus flowed to the west and, 5 to 7 million years later, a second river began to flow east from Tibet—the Brahmaputra. The fount of both these rivers lies close to where the oldest mountain in the High Himalaya, Mount Kailas, rose. The Indus has stayed faithful to its course for the past 18 million years, with little deviation. As it leaves Mount Kailas, the Indus travels west and cuts through volcanic terrain, forming tortuous gorges in Ladakh and Zanskar, before crossing the spectacular Karakoram Range and entering the gently sloping valleys of western Punjab. Here it turns south and travels 1200 kilometres before off-loading its montane sediments into the Arabian Sea. The Brahmaputra and Indus, which originate only a few hundred kilometres from each other, travel in opposite directions, sculpting some of the deepest gorges in the world. These two rivers are about the same length—roughly 2900 kilometres long—and wend their separate ways east and west from Kailas in Tibet for about 1000 kilometres, veering slightly southwards to mark the end of the Himalaya before emptying themselves into the Indian Ocean.

During the second major upheaval, from 13 to 9 million years ago, the Tibetan plateau rose up and dammed the rivers and streams that fed the Indus. The monsoon also grew in intensity at this time and the larger volume of water run-off forced itself through the rocks to create new courses and these became tributaries of the Indus: the Sutlej (which would become its largest tributary), Ravi, Chenab, Jhelum and Beas.

These tributaries pass through Punjab—the land of the five rivers—creating a fertile alluvial plain before joining the Indus in Pakistan. Until this time, the Yamuna had been a major tributary of the Indus but under the new arrangement it parted ways, drifted eastwards and was 'captured' by the Ganga. The Yamuna originates less than 150 kilometres east of the Bhagirathi, a major fount of the Ganga, but runs parallel and south of the Ganga for most of its course, merging only in the holy city of Allahabad in the centre of the Indo-Gangetic plains. The Yamuna is the Ganga's longest and best-known tributary but the crown for the largest tributary in terms of volume goes to the Ghaghra (called the Karnali in Nepal), which carries glacial melt and sediments (including rocks from

The Namche Barwa marks the end of the 1200-kilometre Himalayan range on the eastern side. This photograph, taken from the Chinese side, is of the Tsangpo Valley where the river makes a hairpin bend and enters India. This region and the lower hills below Namche Barwa are off limits for Indians. The Tsangpo becomes the Dihang and 80 kilometres downstream is where it is called the Brahmaputra.

the Himalaya aged between 16 and 5.2 million years), and meets the Ganga near the city of Patna. Older rivers like the Chambal, Betwa and Sone that once flowed into the Tethys from the Vindhya and Satpura mountain ranges also joined the Ganga, and swelled it into the mightiest river of the subcontinent. The course of the Ganga was carved by rivers that merged together between 16 and 11 million years ago.

The Brahmaputra once wandered a lot further east than it does today—about 350 kilometres or so—and followed the course of the present-day Red rivers, like the Irrawaddy in Myanmar and others that flow into the Andaman Sea. But around 18 million years ago, a part of the Irrawaddy broke off to form the Yarlung Tsangpo river that turned inwards into India, taking a hairpin bend at the base of the majestic 25,600-foot Namche Barwa mountain. In India, this river is called the Brahmaputra. The Brahmaputra was gradually deflected west as the Shillong plateau grew and the mountains to its east, along the Indo-Burmese border, began to fold. Flanked by the Brahmaputra, Namche Barwa marks the eastern extremity of the Himalaya. The Brahmaputra is given different names by various local tribes as it travels through Arunachal Pradesh and Assam. After entering the low hills of Assam, it flows southwards for another 600 kilometres before it merges with the Ganga in Bangladesh (where it is called the Meghna) and the two rivers empty their waters and sediments into the Bay of Bengal. The two rivers, the Indus in the west and the Brahmaputra in the east, define the western and eastern extremities of the Himalaya.

Smaller rivers too have played an important role in deciding the fate of larger rivers and are likely to continue to do so. The Arun river rises 150 kilometres north of Mt Everest and flows between Everest and the Kangchenjunga, cutting through a gorge about 5500 metres deep. Although its headwaters lie less than 5 kilometres from the Brahmaputra, these two rivers do not meet. Instead, the Arun meets the Ganga in Katihar district of Bihar. The Arun is a violent river and gathers its force from the steep terrain it traverses. The Arun has slowly but inexorably began cutting its way through the narrow rock that separates it from the Brahmaputra. If it succeeds in cutting through, the Arun will hijack the Brahmaputra's waters, reducing the mighty Brahmaputra to a trickle, and carry its water down the Ganga. If this were to happen, there will be mayhem downstream in Bihar.

As new mountains have risen and new rocks have been forced upwards, the course of some rivers has changed and a few have even been marooned into becoming lakes. Till about 250 years or so ago, the Ganga and Brahmaputra did not meet at all and were more than 200 kilometres apart in present-day Bangladesh. Then for reasons not quite clear, the Brahmaputra changed its course and began to flow east of the Madhupur forests where it joined the Meghna, east of Dhaka, and hijacked the Teesta river (which until then had flowed between the two), thus creating a gaping riverine mouth before flowing into the Bay of Bengal.

Occasionally the intensity with which rivers arrive in the plains is so implacable that they create new paths for themselves. On 18 August 2008, the Kosi—a tributary of the Ganga which careens out of Nepal into Bihar—witnessed one of the greatest fluvial disasters in recent history when it shifted course by nearly 120 kilometres, displacing more than 5 million and killing an unaccounted number of people. The Kosi flood was significant in two ways. First, the river moved eastward from its charted course, breaking the trend of westward migration that Himalayan rivers had followed for the last 200 years. Second, it was the single largest movement of a river recorded in historical times.

But it is not just temperamental rivers that are moving—the tectonic plates that underlie the Himalaya are restless too. The shifting and adjusting of plates cause small and big earthquakes, some of which have resulted in large-scale devastation. At 7.40 p.m. on 15 August 1950, when India was celebrating Independence Day, the town of Pasighat in Arunachal Pradesh was rocked by an earthquake so powerful that it swept away part of the town. The violent earthquake was caused by the sudden shifting of the plates below. The epicentre lay somewhere along the Indo-Chinese border and shocks were felt across India and beyond—from Kolkata and Patna to Thimphu, Dhaka and Yangon. The force of the quake was such that the course of the Brahmaputra was instantly altered, causing flash floods and tremors that killed an estimated 1500 people in what was a sparsely populated region. It was the sixth-largest earthquake of the twentieth century. Vast areas of mountain and valley were elevated or subsided, altering the drainage

The Charbari glacier, moraine and rock deposits in Mahapanth, Kedarnath, 1888. Kedarnath lies within the Great Himalayan range and from amongst its peaks the headwaters of the Ganga rise. Several sacred Hindu shrines are situated along its ice-choked valleys.

of the entire region. The aftermath of the earthquake was even worse. In the weeks following the quake, huge landslides dammed rivers and streams which breached every few days, causing flash floods and obliterating many more villages. Entire rock massifs collapsed into rivers, creating waves as high as 20 metres. The town of Pasighat had to be relocated from the northern to the southern bank of the Siang river with only a memorial left behind to commemorate those who had died.

Increasingly, the behaviour of these mountain rivers has become more and more erratic. Here's why: first, changing climate patterns have affected the volume of meltwater released from glaciers which, coupled with a variable precipitation in each monsoon, has made these rivers extremely unpredictable. Second, the unplanned building of roads, dams and towns with complete disregard for the complex interplay of geology and topography has altered drainage patterns.

One of the worst incidents of devastation in modern times occurred in June 2013 in the Uttarakhand hills. That summer, run-off from snowmelt was at its peak, and when the monsoon arrived early, rivers swelled and began flowing above the danger mark. On the night of 15 June, a Saturday, the rain intensified and on Sunday morning a large part of the Charbari glacier—which over the past decades had become a depleted block of ice—collapsed under the pressure of water, creating a sudden, massive deluge. Within minutes, the Mandakini river became a raging torrent that breached its banks and sent down a crushing wall of water, ice and rock through the pilgrimage town of Kedarnath (which stands on its left bank), wiping out several hundred villages. Houses, schools and hotels toppled into the torrent, and bustling bazaars were swept downstream. Thousands of kilometres of roads were lost in devastating landslides. The media dubbed this the 'Himalayan tsunami'.

The rise of the Himalaya had another crucially important corollary. The wall of high mountains presented a solid barrier to the circulation of winds and clouds and led to the creation of the Asian monsoon. Around 16 to 13 million years ago, the monsoon (from the Arabic and Urdu word *mausum*, which means season) was just a weak and erratic weather phenomenon which brought spotty, intermittent rain to the Indian subcontinent. Around 8 million years ago, after a period of intensive uplifting, this chain of mountains became an imposing barrier for moisture-laden clouds that came from the south, and the monsoon gained steadily in vigour and intensity.

Even before the Himalaya became a wall, peninsular India and the Tibetan plateau trapped a lot of heat in the summer. When the mountains rose, the northern face of the Himalayan ranges prevented cool air from northern Tibet and the Arctic tundra from entering the Indian Peninsula, making the area even hotter and creating a large area of low pressure. The rain in itself is not the monsoon; it is the winds which bring the rain that are called 'monsoon'. These winds arise in the Indian Ocean, and carry moisture for 4000 kilometres, towards the low-pressure area in the Indian Peninsula. After shedding part of their load in southern India and the Western Ghats and northern India these clouds reach the ultimate barrier—the Himalaya. This makes the middle of July in the northern plains muggy and hot, while southern and north-eastern India begin to receive the early monsoonal showers. This pattern is repeated every year. From May to October the monsoon is determined by the intensity of heating of the Himalaya and peninsular India and the evaporation that takes place in the southern Indian Ocean. Even a small degree of change in temperature or wind can make the difference between a plentiful monsoon and a dissipated one, affecting the lives of over 2.5 billion people of the Indian subcontinent who depend on agriculture for their livelihood.

The arrival of the monsoon is a truly spectacular event. 'Upon what a gigantic scale does nature here operate,' exclaimed British botanist Joseph Dalton Hooker, who watched the march of the clouds from the Sikkim Himalaya in 1848. 'Vapours raised from an ocean whose nearest route is 600 kilometres distant are safely transported without the loss of one drop of water to support the rank luxuriance for this far distant region. This and other offices fulfilled, the waste waters are returned by the rivers to the oceans, and again exhaled, exported, recollected and returned.' Now that we understand the monsoons a little better, we realize that Hooker was incorrect to exalt the miraculous powers of monsoons' containment. We now know that there is much leakage on the monsoon's journey from the oceans on land but our admiration of this seasonal phenomenon and its munificence is no less extravagant.

Another distant but potent force that plays a big role in determining the intensity of the monsoon is Antarctica, the world's last remaining wilderness. Its permanent ice cover regulates the world's climate like no other land mass. The South Pole (or Antarctic) is frozen throughout the year, but in winter (which officially lasts from mid-March to mid-September, although reaching the South Pole is almost impossible from mid-February to mid-November due to unpredictable winds and blizzards) it increases in size, accreting nearly 30 per cent more ice along its margins. This vast frozen continent remains plunged in darkness for about six months every year and practically no sunlight reaches its surface during this time. It seems ironic that such a bleak, icy wasteland with so little life should play such a significant role in supporting life on land and sea, by maintaining the balance of energy on Earth.

Although both poles are extremely cold—mostly because not much of the Sun's energy reaches them and the little that does is reflected back by the white ice—on average Antarctica is 6°C colder than the Arctic. The fringes of the Arctic region thaw out in summer and an amazing array of life, ranging from insects to wolves and whales, can be seen there, but Antarctica remains permanently and forbiddingly icebound. The explanation for Antarctica being so much colder than the Arctic lies in its distance from other land masses and the depth of its surrounding ocean. The Arctic is connected to Greenland, Canada and northern Russia by islands and recently sunken land bridges, and the Arctic Ocean is quite shallow at several places. Antarctica, on the other hand, reached its current position around 25 million years ago and its last remaining land link (with South America) was severed around 23 million years ago when the Drake Passage, which joined the Pacific with the Atlantic Ocean, was formed. This gap widened as the sea floor continued to spread.

To understand how Antarctica affects our monsoon, we need to look at how ocean currents 'work' in the southern hemisphere. When water freezes in the oceans that surround Antarctica (which, quite unfairly, does not have an ocean named for itself, unlike the Arctic, and is instead surrounded by the Indian, Atlantic and Pacific Oceans), salt is separated from the seawater and only pure water is frozen as ice. The ice floats on the surface and most of the salt is pushed deep into the ocean. Antarctica is, as a result, surrounded by the saltiest ocean water on Earth! The concentration of salt also makes the water very heavy causing the ocean current to sink and circulate in deep canyons and trenches in the seas. With no land barrier obstructing it, this cold, dense, salty polar water moves like treacle along the bottom of the ocean towards the equator, and to take its place, warm water from mid-depths to the surface travels from the equator towards the South Pole. In this way, sea ice creates the ocean's global 'conveyor-belt' circulation, centred on the powerful Antarctica Circumpolar Current in the southern oceans which gradually mixes with lighter tropical currents. When cold currents from Antarctica meet warm tropical air, it creates an enormous storm system with massive rotating hurricanes. Howling winds (called the 'Screaming Sixties') driven by a wind system called the 'polar jet stream'

sweep across the icy land mass and open seas. The jet rotates clockwise in congruence with the spin of Earth. The strong frigid currents and the ripping winds that originate from Antarctica make the southern Indian Ocean the most dangerous open seas in the world. Antarctica on average is the coldest, driest and windiest continent, and has the highest elevation among all continents.

This is the image of the delta of the Ganga and Brahmaputra at the peak of monsoons, from where the sediment are carried deep into the Bay of Bengal to form the Bengal Fans. Compared to other rivers, sediments deposited by the Bengal Fans are the thickest and largest in size, and extend several thousand kilometres.

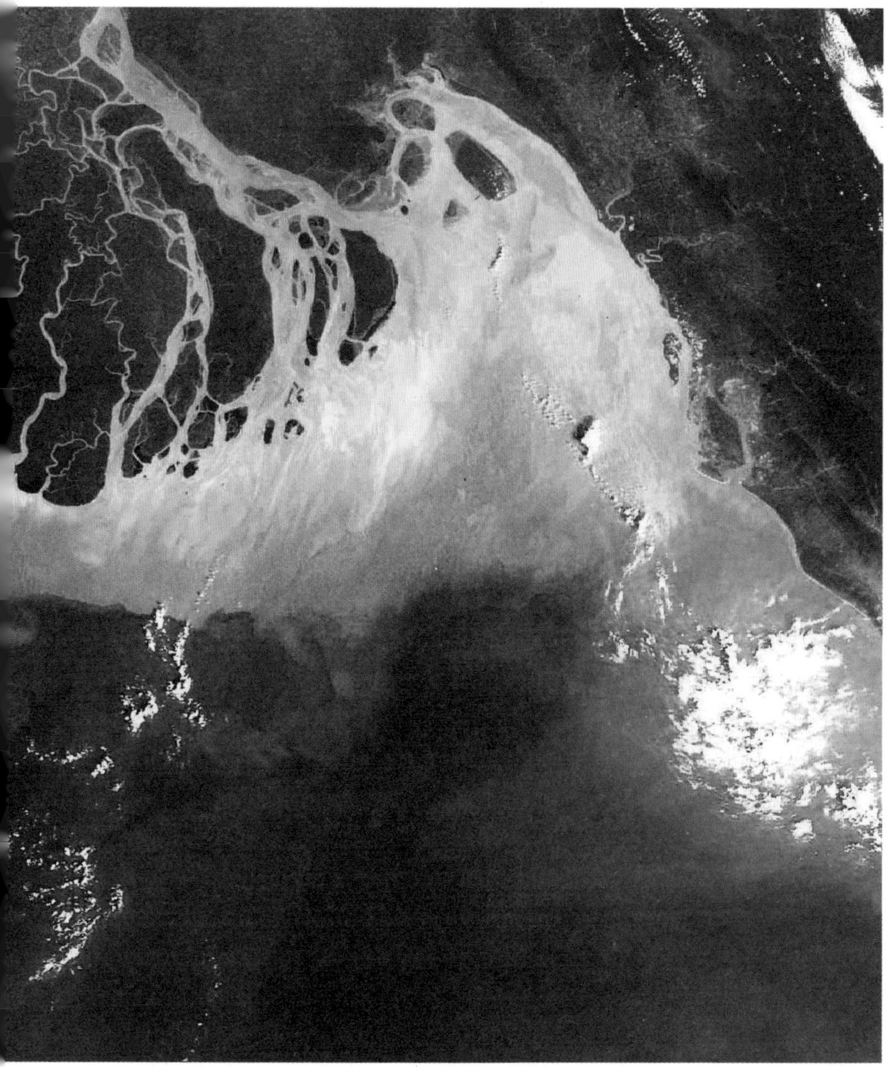

The monsoon depends not only on the intensity of heating of peninsular India or the slopes of the Himalaya and Tibet but also on the intensity of winter in Antarctica. There is an inverse relation between the extent of ice cover in Antarctica in winter of one year and the monsoon rainfall in peninsular India in the following year. When the ice cover is small, more area is available for evaporation and a greater volume of moisture gets across the equator and tropic of Cancer, thus increasing the intensity of the monsoon. More melted ice also dilutes seawater, which also causes water to evaporate more quickly. Together, both these factors work to make a stronger monsoon. In a year, if all these factors play out right, an average monsoon will drop nearly a trillion tonnes of rain on India!

There is another little-known but big corollary to the Himalaya that has to do with its influence on the temperature of our planet. It's not as obvious or dramatic as the stupendous monsoon but it may be just as important. We have already discussed (Chapter 4) how, when Earth's atmosphere was stabilizing, the proportions of oxygen and carbon dioxide had a profound impact upon the temperature on Earth. In the 1980s, scientists discovered that the main component of climate change was a phenomenon called the 'greenhouse effect', a dirty word in contemporary climate change debate with an entirely negative connotation. Very simply, the greenhouse effect works like this: Greenhouse gases (or GHGs—mostly CO_2 and methane) trap heat from the Sun's rays which warms up the planet. In the 1990s scientists discovered the vital role that the Himalaya and its rivers play in moderating this effect. If the level of CO_2 increases, more heat is trapped and the temperature rises. Once in the air, CO_2 combines with rainwater, making a weak (carbonic) acid. About 90 per cent of Earth's crust is made up of silicates—quartz, clays and zeolites and other complex mineral structures. When this mildly acidic rain falls on rocks containing silicates, it reacts with these rocks, eroding them and forming carbonate compounds that use up the CO_2, effectively removing it from the atmosphere. These eroded rocks containing compounds of carbon dioxide are carried by rivers and deposited into the depth of oceans where organisms like molluscs and corals use them to make their shells. When these organisms die, their shells sink to the bottom of the ocean and become part of the ocean floor. This is known as 'carbon burial'. So when you see glistening shells on Indian beaches, remember that their building blocks could well have originated in the high Himalaya! Most of the volume of CO_2 in the atmosphere actually comes from volcanic activity and sea-floor spreading. When sea-floor spreading occurs, sediments on the ocean floor (including these shells) are dragged deep under the ocean floor where they are heated and the trapped CO_2 is released. This is called the 'carbon cycle', and is responsible in very large measure for fluctuations of carbon dioxide (and other carbon compounds). Without it, and other greenhouse gases, Earth would be a frozen world. But increased burning of fossil fuels, forests and biomass and industrial emissions have added nearly 30 per cent more carbon dioxide to the air today than there was about 150 years ago, and Earth is becoming a warmer place. This is why the carbon burial that takes place in the Indian Ocean is a hugely important phenomenon to keep climate altering processes in balance.

When Tibet and Himalayan mountains were uplifted, it caused the monsoons to intensify. The heavy rains caused by the powerful monsoon pummelled the slopes of the Himalaya and increased the rate of erosion on these slopes. The eroded young rocks, in turn, were carried by Himalayan rivers and deposited into the ocean, causing a greater burial of carbon and silica and depletion of carbon levels in the atmosphere. The temperature on Earth decreased significantly and the poles froze. Over millions of years, these rivers that have come down from the mountains have become the most efficient global storehouse for carbon dioxide. If you were to look at the places where these rivers meet the ocean, you would see immense layers of deposited sediments creating a delta and when sediments get carried further into the deep sea they spread out a bit like Chinese hand fans, and are simply called fan.

Sediments from the Indus in the Arabian Sea suggest that the river was formed nearly 45 million years ago, and that it was a mighty river for several million years before its power began to wane gradually about 6 million years ago. The Indus Fan is 10 kilometres thick and spread as far as Kerala, covering a distance of 1600 kilometres in the Arabian Sea! It alone has transported nearly 5 million cubic kilometres of sediment, and this is only about *a third* of the sediments deposited by the giant fan made by the Ganga-Brahmaputra in the Bay of Bengal. The volume deposited by the Ganga and Brahmaputra is so massive that it forms several smaller fans that are collectively called the 'Bengal Fans'. The Bengal Fans covers an area of more than 200,000 square kilometres (slightly less than the area of the state of Uttar Pradesh) and are 22 kilometres thick. Together, the Ganga, Brahmaputra and Indus deposit more sediment than all other rivers of the world combined!

During a short warm period around 4 million years ago, sea levels rose and seawater invaded deep inside the Ganga-Brahmaputra delta. This prevented sediments of the Ganga and Brahmaputra from reaching the sea and, instead, these sediments were deposited around the mouth of the river. The turbid ebb and flow of tides created a forest of mangrove trees here with their unique adaptations of waxy leaves, air-breathing roots, and stilts that prop up their trunks. Mangroves are specially adapted to survive flooding by both fresh water and seawater. This continuous estuarine mangrove forest extends for over 10,000 square kilometres along the delta in West Bengal and Bangladesh and is known as the Sunderbans. This massive swathe of mangrove forest drops its leaves and organic litter directly into the sediments of the rivers, further increasing carbon capture.

The Ganga is more efficient than the Brahmaputra or the Indus in mitigating the greenhouse effect because the clayey soil and silt of the Ganga contains more silicates and therefore captures more CO_2 within it. It is estimated that every year during the monsoon, 1–2 billion tonnes of sediment is transported by the Himalayan rivers that gets buried several thousand metres below the sea in the Bengal Fans. Over the past 15 million years, the Bengal Fans has buried more than 15 per cent of all global carbon. The Himalaya, its rivers and their deltas and fans are, therefore, a key locality for offsetting the world's carbon cycle and some geologists are concerned that pollution, soil erosion and vanishing waterbodies along the Ganga and its tributaries will reduce its potential to capture and bury CO_2, and intensify global warming.

Over the past 50 million years, the continuous mountain-building process has made the Himalaya a place of superlatives. Himalayan ranges are endowed with fourteen of the world's tallest mountains, and Everest (or Chomolangma as the Tibetans call it) rises above all of them at 8848 metres. The Himalaya is the youngest and fastest-growing mountain range in the world—rising nearly 10 millimetres a year at Nanga Parbat. It is estimated that many of its peaks will continue to grow for another 10 million years and add 300 metres more to their height despite having the highest rate of erosion (up to 12 millimetres per year). The slopes of these peaks harbour the largest concentration of glaciers outside of the poles—one reason why they are called 'Himalaya' (Sanskrit for the 'abode of snow'). The Himalaya makes for the biggest climatic barrier and one of the most influential geographic structures to shape the world's climate. For its profound influence and role in regulating the region's water cycle, scientists have designated the Tibetan plateau and the Himalaya the 'Third Pole'.

The Himalaya, Tibet and their rivers have influenced all modern life and the evolution of all modern families of flowering plants and mammals took place along the Tethys. There is another crucial way in which the Tethys had an important impact although this one is from a more narrow human angle. While the supercontinents of Pangæa and Gondwana endowed us with the bounty of coal, it was the Tethys that gave us much of the world's petroleum reserves. Apart from that, along the shallows of the seas (in Kutch, for example) the Tethys created large, flat salt pans containing a variety of salts, like saltpetre, which during medieval times was used to make gunpowder and spurred an interest in chemistry.

The modern geological era in India began with Greater India docking with Eurasia and the subsequent rise of the Himalaya. This affected global climate and transformed the landscape across South East Asia—a change that favoured mammals and they emerged as inheritors of the world. A carnival of mammals arrived from Eurasia into India, and India became a highway for some and a haven for many. Diminutive mammals gradually became bigger and bolder, and the land once again trembled beneath massive grass-eaters that evolved along with vicious carnivores. The golden age of mammals had arrived and unlike their predecessors, the dinosaurs, their ability to adapt and survive ensured that they dominated almost every niche in nature.

Robert

11

THE CARNIVAL OF MAMMALS

The first mammals evolved around 210 million years ago, around the same time or perhaps even a little before the first dinosaurs made their appearance. From a handful of diminutive shrew-like ancestors, mammals have diversified into a staggering 4300 (or so) species, both small and large that live today. There tends to be a certain natural proportion between large and small animals, be it dinosaurs, reptiles or fish, and mammals were no exception. For example, there are only a few species of large-sized mammals such as whales, rhinos or elephants compared to multitudinous smaller ones like mice, mongoose or deer, simply because there are finite limits to what any environment can support. But we are looking here at *survivors* in the great game of evolution, at the way things have played out, not just at nature's own fecundity or inventiveness. It is often the case that groups that are made up of only a handful of surviving species were once a lot more prolific.

The earliest primates to evolve were small and ranged from the size of a thumb to the length of the palm of an adult human. Many primatologists believe that they resembled this creature—the Demidoff's galago which lives on trees in the dense tropical forests of central and western Africa. Demidoff's galagos weigh around 60 grams and feed on insects, tree sap and gum and occasionally on the pollen of nocturnal flowers. They use all four limbs to move through trees and are not adept leapers. They live in small communes and congregate before dawn to rest during the day. A mother galago carries her infants in her mouth when she has to move them. They sleep in a huddle in leaf nests and groom by licking one another. (© Robert Bateman/Boshkung Inc.)

Layers of alluvium, like these along the Markanda river in the Sirmour district that were brought down by massive floods between 17 and 3 million years ago, form the hills of the Lower Siwalik. Within these successive layers of sediments are fossils of large and small mammals. The ideal places for fossil-hunting are streams that cut across these hills and reach the river where large bones, teeth and tusks have often been found.

Take rhinos: there are only five extant species of rhinos in the world today but the fossil record for around 50 million years ago shows that at least 190 different species of rhinos trod earth once.

The class of warm-blooded vertebrates that we call 'mammals' may have arrived early but it came into its own only after the dinosaurs died out and even then, for a short time, had to jockey for pre-eminence with birds and land reptiles. Since then, mammals have diversified into a bewildering variety of body shapes and sizes and have colonized every habitat on Earth with the exception of the deep sea and the continent of Antarctica (which is not home to *any* kind of fully terrestrial vertebrate). Mammals come in all shapes and sizes—from small shrews, marmosets and bats that are only as big as the top of your thumb to whales the size of a town hall. The bounty of mammalian forms is rivalled only by the incredible diversity of insects and fish. With such a broad bandwidth of size, adaptability and temperament, it is no wonder that mammals sit at the top of every food chain in most habitats. The exceptions are parts of the seas and freshwater bodies where fish largely dominate; a few islands and swamps where reptiles like crocodiles and monitor lizards rule; and of course the air, which is the dominion of the birds. It is testimony to the spectacular success of mammals as a group for the last 50 million years that the dominance of fish, birds and reptiles in small, relict environments is seen as an exceptional and surprising fact. The main factor that accounts for the success of mammals is their ability to manipulate the environments they live in.

It is possible, if you have the inclination and the time, to go and see for yourself fragments of the story of mammalian evolution as it played out in India. Your journey will take you first to Ambala in northern Haryana and from there to Nahan in Sirmour district of Himachal Pradesh. As you wind through industrial estates and a maze of sputtering, overladen trucks, you will eventually come to ravines and hills made of monotonous alluvium. These hills were formed between 30 and 4 million years ago when the Himalaya was still young and rushing montane streams brought down huge loads of sediment every season. Gradually, smaller streams merged to become gushing rivers that cut their way through to carve out a narrow V-shaped valley. But your reason for coming here is simply this: these alluvial hills are a rich repository of giant mammalian bones. Local farmers frequently come across immense tusks sticking out of alluvial beds, sometimes 1–2 metres long. It is by no means unusual for a farmer tilling his land to accidentally find rocks that are actually the femur and hips of rhinos or elephants. This place is a fossil hunter's paradise but, of course, you need to know what to look for.

A good place to begin would be slightly off the road 60 kilometres north of Ambala, and 20 kilometres before you reach Nahan, at a town along the Markanda river called Kala-amb (black mango) where the Siwalik Hills begin to rise. To reach a particularly good site for fossil-hunting, just before you reach the *chungi-chowk* (a toll crossroads) that marks an entry point into Himachal Pradesh, take a short detour on a dirt track that leads to Suketi Fossil Park. The fossil park museum will give you a glimpse of what scientists have found here but the thrill of finding your very own fossil is something else!

You now need to travel further north on the dirt track and watch out on the right of the road for stream beds of seasonal rivers which cut the ravines. These beds remain dry most of the year, filling up only during the monsoons, and you will find a trail of uneven-sized stones and rounded boulders in them. Walk along the stream beds looking for odd-shaped rocks that resemble bones and you are very likely to find fossilized bones of a giant herbivore. Locals who bring their sheep here to graze are inured to the sight of people trudging along these inclines. They know what curiosities drive palæontologists and students from universities to these places and are often quite happy to direct you to a few good sites that have been exposed in the previous rains or point you to the place where a recent batch of visiting palæontologists looked particularly excited!

Between 50 and 38 million years ago after the Tethys Sea had drained to the west, this part of what is now the Siwalik Hills would have been a fertile plain. The land was dominated by grasses and isolated clumps of trees and although Greater India had not yet docked with Eurasia, their

proximity was an opportunity for animals to cross over. It wasn't until 18 million years ago that the first mountains began to rise. But this suture of the two land masses would witness an upheaval of another kind—an epic war that raged between those mammals who developed a fondness for grass and those who developed a taste for grass-eaters.

About 35 million years ago, life in the undergrowth saw a proliferation of several small browsing mammals. Watching them were the earliest carnivores which were about the size of foxes. Having adapted to life in and around trees and feeding on insects, they quickly developed a taste for flesh and their cousins who had stayed vegetarian were among their earliest prey.

An area stretching from Kurukshetra in Haryana till the outskirts of Chandigarh forms the base of the Siwalik Hills (Siwalik: the land of Shiva) which are a disjointed range of low hills. The Siwalik is the youngest part of Himalaya that buckled and folded during the last major thrust of continental compression. These plains witnessed an explosion of diverse and giant mammals between 30 and 8 million years ago.

For more than 15 million years—from 65 to 50 million years ago—mammals did not take to grazing, a niche that had remained unoccupied ever since large dinosaur grazers had died out. The trees of the rainforests provided ample fruit and foliage, and mammals did not have enough reason to develop a taste for grass which had evolved with a number of features to discourage grazing.

No one quite knows exactly when it happened but eventually a few tree-dwelling mammals that lived off fruit, seeds and nuts, came down from the trees and, over time, took a liking to grass and began to live on the ground. Climatic fluctuations were helping grasses spread during prolonged dry spells and a few ground-dwelling mammals probably found succulent grasses an inviting option, especially when fruit and foliage from trees became scarce. These new grazing mammals grew larger than their arboreal cousins, and over the next 3 million years, grazing turned from a fashion to a craze, and grazing mammals grew even larger. Around 44 million years ago, there would have been as many pig-sized herbivores on the ground as there were fruit- and insect-eating mammals on trees. Tree dwellers remained small but gradually developed a fondness for insects and the occasional lizard that visited the trees. This diet was a rich and easy source of protein and some mammals, more than others, acquired a taste for it.

About 40 million years ago, the earliest true carnivores began to evolve from tree-dwelling mammals. They were slender, agile and in the earliest stage still tree-dwelling, and probably looked a lot like the modern linsang, an elegant hunter of the night. When more and more tree dwellers began adapting to life on the ground, the linsang-like carnivores followed suit. The first herbivores had an early start and grew so large that many would not have felt threatened by the small carnivores. Most carnivores of this period were about the size and shape of modern-day civets and toddy cats. These small hunters retained their arboreal

Although camel-like in appearance, the largest land mammal, Paraceratherium, *belonged to the families of rhinos. After its discovery in 1908 in the Bugti Hills of eastern Baluchistan, nobody could conduct further investigations in the region because of incessant tribal clashes. It was only in 1985 that a French-Pakistani team managed to undertake a short expedition in this region. More research is needed to date the formation where the fossils were found for palæontologists to ascertain the time period when this giant ruled.*

sure-footedness while gaining stealth and agility on the ground. The first true cats evolved from here, and with their sharp teeth and retractable claws they became adept at hunting on trees, in the undergrowth and in the open savannah. Over time they developed longer limbs to run short distances at high speeds in pursuit of their prey. The ambushing cats with specialized canines could bring down prey much larger than themselves with a single decisive bite on the throat or nape.

With an abundance of grass as it spread and conquered new habitats, and with improvements in physiological adaptations that enabled them to digest grass, herbivores began to grow larger still. To derive more nutrition from grass and leaves, grazing animals needed a larger gut and the way to accommodate larger guts was to expand the body. Herbivores travelled long distances to consume copious quantities of grass and helped disperse grasses to new places. Grazers developed sociability and the instinct to stay together in small herds, often with different species mixed together, which was another important step in mastering the open savannah. With imposing numbers and impressive mass, a marching or charging herd could intimidate early, small-sized carnivores.

In the next 5 million years, from 40 to 35 million years ago, the ancestors of pachyderms travelled from the Americas and reached Africa where they attained 'elephantine' proportions, while giraffes and rhinos crossed from Africa into Eurasia and India around 22 million years ago. Some of these giraffes were roughly three times larger than their modern counterparts. Not since the dinosaurs had such immensely proportioned giants walked on Earth. This was the age of extremes.

Some mammals remained diminutive furry creatures that scurried in the undergrowth just as they had done in the time of the dinosaurs. Bats evolved to take flight and a few squirrels were able to glide between trees. On the other hand, some herbivores became so large that palæontologists were forced to invent a separate word for them: Megaherbivores. Among them were rhinos, which evolved from small pig-like browsers to become the largest mammals ever to roam on land. One species of rhino, called *Paraceratherium* (or *Indricotherium*, from Russian, *indrik*: fabulous beast in Russian folklore; and Greek, *-therion*: wild beast, also called *Baluchitherium* because the fossil was first found in the province of Baluchistan), resembled a giraffe on steroids. It reached a height of 5.5 metres at the shoulder and weighed in at 15 to 20 tonnes. *Paraceratherium* lived between 35 and 17 million years ago, and its fossils have been found from Baluchistan through the Siwalik in Jammu to Bengal, suggesting that all of northern India at the time was a dry tropical forest with grassland. This massive browser stripped leaves from trees with its downward-curving, conical upper teeth which looked like tusks, and these matched its forward-pointing lower teeth which met with the upper teeth like a pair of giant forceps. *Paraceratherium*'s upper prehensile lip was thick and muscular, useful for grasping leaves and shoots. Its neck was very long, the body robust and the limbs long, thick and pillar-like. Unlike modern rhinos, these giants had no horns. The male *Paraceratherium* was massive and intimidating for all other animals. It had larger incisors and browsed on treetops, while the female was more graceful and slightly smaller, and browsed on trees and shrubs.

A day in the life of *Paraceratherium* would have been spent moving from tree to tree and grove to grove in search of fresh leaves and shoots. Having access to a large territory with lots of foliage trees was critical for its survival. However, as the climate fluctuated, there were increasingly long summers during which trees produced modest flushes of leaves, and few new shoots germinated. In such a situation, it became increasingly important for these browsers to guard their treetops from competitors like *Giraffokeryx*, an ancestor of the giraffe. A dominant male *Paraceratherium* with one or two females would lord over a large area. The arrival of rain (the monsoon had not yet become a regular feature of the Indian subcontinent) would provide respite and this was when a single calf would be born to a female. Though the calf would have been too large for a single carnivore to take on, it would have been easy prey for a pack of megadogs or the meanest hyenas. This would be a time of great vulnerability for mother and calf. The docile mother would turn aggressor and exile the dominant male to the periphery of his own territory, so that the newborn would not be intimidated by an aggressive male. Mother and calf would have to fend for themselves and keep their predators at bay. And with sheer grit, the *Paraceratherium* mother would probably prevail over these dangers. If the rains were sufficient, the trees would offer a leafy banquet for the nursing mother. If the rains failed, the calf would survive but only just, and if the water dried out before the long summers, the calf would surely die as the mother would not be able to provide enough milk to her demanding calf.

How could such a large animal obtain enough food from an increasingly desiccating landscape? A possible answer could be that giant herbivores like *Paraceratherium* possessed greater gut efficiency. They had to extract all the nutrition they needed from the leaves of trees. A combination of factors would have worked in tandem for the *Paraceratherium*—strong mechanical muscles in the jaw and the gut to grind up cellulose and fibre; juices and microbes to assist digestion; and a long gut to intercept and absorb every available nutrient. It is also possible that in those times there was greater diversity of trees, many of which were large, like the animals themselves, and that their leaves and shoots were more nutritious. As opposed to this, the modern elephant's digestive tract processes food quickly and it excretes about three-fifths of the mass it consumes; what passes out contains a large volume of undigested and unassimilated nutrients.

Although *Paraceratherium* had efficient digestion, its method of browsing treetops eventually contributed to its downfall. The reason why we still have large elephants but no enormous rhinos is because of the design of the elephants of those times. Elephants had an extendable trunk and their head and teeth enabled them to feed both off the trees and on the ground, allowing them a more cosmopolitan diet than *Paraceratherium*. About 22 million years ago, several species of archaic elephants moved into the habitat of *Paraceratherium*. Elephants originated in North Africa from a pig-like ancestor that reached India through Eurasia. As they grew in size, elephants like *Gomphotherium* and *Deinotherium* and later *Stegodon* changed the playing

A tusk of the extinct elephant Stegodon ganesa *that was exposed on the steep alluvial banks of the Markanda river, between Ambala and Nahan.* Stegodon *was among the last of the great elephants which reigned from 11 million years ago to about 4100 years ago. Modern elephants arose only about 4 million years ago and continue to prevail.*

field slowly by stripping and uprooting the trees that the huge rhinos depended upon. *Gomphotherium* and other elephants had arrived in India from the west, from Eurasia through Arabia. As permanent ice began to form over Antarctica, the seawater receded, land bridges became a permanent feature and these narrow strips were covered in lush tropical forests that became a highway for mammals. As the climate began to cool again around 21 and 19 million years ago, the number of tall forest trees dwindled and conditions swung in favour of the tougher grasses that elephants could feed on.

Paraceratherium were unable to bend down to reach grasses and low trees, and around 18 million years ago, this limitation was the most likely factor in driving these giant rhinos to extinction.

After *Paraceratherium* vanished off the face of Earth, barring elephants, there were few herbivores big enough to intimidate the carnivores, and everything that fed on grass became 'easy meat'. Since then, for every grass eater there is one or more predator (although herbivores are found in large numbers, there are often fewer species of herbivores than those who prey upon them). The evolution of prey and predator is dictated by their constant effort to outdo one another. With every major climatic shift, grasses encroached on forests and flowering plants evolved newer species and new strategies of survival and success. In response, herbivores too were forced to adapt rapidly into several different types, changing their diets and body shapes as they diversified into every available niche. For carnivores, plant eaters were a factory that converted grass into flesh. Unlike grass, flesh provides a rich source of muscle-building protein and fat for energy, and needs to be eaten only every now and then.

Being a carnivore was, therefore, a high-risk but also a high-reward life (in terms of a high-protein diet), and if a mammal had developed a taste for meat and was also fast and ferocious, it only made sense for it to evolve into a meat eater. The earliest carnivores evolved from omnivores and took to eating meat only after they developed specialized teeth. Meat eaters also developed a strong sense of smell, a battery of lethal weaponry, highly developed senses like stereoscopic sight (although some non-meat eaters also possess it), ultrasensitive hearing, and the ability to stalk with guile and stealth and to pounce, chase and

A philatelic tribute by the Government of Nepal to extinct elephants.

kill while minimizing the chances of becoming prey themselves. All of these abilities—both physical and mental—allowed them to follow and live off herbivores wherever they roamed.

But evolution is never straightforward and nature provides little quirks to keep things interesting! Over time, a few species of carnivores drifted towards the edges of their specialization to become exceptions. These were carnivores that abandoned their instinct for eating meat altogether and took to eating other specialized foods. Pandas, like the red panda of Sikkim, are bears that feed exclusively on shoots and leaves. They evolved a sixth digit on their paws that made them adept

An elephantine blunder? To mark the centenary of the Geological Survey of India, the Post & Telegraph Department issued a stamp depicting Stegodon ganesa. *The representation of the trunk dangling over the giant tusks is inaccurate.*

at climbing trees and plucking their favourite food. The sloth bear is another example: it rarely eats meat and lives mostly on a diet of honey, fruit, termites and other insects. In Africa, there is a species of hyena and a type of fox that feed mostly on termites. Interestingly, mammals with a similar diet do not always have similar ways of eating. The red panda (*Ailurus fulgens*) of Sikkim and the giant panda (*Ailuropoda melanoleuca*) of eastern China are mammalian carnivores separated from each other by about 40 million years of evolution. The red panda is most closely related to animals like raccoons and weasels and weighs about 6 kilograms and feeds largely on bamboo and occasionally on insects. The giant panda is a member of the bear family and is at least fifteen times bigger than the red panda. Giant pandas probably evolved in the bamboo forests of Indo-China but they use a different strategy to eat their favourite shoots of bamboo. The giant panda uses its strong skull and jaws to chew tough bamboo shoots, swallows the coarse cud and lets its tough stomach juices do the digesting. The red panda, on the other hand, is a more refined eater. It uses its slender jaws more elegantly and efficiently by distributing the force evenly and masticates the cud which makes it easier for its stomach to digest it.

Cats evolved as specialized hunters of large herbivores. Like the small early tree carnivores, cats retained their special ability of bringing down prey bigger than themselves. In fact, they developed a unique method of killing their prey for which they even evolved special teeth. Instead of shearing their large teeth through the spine or the back of the neck of thick-skinned herbivores, cats perfected the technique of suffocating their panicking prey to death. The second mammalian revolution occurred when giant cats began to dominate and rule the grasslands and scrub forests. Within the family of cats, one group that we call sabre-toothed cats developed bizarre and dagger-like front teeth. *Homotherium* was among the longest surviving sabre-toothed cats that lived from about 3.5 million years ago to about 300,000 years ago. They were among the chief nemeses of our human ancestors, *Homo erectus*. They were small, about the size of a male African lion, but more agile than other cats of this time. Their long upper canines were flat and finely toothed, which gave them a powerful puncturing bite and grip. Their molars were weak, which suggests that they did not crush bones and therefore did not feed on bone marrow, like the hyena or our own human ancestors did. Compared to other cats, the limbs of *Homotherium* were disproportionate—its forelegs were elongated and its rear legs small. Its appearance was perhaps hyena-like, though its skull suggests that it had large nares (openings of the nostrils), much like those of the modern cheetah, which enabled it to breathe more efficiently especially when it was chasing its prey. *Homotherium* preferred feeding on thick-skinned herbivores like elephants and rhinos. They probably hunted in packs, dragging away any large animals they managed to bring down.

Like *Homotherium*, *Megantereon* (or giant beast) and a smaller sabre-tooth, *Vishnufelis* (Vishnu's cat), were other species related to the sabre-toothed cats and lived between 3 million to 500,000 years ago. *Megantereon* was built like a stocky leopard with strong forelimbs,

One of the largest predators of the plains and low hills was Megantereon *and it was possibly the longest ruling among all predators (from 9.5 to 2.5 million years ago); other smaller cats and social dogs and hyenas lived alongside.*

weighed about 90 to 150 kilograms and was 1.5 metres long. *Vishnufelis* was smaller, only about 70 to 110 kilograms. Both these sabre-toothed cats had large neck muscles designed to deliver a powerful bite with their long canines. The angle and power of their teeth ensured that they could pierce and also shear through the neck muscles of large prey and rupture their neck vertebrae, major nerves and blood vessels, killing them quickly. Their favourite prey included small horses, the young of rhinos and elephants, and antelopes. *Megantereon* was not built for speed and was almost certainly not equipped to sprint after its prey across open savannah. It was more likely an ambush hunter that waited for its prey to get close before pouncing from a rock or a tree. Like all sabre-tooths, *Megantereon* used its enlarged canines to inflict a

Between 11 and 7 million years ago, seasons became distinct, and dry grasslands began to create gaps in forests. The Siwalik, about 10 million years ago, would have resembled the scene above, with ancestors of horses, antelopes, elephants and camels feasting on a vast spread of grass.

single devastating bite and then backed off while the prey died. Yet, despite such fantastic adaptation, adult rhinos and elephants would have been a handful for these sabre-toothed cats.

Hyenas were vicious competitors for all carnivores, including our own early ancestors. Among these, *Pachycrocuta* (literally, 'thick hyena') was the menace-in-chief. It weighed about 130 kilograms and was the size of a modern African lioness, making it the largest known hyena to have been discovered. *Pachycrocuta* appeared on the scene about 3 million years ago and became extinct around 400,000 years ago.

It hunted in small packs and brought down large deer and horses. Eventually, large beasts like *Pachycrocuta* disappeared because large herbivores died out and they could no longer compete with the more agile cats. It is more than likely that the ancestors of our own species, *Homo erectus*, played a role in their extinction, too.

The decline of the large sabre-toothed cats and hyenas paved the way for smaller fleet-footed hunters like dogs and smaller cats which hunted deer and small forms of horses, pigs and hares. A common prey for these predators was a giant hare (*Caprolagus*) which survives in a miniaturized version. This is the hispid hare, rare today, but it once ranged in submontane savannah from Uttar Pradesh to Assam. Their original range extended from the Levant to China, but as the Himalaya rose, the population of hispid hare got isolated and they survive only in the lower hills of the Himalaya and the terai.

Hunting dogs emerged from an entirely different lineage than that of cats. About 38 million years ago, dog-like hunters known as 'bear dogs' arose. These giant carnivores were capable of running great distances on

This scene from about 20 million years ago shows a pack of bear dogs (Amphicyon) *ambush and attack a solitary* Entelodont, *an ancestor of the pigs. The bear dogs had powerful front limbs (which were probably used to grasp their prey) and reduced shearing canines (or carnassials) but retained a long tail and shorter hind legs with five toes like their ancestors. They became more and more like bears in their habits but were not at all like them in proportion. Bear dogs died out around 9 million years ago in Europe and America, but survived longer in the Siwalik of India and Pakistan, until around 7 million years ago.* Entelodont *were vicious large pigs with a nasty bite and a taste for carrion. They lived well until the arrival of an ancestor of the humans* (Homo erectus).

open grassland and were the dominant hunters of Eurasia for 10 million years. As their name suggests, bear dogs had attributes in common with both dogs and bears. They resembled large bears with flat faces and large teeth. They were strong, lone hunters that ran after their prey till they brought it down. Over time, bear dogs were outcompeted by hyenas, cats and dogs (which diverged early from bear dogs). Genetic studies show that from the primitive line of bear dogs, only two lines survived. One lineage developed into bears and became reclusive, feeding off fruit, tubers, insects and only the occasional portion of meat. They evolved paws that helped them dig and climb trees. The omnivorous bears traded the large slicing and crunching teeth they inherited from bear dogs for broad, flat, grinding surfaces. Most bears that live in open forests, like our own sloth bear, have naked soles on their feet and curved claws to help them dig into termite mounds and climb trees. Larger bears forage and steal food, or anything that comes their way. Child-rearing for big animals with slow-growing babies is a major investment of time and energy, and for bears a cosmopolitan diet is an asset. Yet, even today, some dog-like traits can still be seen among the larger bears. Occasionally, a lone bear goes out into the open to chase and successfully bring down prey much like a solitary jackal or a wolf does.

The second lineage that diverged from the bear dog family underwent a complete makeover between 32 and 28 million years ago when they shed their bulk to become long and slender so they could enter and hunt in holes and burrows. These were sinister long, thin killers, early prototypes of minks and otters in water, and martens and toddy cats in trees. Around 22 million years ago, some even took to living underground. With abundant rodents, insects, fruits and nuts, these carnivores became extremely successful. Grazing herbivores also provided an incentive for some other descendants of bear dogs to stay close to the ground but for a different reason. Grass eaters produced copious amounts of dung, which attracted several types of beetles and flies, and many of these laid their eggs within this predigested and decaying cellulosic mass. The descendants of the bear dogs that were enticed by the insects and their larvae in the dung, evolved into ground-living mammals, the ancestor of the modern mongoose.

The carnivorous mammal survived by adapting its diet, and to do this it needed to constantly evolve new kinds of teeth that would enable it to eat new kinds of food. Teeth, therefore, are a good way to determine the durability of mammalian evolutionary lines. Although sabre-toothed cats are extinct, cats retain the extravagantly sharp incisors of their ancestors, located in the front of the jaw and used mainly to puncture prey in their neck. In the rear of the lower and upper jaws are scissor-like teeth which shear chunks of flesh from their large kills. Dogs' teeth by comparison are smaller and less effective as killing fangs. But dogs (and even bears and some smaller carnivores like civets) are armed with a set of grinding teeth which help them eat a variety of coarser food, from nuts, insects, bones and carrion to fresh meat. Unlike cats, dogs like the wolf, jackal and wild dog hunt in packs and bring down prey larger than themselves by cooperative effort. They depend on endurance to outrun and outsmart their prey, but lack innate strength to do so alone. Dogs are unable to wrestle and overpower their prey or bring them down easily. Social dogs communicate efficiently through whistles, barks and scents and form packs to draw strength and drive away a large cat. A pack of dogs has been known to occasionally bully a solitary cat or even a pride of lions and steal their kill. While cats are the supreme killers, dogs are tactically superior and efficient to boot, making up in numbers what they lack in individual strength.

As the Himalaya rose higher and the climate became more variable, some herbivores like the ancestors of the mountain goat moved in search of new grazing grounds and their contemporary carnivores followed them. Over time, small populations of these herbivores stayed put while others migrated further south through the Eastern Ghats. Those who stayed back adapted to feeding in high altitudes in summer when fresh flushes of grass appeared, and descended to lower slopes in winter. These evolved into the Himalayan tahr. A few herds of mountain goats travelled along the Eastern Ghats and reached the lush Nilgiri Hills, and over the next 9 million years, the tahr here shrank in size and developed a darker coat and evolved into Nilgiri tahr. Several other plants and animals travelled into and out of India using Himalayan corridors. Around 12 million years ago, many small mammals like the marten, civet, giant and flying squirrels and the cats that arrived from South East Asia through the eastern Himalayan corridor followed flowering plants like the rhododendrons and orchids and migrated to the cool southern hills of India, travelling more than 2200 kilometres in all! Some of these creatures crossed into Sri Lanka which was then connected to India by a causeway of rock, and some undertook a journey along the Western Ghats and went further along the ancient rivers of the Narmada and Tapti and the mountain ranges, the Vindhyas and Satpuras, to colonize central India. While India received this bounty of animals from Eurasia, in exchange it contributed a tree that makes South East Asian forests distinct. India provided the region the ancestors of the dipterocarps (*di*: two, *-ptero*: wing, *-karpos*: fruit), the most handsome

Clouded leopards are nocturnal, arboreal, secretive and solitary. Although they resemble leopards, they are smaller and their spots are patches of pale yellow which are mottled with grey, which makes them look like a cloud, hence their name. Clouded leopards are the most arboreal of all big cats. They have broad, soft feet for leaping, very long tails for balance and hind ankles that roll outwards when descending trees head first. They have the longest canine teeth in relation to the skull size of all big cats. They also have the widest gape among all cats (100 degrees compared with 60 degrees in other large cats). Clouded leopards also hiss rather than roar, a characteristic that is only seen among the group that comprises the roaring and purring cats.

representative of which is the sal tree. In addition, a variety of frogs, snakes and mammals, orchids, bananas, grasses and bamboos, crossed over into South East Asia.

The last upheaval of the Himalaya between 3.7 million years ago and 300,000 years ago was the critical driver for the evolution of a number of modern mammal species. The thirty-seven species of cats alive today all emerged around 11 million years ago from a single panther-like ancestor that lived somewhere in South East Asia. With new rivers being created, this ancestor was isolated and diverged into three lineages of cats around 9 million years ago—one founding the family that now consists of the clouded leopard (found from Sikkim to Arunachal Pradesh), the second the 'great roaring cats' of the genus *Panthera* comprising all other cats including the lion and the tiger, and the third diverged into smaller purring cats that eventually led to the evolution of the domesticated cat. The relationship between large cats is entangled and there were perhaps several instances of cross-breeding among cousins before they became distinct species. The ancestor of tigers and snow leopards was the first to branch off, around 3.9 million years ago while lions and leopards split from one another about 3.1 to 1.95 million years ago. The clouded leopard (*Neofelis nebulosa*, or 'new cat with cloud markings') actually comes from the same founder as other cat species (*Panthera and Felis*) and therefore its genus name is inaccurate and needs revision.

Experts believe that tigers arrived into India from South East Asia through the low hills and along the coastal route around 1 million years ago and later spread across India. Once here, this uniform population became isolated and then evolved into several species, relatively recently. The junction between India and Myanmar is perhaps the only place in the world where four great cats—the tiger, snow leopard, clouded leopard and leopard—exist in close quarters. In a particular place, about 90 kilometres south-east of Namche Barwa, along the border with Myanmar, called the Namdapha forest reserve in Arunachal Pradesh where snow-covered peaks descend precipitously to reach well-watered plains, these four great cats converge. To the north of Namdapha are mountains that are nearly 4500 metres high and are home to the snow leopard. The snow diminishes as one descends to lower altitudes where dense tropical forests become the home of the clouded leopard. In the jungles and grasslands along the banks of the Brahmaputra, leopards and tigers rule the plains and riverine forests, much as they do in the national parks of Corbett, Dudhwa and Chitwan in the terai. The four great cats seldom encounter each other since their range is clearly defined by their habitat but this makes Namdapha the only place in the world where four large cats can be found within an area of 60 square kilometres.

Lions, on the other hand, migrated from Europe and West Asia, entered and eventually evolved in Africa and then re-entered Asia, where they gradually became extinct (largely due to hunting by humans till as recently as the sixteenth century) with only one small relict population surviving in Sasan Gir in Gujarat.

Isolation did not just produce new species of large mammals like the cats; new species of other mammals evolved too, some relatively recently. About 9 million years ago, the Brahmaputra and its tributaries in the east isolated the ancestors of grey langurs between two fast- flowing mountain rivers, the Sunkosh in Bhutan and Maanas in Assam.

Gainda from Gujarat! This engraving of the rhino (or ganda *as it was titled) from 1515 suggests that rhinos perhaps lived at least as far south as Gujarat and as far west as the Indus. It is well accepted that rhinos lived across the Siwalik range—from western Pakistan through Assam and beyond into Myanmar.*

The two isolated populations evolved into the golden langur and the capped langur which are quite distinct from their ancestor, the grey langur.

Unlike social dogs which could run long distances and hunt in packs, solitary cats became specialized ambush hunters, while smaller omnivores like civets, mongooses and polecats perfected the art of hunting in holes, on trees and in the undergrowth. Dogs and their descendants—hyenas, foxes and wolves—have retained the instinct for strong social cohesion. Cats on the other hand are mostly solitary and, with lions as the outstanding exception, don't hunt together. In the last 2 million years, only one carnivore has developed the design and skill to outrun the fleetest herbivore. This is the cheetah (from *chitta*, Hindi for spotted), the fastest-sprinting animal in the world. Because young cheetahs are easy to tame, they were used as a hunting aid by Mughal emperors and Rajput kings. These rulers ordered the capture of live cheetahs, which often involved taking the young away from their parents and killing the mother. Over 200 years, their numbers began to decline in the wild. The last cheetahs were reportedly shot in Sarguja and Baroda in 1947, and the species was declared extinct in 1952. The cheetah once roamed across North Africa, from Arabia into India and southern China. In India, they ranged from the Punjab in the north and west to the Chota Nagpur plateau, and covered the monsoon forests and grasslands till central Tamil Nadu.

Predators like social dogs and solitary cats followed large herbivores across grasslands and scrub forests, therefore spreading far and wide. Elephants in particular seized their opportunity of growing very large. A 13-foot thigh bone of *Palaeoloxodon namadicus* (or 'old elephant of the Narmada') discovered in Narsinghpur, Madhya Pradesh, suggests that some elephants grew very large, perhaps even matching the size of the giant *Paraceratherium*! Rhinos inhabited grasslands and marshy wetlands in the Indian subcontinent and have been found as far south as Tamil Nadu as recently as the early seventeenth century. The first live Indian rhinoceros to have been seen in Europe was a present from Muzaffar Shah, the second sultan of Cambay, which was shipped to Emmanuel, king of Portugal, in 1513. The *gainda* (or *ganda*) became the toast of the court, royal fair and festivals of Lisbon, and tales of this fabulous beast reached the ears of the great German printmaker

Albrecht Dürer. Dürer did not see the Indian one-horned rhino himself but made a celebrated (and somewhat fanciful) engraving which became the most reproduced sketch of the sixteenth and seventeenth centuries (at least eighty versions are known to exist) and appeared in several books on natural history in Europe. Sadly, rhinos and several other mammals, large and small, were beginning to get severely depleted in numbers during the Little Ice Age which began late in the sixteenth century and lasted until about 1850. The arrival of Europeans in India intensified hunting with guns, and the ever-obliging native rulers abetted this, which together hastened the demise of many wonderful wild animals.

Over the 40 million years of their evolution (from 55 to 15 million years ago), mammalian carnivores have taught us a few rules of thumb. One is that Generalists are more likely to succeed than Super-Specialists. The early carnivores that evolved from tree dwellers and ungulates were Generalists, and in a purely evolutionary sense, showed their resilience and adaptability. They were jacks of all trade and masters of most! Generalists tend to be smart and adaptable and have a catholic taste for food and a natural curiosity to learn about their changing environment. Together, these qualities improved their chances of survival and probably abetted their domination when Super-Specialists died out. Extrapolating from the fossil record, it's possible to say that Generalists are the drawing board upon which evolution fashions new innovations as Specialists. While the design of many Specialists became obsolete over time, the body plan of Generalists has hardly changed over millions of years. The last lineage to emerge from the bear dogs was the bears who became exclusively Generalists. And being Generalists over several million years paid dividends for more successful mammals like polecats, civets, martens and foxes, who have found no reason to change their diets in any significant way.

Around 20 million years ago, tropical forests had reached their maximum diversity and spread, and many specialized tree dwellers like primates had evolved in Eurasia and India. Gradually, as the climate cooled, forests began to recede and life on trees for most animals became difficult. The early primates were social and lived on trees for the fleshy fruits and leaves they provided, disseminating the seeds of their favourite fruits far and wide. On the ground, however, were rodents who had developed a particular fondness for seeds and nuts and their consumption constricted the expansion of some tropical fruit-bearing tree species. As the climate became colder and drier, tropical forests became deciduous, and large swathes of forests were reduced to isolated clumps of trees. Over the next few million years, the changing climate and evolving rivalries with rodents and other mammals would compel the primates to adapt to these changing times, and one family of primates would evolve into apes. And sometime in the next 3 million years, some apes would venture on to the ground to find new sources of food.

About 18 million years ago, the giant *Paraceratherium* and many large mammals had died out and their place was taken by smaller and more efficient mammals. The 'push' was provided by the decline of tropical forests which forced several mammals to invent new means of survival. At least 120 species of mammals have been recorded to have existed in the Siwalik and lower Himalayan region and the northern plains of India between 18 and 2 million years ago. Even though the giants had died out, this was a time of greater diversity within all the mammalian families than ever before. There were at least twelve species of rhinos, eight of elephants, four of giraffes, and five of primates living in the Siwalik Hills and valleys. By 1 million years ago, after all large mammals had died out, the ancestors of modern mammals had emerged.

The evolution of modern mammals—including us humans—is linked with tectonic upheavals in the Himalaya and beyond, and with concomitant climate, sea-level and vegetation changes. It is a complicated story but here's a short version.

The evolution of modern mammals occurred in seven short, but intense, pulses from 23 to 22 million years ago, and then 20-18, 15-14, 11-10, 9-8, 3-2, and the most recent around 1 million years ago. We have been able to establish these dates from the fossil record of immigrant mammals that can be traced from Africa in the west, into West Asia and Europe along ancient coastal routes (such as Kargil and Kutch between 23 and 20 million years ago), eventually leading into India. A less intense but sustained immigration of smaller mammals took place from the east (Myanmar, Thailand and South China).

Between 22 and 18 million years ago, giant elephants like *Gomphotherium* left Africa and travelled through Arabia and Eurasia over a newly created land bridge (named the '*Gomphotherium* land bridge') and entered India to displace the ruling *Paraceratherium* and other herbivores. Several other types of mammals followed in the footsteps of the giant elephants. Creodonts (a group of vicious carnivores that is now extinct) and smaller mammals like rodents were the first to take this trail. Rhinos, chalicotheres (odd-looking, large sloth-bear-like herbivores who sat on their hind legs and munched leaves from trees, and are now extinct), the bear dog (like *Amphicyon*) and a variety of pigs left Eurasia and entered Africa between 20 and 18 million years ago perhaps through this corridor. From the east, smaller mammals like squirrels arrived into India. Over the next 2 million years or so, several such exchanges took place, predominantly from the west.

Between 15 and 14 million years ago, a new lot of immigrants arrived from Africa and Eurasia which included apes, Eurasian buffaloes and perhaps the first otters. From the east arrived three species of slightly smaller buffaloes, more pigs and gazelles. A third wave between 11 and 5 million years ago occurred when the monsoon intensified and grasses seized the opportunity to expand their dominion. A cavalcade of horses and camels entered China from North America travelling over a land bridge to (eventually) reach the Siwalik around 10.7 million years ago. The ancestor of the langurs (*Presbytis*) also arrived in the Siwalik foothills around 6.3 million years ago. Between 5 and 4 million years ago the climate again became warm and humid and the landscape was dominated by wooded grasslands and scrub. The ancestors of the modern family of mice and rodents travelled along the grassy corridor, and small hippos (*Hexaprotodon*) and more elephants marched into the Siwalik from Africa. However, two elephants, *Anancus* with giant tusks and *Stegodon*, evolved in the Siwalik and migrated and dispersed widely into Africa, Europe and North Asia. This period between 5 and 3.5 million years ago witnessed a decline in the variety of plant eaters (elephants, giraffes, deer, horses and a few primates) and the extinction of the giant rhinos and several other large mammals, most of whom left behind no descendants. The animals to flourish in these drier conditions were rodents like rats, mice and the hyrax. The next million years (from

Around 7 million years ago, in what is today Abu Dhabi, a variety of mammals of various sizes ambled across a stretch of very shallow water. A narrow land bridge connected Africa with Eurasia which facilitated the exchange of an amazing variety of plants and animals—elephants, camels, geckos, grasses and rats. For about 3 million years (from about 8 to 6 million years ago) Arabia was covered with trees and savannah-like vegetation, where a criss-cross of rivers cut through. As the animals left and entered Eurasia, they made tracks on the soft river mud and some of these footprints got preserved in the sand over time. This track of circular footprints extends for about 250 metres on a salt pan in Mleisa, 180 kilometres west of Abu Dhabi. Palæontologists believe that this track was made by a four-tusked elephant known as Stegotetrabelodon syrticus. *Alongside this trackway are footprints of twelve other elephants, and in the layers of rocks that surround these mud flats are more trackways and fossils of a variety of other mammals from these times.*

3.6 to 2.6 million years ago) witnessed cooler and even drier conditions. Africa's grasslands received less rain, and Arabia and India offered the grazers greener pastures. This phase therefore saw more African antelopes, Eurasian deer and rodents enter Kashmir and the Siwalik, from where they travelled further inland towards central India as the monsoons began to intensify around 2 million years or so ago. In the final wave, between 2.6 million years ago and 780,000 years ago, large antelopes and fierce hyenas arrived from Africa, and giant cats made their way from Eurasia to the Pinjore Hills around Chandigarh. Around this time, the many mammals that entered the Indian subcontinent from the west continued their journey along the Siwalik from Assam into Myanmar and onwards to Java, and the same path was likely to have been taken by *Homo erectus*. By this time, the last giant elephants had been replaced by the smaller African immigrant (*Elephas*) and the long-reigning rhinos by modern varieties. In just 1.5 million years (around 7 million years ago), four species of rhino that dominated the Siwalik Hills and plains of India perished, leaving just one species to rule the grasslands and forests. Modern cats, bears, social dogs, hyenas and smaller carnivores which came into existence around 3 million years ago began to dominate over their ancestral cousins.

Meanwhile, another momentous story was unfolding in the background. As some modest mammals reached near-dinosaurian proportions, with viciousness to match, some mammals were still struggling, somewhat stranded in their evolutionary journey. Not every animal could become large or stay small to escape competition. The shores of the Tethys, the narrow lush valley that ran from Himachal Pradesh to Kutch, became a laboratory for mammalian tinkering and innovation, and a few mammals found an entirely new solution: a return to life in water. This began with a few of them foraging in water reeds, floating watercress and succulent plants in and around lakes and rivers. The arch enemies of mammals—crocodiles and other reptiles and freshwater sharks—did not make these attempts any easier. But by sheer doggedness and a series of adaptations, some mammals took to life in water and, in just 3 million years, became fully aquatic and were able to defend themselves against domination. Over the next 20 million years, this evolutionary twist ushered in the age of the whales.

12

BIRTH OF THE WHALES

Late in the summer of 1975, two Indian scientists stood ankle-deep in a ditch, excavating soft chocolate-coloured limestone on the outskirts of Harudi village (Lat: 23°15'N, Long: 69°79'E) in Kutch district of Gujarat. Their interest had been aroused by some unusually interesting bone fossils that had been found at this spot. The scientists poked around in the limestone for several weeks under a harsh sun and intermittent rain and unearthed several more bones. After a few months of careful analysis in laboratories in Chandigarh and Lucknow, the various bones were identified as belonging to a sea cow, an early species of elephant and three kinds of primitive whales. Across the border in Pakistan too, bones of whales were being discovered along with fossils of rhinos, pigs and a tapir-like creature. In the next two decades, additional fossil evidence turned up from the Punjab and Sindh regions of Pakistan, in Kutch, the Siwalik Hills and Assam, and pieces of the puzzle slowly began to come together.

What were the bones of primitive whales doing alongside the bones of elephants and other grazers? Scientists already knew from the fossil

The earliest ancestors of whales evolved from a mouse deer-like ancestor, the Indohyus, *which possessed a bony ear structure similar to that in modern whales. Some of these ancestors branched off to create the family of horses, hippos, camels and their very distant descendants became the founders of the family of whales.*

record that whales had evolved from a land-dwelling ancestor they shared with hippos, cows and sheep—a group of mammals classified as artiodactyls (even-toed grazers). Their common ancestor, the progenitor of the family, was a wolf-like sheep that sprinted on hoofed feet, was armed with teeth, had a short furry coat and a ravenous appetite for meat.

The corridor extending from Jammu all the way south to Kutch in the western margin of India and also in neighbouring Pakistan has been the site of a rich haul where scientists have found a graveyard of whales that evolved during the narrowing of the Tethys Sea. Since 1975, long bones and fossils of at least eighteen different species of ancestral

whales have been discovered in this territory. The oldest ancestor of the whale to have been found in India is *Himalayacetus* (the Himalayan whale) which lived 53.5 million years ago and whose fossils were found close to the Simla Hills in Himachal Pradesh. *Himalayacetus* was closely followed by two other amphibious carnivorous relatives: *Nalacetus* (Hindustani, *nala*: stream) discovered in the province of Punjab in

Oddly, the first animal to be labelled as a whale lived on land and hunted for prey around the margins of large shallow lakes along the Tethys Sea. This wolf-sized, fish-eating mammal named Pakicetus *had the body of a land-dwelling animal, though its skull had a distinctive long shape, similar to that of a whale's.*

Pakistan and *Indocetus* (India's whale) discovered by Professors Ashok Sahni and Vijay Prakash Mishra in Harudi in 1975. These were the two scientists digging in the mud with whom we began the story of whales! These proved to be early 'experiments' by mammals that lived on land but depended on water for food and to escape from deadly predators on land. Even though they were quite small at this stage, they had already developed some anatomical features that were whale-like. By about 47 million years ago these wolf-sized animals had developed webbed hind legs to propel them in shallow water. Their head and body resembled a wolf or fox but their teeth, the structure of their ears and the physiology of their bones were all unmistakably akin to those of modern whales.

Ambulocetus *scoured coasts and banks of brackish rivers between 49 to 40 million years ago. This amphibious land-living whale had short limbs and large feet which it used for swimming. Because it lacked a tail fluke and its pelvis was still attached to its spine, like in land mammals, it was an inefficient swimmer. It had a long snout armed with pointed teeth and a strong jaw, and presumably it thrashed its prey violently over water before swallowing it.*

They were also differentiated from most other land mammals in that their eyes were set close on a long snout and they had long, muscular tails.

Even more clinching evidence of early mammals taking to water came from a fantastic array of fossils discovered in Pakistan. These belonged to *Pakicetus*, a wolf-like aquatic animal with dense bones that was adapted for swimming or wading in shallow water. *Pakicetus* would have paddled like a dog in water and would have looked distinctly clumsy compared to its sleek contemporary predators like sharks and crocodiles that dominated the water. *Pakicetus*'s search for food in or around water would have been more opportunistic than the targeted hunting that sharks, for example, were already so good at. Perhaps *Pakicetus* would have hidden behind rocks underwater or in the shallows, so that it could dart out to snap up unsuspecting small prey. Not surprisingly, *Pakicetus* and its immediate descendants were not too successful at thriving in water, and their kind had to constantly reinvent themselves. Their success, however, should not be measured by how long they endured, but in the small but decisive innovation of taking to water—a small step that proved to be a giant leap for mammal-kind!

The mammal that we think evolved directly from *Pakicetus*—because it resembled it so closely—showed a few crucial modifications in its body. The design of its feet became more like those of present-day wading mammals. It developed a more muscular and flattened tail like that of modern otters, and its hips and rear legs transformed into flippers. Even more importantly, its entire body became more spindle-shaped and streamlined, all the better to swim with. This new species, named *Ambulocetus* (or the walking whale), had a crocodile-like head mounted on a hyena-like body, with its feet pointing outwards. *Ambulocetus* lived primarily on land and close to rivers and lakes when it was still young but took to the seas as an adult when its teeth became fully formed. Compared with *Pakicetus, Ambulocetus* had better and stronger lungs, and nostrils which enabled it to stay underwater longer. But it had not yet developed the essential physiological adaptation to tolerate sea salt and was therefore forced to live (much like *Pakicetus*) at the cusp where fresh water met the sea. *Ambulocetus* still did not have the physiology to outpace predators or hunt efficiently in water. Surprise and stealth remained its primary strategy as it hid behind rocks and, camouflaged

by its sand-coloured coat, stayed still until it could lunge at fish or turtles that came within reach, or at any animal coming for a drink at the water's edge. On land, *Ambulocetus*'s movements were clumsy because of its flipper-like limbs and it ambled and dragged itself on the ground quite like modern seals and walruses. Like *Pakicetus*, *Ambulocetus* too did not endure but it put into place an improved design that was better suited for success in water than any of its predecessors.

Around 50 million years ago, as the northern margins of India came close to docking with Asia, the monumental forces of a continent in motion were already folding the land into mountains. The lakes were now drying and were perhaps connected through seasonal rivers to the rapidly closing Tethys Sea. The next prototype of a water-living mammal that we have found was the first fully aquatic ancestor of the modern whale—*Rodhocetus* (*rodho*: a gentle fold in the hill, where the fossil

Kutchicetus was a small otter-like animal which lived in the shallow tropical seas and islands which were formed when the Tethys began to close. It had smaller hind legs compared to the earlier whales but a powerful tail which helped propel it underwater. This made it a better diver. It was perhaps also endowed with blubber which provided insulation underwater and gave it a more streamlined body.

was found; *-cetus*: whale)—which evolved around 3 million years after *Pakicetus*. It was 3 metres long with a streamlined body and a shorter neck. It had better jaws, teeth and an ear design that equipped it for life in the sea. *Rodhocetus* perhaps also evolved kidneys that could maintain its water balance in the sea and fur that could tolerate salt water. Scientists believe that the ability of *Rodhocetus* and other ancestors of whales to tolerate salt developed very slowly, as freshwater lakes and lagoons mingled with seawater giving them the opportunity to adapt gradually to life in the sea. They must also have been helped by the fact that the sea at this time was not as salty as it is today. The build-up of salt in the seas too has been gradual. Clouds that form in the oceans shed rain on land, and when this water seeps through layers of soil and rocks, it etches away salts from Earth and deposits them into the ocean. As the erosion of rocks on land increased over time, salt levels too increased in the seas.

Other improvements in the design of *Rodhocetus* included hollow bones with hardened exteriors, strengthened with extra calcium. Its toes were webbed, its limbs were flipper-like and its tail became flatter and muscular. But there was one more crucial adaptation of an organ that—unlikely as it seems—gave the ancient whales an important fillip towards thriving in water. It was their ears!

The mammalian ear is more than just an organ of hearing. It is the primary organ that maintains balance and relays this information to the brain. The ear was built for hearing on land and many of its functions, especially to do with balance, needed to be tinkered with to enable whales to live in water. All whales have a special organ of balance inside their inner ear. This special organ was essential not only to maintain their balance in water but to endow early whales with the ability to turn and twist at higher speeds, enabling them to outmanoeuvre predators but also to communicate with each other.

The structure of the ear and design of its bones has helped palæontologists identify another mammal that (along with *Pakicetus*) was probably among the earliest known mammals to have foraged along riverbeds even while it still lived on land. This was *Indohyus* (literally, India's pig), a complete skeleton of which was found in a rocky outcrop in Kashmir (on both the Indian and Pakistani sides). Scientists believe that *Indohyus* may well be the 'missing link' between the land-living and sea-dwelling ancestors of the whales. In appearance, this 'pig' would have resembled the modern-day water chevrotain. While chomping on succulent reeds and water weeds, it probably ducked into water to escape predators and scrambled back to safety on land, much like African water chevrotains do today. It had ears that were effective on land but useful underwater as well and resembled the ears embedded in the skulls of whales and the earlier land creatures like *Pakicetus*. The insides of its ears were thickened like those of contemporary whales. Its bones too were very thick, like those of the hippopotamus, that lives almost entirely in water and forages at the bottom of rivers and lakes. But a cross section of the bones of *Indohyus* showed that they were almost completely ossified with very little marrow space. The organ of balance too was poorly developed in these animals, suggesting that it probably spent a fair amount of time on land, too.

Around 39 million years ago, the constant adaptations at last threw up a design of a marine mammal that could match up to reptiles and giant fish. This was a 15-metre-long predator that was dubbed *Basilosaurus* (king of lizards). When the first fossils of *Basilosaurus* were discovered, it was mistaken for a giant lizard (hence the name) but later fossil discoveries confirmed that it was indeed a mammal. Although in India and Pakistan only incomplete fossils of *Basilosaurus* have been discovered, complete skeletons have been unearthed in Egypt. *Basilosaurus* was truly king of the seas. Looking at its majestic fossil it is difficult to imagine that just 25 million years earlier, the shrew-like ancestors of this imposing beast were living in trees and scampering about in the undergrowth! *Basilosaurus* had a fluke-like tail that went up and down (instead of side to side) aided by a flexible lower back that enabled it to move forward. It swam like its modern cousins, the dolphins, and had developed the ability to stay underwater for much longer than any of its mammalian predecessors. It had exceptional eyesight, improved hearing and a slender, serpentine body with small vestigial limbs that allowed males to attach themselves to a female during mating. Its flexible ribs allowed the lungs to collapse and thickened tissues in its middle ear enabled them to withstand tremendous pressure. *Basilosaurus* possessed most of the instincts—such as being doting mothers—that are considered mammalian. It had developed a special talent to dive deep and for a prolonged time as it hunted for prey or escaped predators. One exceptionally well-preserved skeleton of *Basilosaurus* from Egypt was discovered with two small sharks in its stomach. The hunted was now the hunter. Other smaller whales (like *Dorudon*, meaning spear tooth) also evolved in these years of plenty, and each specialized in diverse habitats of the seas and giant lakes, developing better social instincts and improving their design to ensure better survival.

But these idyllic times would soon change. When India fused with Eurasia and the Tethys closed, Australia and Antarctica too began to separate, creating new ocean currents. Around 23 million years ago when South

Basilosaurus *were among the first fully aquatic whales that reigned the seas from 41 to 34 million years ago. What made them successful was the refinement of their anatomy over their immediate ancestors. The nostrils, or blowhole, had moved towards the top of the head, their ear bones suggest that they could hear well underwater, their forelimbs had become paddle-like flippers, while their hindlimbs were rudimentary and their pelvis has detached from the spinal column, freeing up the lower spine to power their fluke-like tail. They eventually covered the entire tropical seas—from Alabama in the US to Egypt and Pakistan.*

America broke off from Antarctica and created the Drake Passage, Antarctica developed an ice cap that became permanent only about 11.5 million years ago. As ice caps formed at both the poles, sea levels dropped. Microbes, plankton and plants began to perish around the poles and the bounty of fish in the seas waned. Marine predators were hit particularly hard. For a short while, whales persevered with the help of the energy of their fat reserves, while larger predators like sharks began to die out at the poles. The eel-bodied *Basilosaurus* was ill-suited for these changes in the deep seas because, in proportion to its size, it had lower fat reserves. The rapid cooling of the seas and land began to take its toll on several species of mammals—the largest extinction since the dinosaurs. Nearly a fifth of all mammal species died out. Some whales did manage to survive in the colder oceans and seas, but *Basilosaurus* and its cousin *Dorudon* were not among them.

The tropical latitudes were a refuge for many marine creatures including some struggling species of whales and all sharks. But the tropics were not without their own hazards. Whales and dolphins had to contend with new monstrous shark predators such as *Megaselachus* in the Indian Ocean and *Megalodon* in the western seas. This was nature's twisted video game where old enemies, whales and sharks, rearmed themselves in an ever-escalating competition to gain dominance. *Megaselachus* ruled the deep waters of the tropical seas and preyed on young whale calves. As tropical waters became much too dangerous for whales and escape seemed a better strategy than staying and fighting, they began to spend more time closer to the fringes where warm tropical water met cold water. This water, fortunately for whales, was too cold for sharks. Gradually, whales migrated to colder waters of Antarctica. Because they are warm-blooded and possess the physiology to adapt to living in colder water, they were safe in the frigid polar seas. Whales can also control the flow of blood to their heart and brain and as a result, they did not suffer lack of oxygen in these vital organs in deeper waters. They developed kidneys which had a number of lobes and were efficient at concentrating urine. Whales also evolved special proteins in their red blood cells that helped them store higher concentrations of oxygen in their muscles and this enabled them to dive deeper and stay underwater for longer. Traditional whaling communities like the

Nordic people, Greenlanders, the Inuit and Japanese have been eating this deep-maroon whale meat for centuries, although now, because of the declining numbers of whales, whaling has been restricted. For cold-blooded sharks, staying in cold water was not an option for it would shut down their vital life-support systems.

On the poles, microscopic life adapted quickly to cold water, just as it had 700 million years ago. Small crustaceans and fish that had evolved to feed on microbes in cold water became fodder for the first whales. But visibility in these sun-starved waters was poor and whales needed to come up with a way to hunt for prey that they could hardly see. The ancestors of whales were bestowed with one last adaptation around this time that they retain to this day: they developed an apparatus for

Massive sharks like Megalodon *lived 19 to 3 million years ago and were the most powerful sharks in history. Large sharks like this, and* Megaselachus, *challenged the dominance of whales and marine reptiles during their long reign of the tropical seas. Based on the size of the teeth of* Megalodon, *scientists estimate that these sharks could have attained nearly 18 metres in size and weighed nearly 40 tonnes (less than a third of a blue whale). Many shark enthusiasts believe that* Megalodon *still lurks in the deep!*

echolocation, using sound to assess the size of their prey as well as to calculate its position and the speed at which it was swimming. Whale 'songs' consist of distinct sequences of clicks, harrumphs, snorts, roars and high-pitched squeals that may last up to ten minutes before the whale surfaces or dives deep into the water, at which time the sound diminishes. Scientists believe that apart from aiding the hunt for prey, these sounds are also used for navigation, identifying other individuals, communicating over long distances and to warn other whales of threats.

All these developments ensured that while sharks continued to dominate warm tropical waters, the cooler and frigid waters around the poles became a sanctuary for the whales. *Basilosaurus* were the true ancestor of the two families of whales that exist today, and they developed a few modifications that are still found in the whales. Their external nares or nostrils migrated to the top of their skulls, between or behind the eye sockets and their facial bones were reorganized to improve echolocation and communication.

In their own right, sharks too have been amazing escape artists, having survived multiple extinctions for over 470 million years since they first emerged. Incredible varieties of sharks have developed which live in the deep sea, in shallow lagoons and in rivers. Sharks rapidly diversified between 24 and 11 million years ago to become truly cosmopolitan along tropical seas and rivers. The layer of rocks and sediment around the sleepy town of Baripada in Odisha is a prized site for finding the fossilized teeth of extinct sharks. About 10 million years ago, the eastern coastline of India lay 60 kilometres inland and as the sea receded over the next 6 million years, it left many sharks stranded in inland lagoons. In 2001, during digging to build a road linking Cuttack with Kolkata, a cache of teeth of nearly forty different types of sharks was exposed in one stretch in Baripada alone. Sharks are cartilaginous and cartilage seldom gets preserved as fossils, so it is exceptionally rare to find a complete shark fossil. Serrated teeth more than 10 centimetres long, belonging to monster sharks like *Megaselachus*, have also been found in a number of places in Kutch and Mizoram.

Once whales grew large enough to challenge the might of the sharks, they were able to extend their territory deep into tropical waters once again. The Indian Ocean became a battleground where the last marine reptiles like massive sea crocodiles and monstrous sharks battled it out with whales. These internecine wars for dominance would push a number of species towards extinction, but inevitably some fared better than others and recovered from their battles to establish their kind. When the global climate cooled between 11 and 8 million years ago and then again between 3 and 1 million years ago, and tropical oceans also became cooler, marine reptiles like crocs were outcompeted by whales and sharks and they began to perish. Just one species continues to venture into seas—the saltwater crocodile—that is found from Eastern India to north-eastern Australia and across the Pacific Islands till the Solomon Islands, but it hardly ever confronts the sharks or the whales, not like it used to in the olden days! However, crocodiles remain the largest predators in freshwater bodies.

Tropical waters provided new varieties of seasonal food like plankton and krill and some whales began to undertake an annual migration from the poles to the tropics. This ritual of migration continues till today. A fourth of all sea mammals come to the coast of India at some time. In June every year, some of the largest and the most magnificent whales visit tropical waters from Antarctica in search of plankton or fish and to find sanctuary in the Bay of Bengal and the Arabian Sea. In the summer of 2015, a blue whale was sighted off the coast of Alibaug in Maharashtra and some whales and dolphins were found stranded along other beaches of India.

A shy and less spectacular cousin of the whales lives in the slow-moving rivers of the Himalaya (the Ganga, Brahmaputra, Indus and all their tributaries). This is the river dolphin. Some river dolphins, like the Gangetic dolphin, have become blind as they do not need eyes to see in the murky water. Instead they have highly sensitive echolocating abilities that help them scour the water for fish, crabs and molluscs. River dolphins are shy aquatic creatures but, tragically, nobody seems to care about them any more and they have been largely forgotten. Their habitats have been broken by dams and constructions along the riverbanks which, in addition to restricting their movement, has also reduced the stock of fish that they feed upon. Poisoning of the water with sewage and industrial and agrochemical waste is also wiping them out in large numbers. These creatures that evolved over 6 million years ago and survived two waves of extinction of mammals are unlikely to be able to overcome the anthropogenic crisis they face today because, despite their evolutionary adaptations and their hardy genes, nothing has prepared them for this kind of rapid and irreversible change.

In a span of 40 million years, mammals undertook an incredible journey of evolution and transformed from small arboreal shrew-like ancestors to water-dwelling giant whales. The story of whales tells us that evolution is not just about surviving but also about making several starts. After whales, mammals made three other attempts to make a living from water. The first to follow whales were the gentle, seagrass-grazing dugongs and manatees that emerged on the scene around 28 million years ago in the Pacific north-west, a region around the evergreen temperate forests between Oregon and Washington state in the US. Dugongs and manatees trace their descent to the same ancestor from whom elephants (*Moeritherium*) emerged around 50 million years ago. Although fossils of as many as fifty-two different types of these early aquatic or marine animals have been found, only five species of dugongs and manatees survive today. Several species have become extinct because of overhunting. Their name (order *Sirenia*) possibly originates from Homer's Odyssey in which the Sirens, the half-woman sea creatures, lured seafarers and the hero, Odysseus, and his ship on to their island with sweet songs of love. While recording the sea stories of sailors, some scribes confused Sirens with mermaids. It is said that sailors who had been at sea for long pined for attention from the fairer sex. When they heard cooing and singing in placid seas and saw a suckling calf feeding from the pectoral breasts of a mother manatee, they imagined these manatees to be women of the sea, which is probably the root of the legend of mermaids. In 1493, Christopher Columbus wrote that the 'mermaids' he had seen in the Caribbean were not as lovely as he had expected. So the next time you call a Bollywood starlet or a college heart-throb a 'siren', remember the story of these highly threatened sea mammals!

After the dugongs and manatees, a group of semiaquatic sloths from South America emerged 4 million years ago but they perished when lakes and vegetation shrunk, and caused these giants to starve to death.

The final attempt to live in water was made by modern-day polar bears and otters, who took to water because they developed a taste for the creatures that lived in it. Though there are no significant fossil records to show this, it is quite certain that they were the most recent mammals to enter water and have happily lived an amphibious life—that is, they live and breed on land but prefer to hunt in the water.

The transformation of the fierce wolf-like ancestors of sheep into giant whales and graceful dolphins is the Cinderella story of the animal world. But there were other mammals too which set off on their own evolutionary trajectories and among these were the mammals living in the trees of the Siwalik Hills. Although tree-dwelling primates initially remained largely unperturbed by the power struggles between the ever-growing grazers and vicious meat eaters that were taking place on the ground, the changing vegetation and the decreasing number of trees meant that they had to leave the trees to look for more options to eat on the ground. The same evolutionary forces that compelled *Indohyus* and its kind to test the margins of water began driving tree-dwelling mammals like the primates to explore new sources of food. Primates, who had so far lived entirely off trees, between 28 and 9 million years ago, evolved into several new types, adapting and surviving first the moist forests and then the arid habitats that appeared rather suddenly on the scene. When the Tethys drained completely, India witnessed the arrival of strange new beasts from Eurasia and Africa, and these, over the next 11 million years, evolved into the ancestors of modern mammals, including our own species.

13

HOW TO MAKE A MAN

Between 60 and 12 million years ago—a tiny slice of time, relatively speaking—an impressively large number of mammal species descended from trees to occupy an array of different niches and habitats, developed a taste for new kinds of foliage and flesh, and adapted into myriad shapes and sizes to suit their new lifestyles. A few, however, remained in the trees. It is these mammals—the least enterprising of all mammals, in a sense, because they stayed up in the trees—that are of particular interest to us, because among them were the early primates, from whom arose the ancestors of humankind.

Initially, the mammals that stayed back on trees remained small and were of little consequence to other creatures. Forests and grasslands were dominated by mega-herbivores and fierce meat eaters, while the first whales were beginning to test the waters of the sea. These creatures would have taken scant interest in tree-dwelling mammals who posed no threat and were hardly desirable prey. But around 20 million years ago, a revolution occurred in the realm of flowering plants that saw the development of modern species of fruit and foliage. This, in effect, led to the growth, both in size and importance, of primates.

Hispanopithecus *were small, slow-moving tree dwellers with prominent ape-like canines, very slender ankles, and ape-like big toes and were therefore not really bipedal walkers.*

'Primate' is an overarching term for a group of animals (technically, an 'order') that encompasses lemurs, monkeys and apes. The order Primate is divided into two branches—the prosimians, which includes lemurs and tarsiers and is considered the more 'primitive' group, and the anthropoids, which is further divided into three subgroups: monkeys, apes and all 'hominins', including humans and all of our immediate ancestral hominids (a term covering *all* members of the genus *Homo*).

The monkey subgroup includes Indian langurs, African baboons and all the macaques, making up over 220 different species in all. Around 40 million years ago, while they were still quite small, a few species of monkeys left Asia, reached Africa and then crossed over into South America by land bridges. These pioneers laid the foundation of a new family of monkeys that we call 'New World monkeys', which includes tamarins, marmosets and capuchins. New World monkeys differ from those of the Old World mainly by having flatter noses and, longer prehensile tails, and by being colour-blind (the howler monkey is a singular exception). Old World monkeys, in comparison, have prominent nostrils, shorter non-grasping tails and opposable thumbs, and can see in colour. It is the Old World monkeys who provided the genetic platform for all the species of monkeys and apes that evolved subsequently. Genetic studies tell us that a common ancestor of an Old World monkey that lived in Asia was the progenitor of all modern great apes (the larger apes that include orangutans, gorillas, chimpanzees and humans) and lesser apes (gibbons) that evolved over the next 18 million years or so. Humans, gorillas and chimpanzees all came from this common ancestor but then evolved in different ways and under different circumstances. The ancestors that led to the modern gorilla had diverged about 12.5 million years ago from the common ancestor of chimpanzees and humans to form a distinct lineage of their own. Our human ancestors diverged from the chimpanzees between 8 and 5 million years ago. After several criss-crossing evolutionary pathways, the ancestors of the apes and the ancestors of humans—many of whom would have lived alongside each other and might even have procreated with one another—emerged and created separate lineages but with a diverse genetic pool.

One thing is certain—humans did not evolve *from* monkeys. Humans and monkeys evolved from a common ancestor that is now long extinct.

But human evolution is complex and politically and culturally sensitive and there are no easy, straightforward explanations or answers that rise unequivocally above controversy. Every year, discoveries of new fossils, sites and sophisticated new genetic analyses help recalibrate the timing and improve our understanding of our origins. But we are still some way away from a clear statement of origins uncluttered by other biases and considerations.

So how did we get here? The first question to ask is where primates originated from. The examination of a rather awkward part of the human body proves surprisingly to be an elegant place to start. This is the testes, the external gonads that hang in a sac of skin between the legs of human males. The word *testis* in Latin means 'witness' or 'spectator', as in the words 'testify' and 'testimony', referring perhaps to maleness or virility. Fish, reptiles, birds and many mammals like elephants and whales have internal or undescended testes, i.e. their testes are not suspended between their legs but are enclosed *within* the abdominal cavity. For animals that fly, swim, leap or run, it is easy to see how dangling testes could be an inconvenience. Many warm-blooded animals like birds, which have internal gonads and a higher body temperature than humans, have special adaptations to cool their sacs and prevent the destruction of their sperm. Whales, seals and walruses too use blubber and a special circulatory system to protect their gonads from freezing. In humans, the external or descended testes are primarily a mechanism for keeping them cool and preventing body heat from destroying delicate sperm.

It is instructive to see how external testes are distributed among mammals because it points to evolutionary cleavages. Mammals whose testes are contained within their abdominal space belong to a class of mammals that originated in Africa (moles, aardvarks, hyraxes and elephants) and South America (armadillos, sloths and anteaters). By contrast, the dominant mammals that evolved from Asia and Eurasia (*Euarchontoglires* which comprises primates, squirrels, some marsupials and small bats; and *Laurasiatheria* under which fall shrews, horses, rhinos, cats and dogs) have external testes. There are exceptions: whales, for instance, originated in South Asia and have internal testes, but the receding of their gonads was probably a late adaptation to life in (cold) water.

Nevertheless, it is generally true that nearly all the mammals that originated in Eurasia have external testes. The presence of external testes in primates is probable confirmation that they did not originate in Africa, and in all likelihood entered Africa from Eurasia.

Between 22 and 5.5 million years ago, a time known as the Miocene epoch, apes dominated the primate world. More than 100 species ranged throughout Eurasia, Africa and India. Out of these, at least nine species of primates lived between 15 and 9 million years ago in the tropical forests of the eastern part of Eurasia but two of these were dominant, especially from the perspective of the future of ape and human evolution. The largest among the apes was *Gigantopithecus* (Greek, *gigas*: giant, *-pithekos*: ape) that lived from around 9 million to about 120,000 years ago in China, India, Pakistan, Nepal and Vietnam. *Gigantopithecus* was the largest ape ever to have lived. A fully grown adult would have stood 3 metres (nearly 10 feet) tall and weighed in excess of 500 kilograms. The species found in India (*Gigantopithecus bilaspurensis*, named after the first fossils found in Bilaspur in Himachal Pradesh, although more fossils were found in other parts of the Siwalik Hills of Himachal and Punjab later) lived about 9 to 6 million years ago. This species became extinct long before several allied species found in China and other parts of South East Asia and the reason for this remains something of a mystery. Some scientists speculate that *Gigantopithecus bilaspurensis*'s disappearance may have coincided with a time when forests began to wane and grasslands expanded, shrinking the options for a species whose diet was based on foliage and fruit.

The dominant species in South Asia was *Sivapithecus*—ape of the Siwaliks. Unlike the bones of giant rhinos and elephants, the fossil remains of *Sivapithecus* are not at all easy to find. *Sivapithecus* lived like a chimpanzee in the trees, along with other species of primates. Fossils of *Sivapithecus* have, until now, been found only at three sites—in Haritalyangar (the hillock and a stream which exposes the Dhok Pathan formation behind the public works department guest house is a particularly good place to start; Lat: 31°51'N, Long: 76°63'E), about 30 kilometres north-west from Bilaspur town; in Chakrana (Lat: 31°52'N, Long: 76°67'E), around 6 kilometres from Haritalyangar; and in the Chinji formation which is best exposed along an 8-kilometre strip beside the road between Kulwanta (Lat: 32°79'N, Long: 75°28'E) and Ramchand on the outskirts of the town of Ramnagar in Udhampur district, 90 kilometres east-north-east of Jammu. *Sivapithecus* was much like an adult orangutan in size (about 1.5 metres tall) as well as in its facial features. In western Eurasia, from France to Hungary, the dominant ape species was *Dryopithecus* (from Greek, *drus*: oak tree, *-pithekos*: ape) and it lived alongside another ape called *Hispanopithecus* (*Hispano*: from Spain, *-pithekos*: ape). *Hispanopithecus* was about 1.2 metres tall and resembled a chimpanzee. *Dryopithecus* and *Hispanopithecus* spent most of their time on trees, swinging from branch to branch like the modern orangutan. They had distinctively different jaws and teeth but shared similar tastes in fruits and leaves. Both these species had large brains. Around 11 million years ago, either *Dryopithecus* and *Hispanopithecus* or *Sivapithecus*—or perhaps all three—reached North Africa from Eurasia or West Asia. *Dryopithecus* and *Hispanopithecus* moved southwards towards a more tropical part of Africa and coped well in the changing climate and evolved into the earliest ancestors of a lineage of Africa's great apes and the hominids. Although *Sivapithecus* may have moved into North Africa, it preferred living in the more stable tropical jungles that extended from northern India to South East Asia, whose salubrious environments did not push it to take the evolutionary strides that *Hispanopithecus* took. About 10 million years ago, the lineage of *Sivapithecus* led to the emergence of the modern gibbon and the orangutan lineages and they migrated eastwards into South East Asia and India. The hoolock gibbon is the only surviving ape in India (other than us humans!) and is found east of the Brahmaputra and Dibang rivers.

Around 11 million years ago, some apes in Africa would have begun walking upright but only with the support of overhanging branches. They still needed props to be able to stand erect on two feet. The earliest attempt at bipedal movement would probably have been made not on the ground but along large branches of trees. Around 7 million years ago, as the savannah spread, the corridor of trees broke and trees began to seem like little islands of isolation in a sea of grass. A few apes began venturing on to the ground to pick up fallen fruit, have a drink of water from a pool below or to feed on termites, perhaps even to defecate. Once on the ground, they needed to raise their heads to keep a lookout for

Like all apes, hoolock gibbons' arms are nearly double the length of their legs which is useful for balancing when they swing from branch to branch (called brachiation) at high speeds (sometimes covering as much as 6 metres in one swing). Hoolock gibbons are almost entirely arboreal, coming to the ground only in exceptional circumstances. Only 5 per cent of their locomotion is bipedal walking using overhead branches of trees. Their name is derived from the Assamese word 'uloock' meaning the loud call or howl of the gibbon. Their emotive calls echo across long distances in the forest which communicate with other individuals of the troupe or with other hoolock communities. It is said that the calls are so regular that in the low hills of Nagaland and Assam local people use these calls to synchronize their days' routine.

Australopithecus *was perhaps the earliest ape to crack rocks and create crafted tools and was probably among the earliest to share traits with modern apes and humans. Sophisticated genetic research has found that these gracile apes may have been the first to carry a duplicate copy of a gene called* SRGAP2A *that led to the increase in the length of neurons and made the nerves more dense, creating a more advanced neural network leading to and within the brain. This gene is one of the twenty-three genes that got duplicated in the genetic structure of early human ancestors, and it is the duplication of this gene which, by increasing the complexity of neural networks, enhanced the capacity of the brain to perform more complex tasks. Genetic studies have found that the gene duplicated three times—3.4 million years ago (the gene became* SRGAP2B*); 2.4 million years ago (*SRGAP2C*); and about 1 million years ago (to create* SRGAP2D*).*

signs of danger, and as a consequence took to knuckle-walking to move over short distances. Some apes evolved knuckle-walking after they split off from our human ancestors nearly 6 million years ago, while the first bipedal ancestor of humans did not knuckle-walk at all. Surprisingly, among all the apes, gibbons display the least amount of cognitive skills but are the most bipedal of all non-human primates. In contrast, African apes like chimpanzees and bonobos are more closely related to humans but rarely use bipedalism. So bipedalism is not necessarily associated with being 'modern' nor is knuckle-walking a sign of being primitive, but knuckle-walking did indeed play a significant role in the future progress of apes, especially the ancestors of the humans.

Bipedalism offered several advantages to apes: it freed their hands, enabled them to walk tall and allowed them to keep a lookout for potential predators in the savannah. The transition from an arboreal life also set off changes in many of their organs, like the rearrangement of the ear bones, the placement of the eyes in the skull, the redesign of joints in their arms and legs to improve locomotion on the ground, the decrease in the size of hips and pelvis, and the reshaping of the spine to give it the suspension that would allow apes to walk erect.

Around 4 million years ago, a second intensive uplift of the Himalaya further intensified the Indian monsoon, and this had a knock-on effect on the regional climate of East Africa as well. In some years, there was plenty of rain to fill up lakes and even create marshy conditions, but at other times this region received virtually no rain. The vagaries of climate transformed the region, and the deep, freshwater lakes that had existed around 2.6 million, 1.8 million and 1 million years ago also drained out in cycles lasting for almost 20,000 years before drying out altogether. These were key dates in human evolutionary history, with East Africa and southern Africa set to become fecund incubators of human evolution. The development and intensification of the Indian monsoon coincides with the split between Asian and African apes, the former evolving into the apes in Asia and the latter eventually evolving into African apes, early humans and us (the *Homo* species).

Among the earliest ancestors of humans was a group of bipedal apes called *Australopithecus* (meaning 'southern ape', because the first specimen was found in southern Africa). At least four species of

Australopithecus (some believe there were seven) lived throughout eastern and northern Africa between 3.9 and 1.9 million years ago, and one of these eventually became the ancestor of *Homo*. Members of this genus resembled a flat-faced gorilla, with short limbs, wide hips and a broad ribcage, and only a small ape-sized brain. An adult would grow no more than 1 metre tall. *Australopithecus* had no grasping big toe on its feet—a clear sign that it was by now adapted to walking upright. Males were a lot larger than females. Their jaws had large cheek teeth with very thick tooth enamel. They had a short lifespan: they grew fast and died young, like African apes today. *Australopithecus* was a more adept walker and climber than a long-distance strider and runner. About 3 million years ago, the *Australopithecus* lineage split into two—*Australopithecus* remained and an offshoot led to the creation of the genus *Homo*.

The ability to walk upright was not the sole criterion for being labelled an ancestor of humans. In order to qualify, a fossil needs to possess any one of the following attributes: its premolar teeth should have two cusps (raised bumps on the chewing surface), or the base of the skull, the backbone, pelvis, legs, feet and toes of the skeleton should all have been constructed for walking on two legs.

The earliest species of *Homo* to emerge was called *Homo habilis* ('handy' or 'skilled human'; I will refer to this species as '*Habilis*' from here on) which evolved and flourished between 2.3 and 1.4 million years ago. *Habilis* was still chimp-like at about 1.2 metres tall. But their hands were more like those of modern humans, which meant that they possessed the ability to shape rocks. Clear evidence that *Habilis* had the ability to craft tools comes from the anatomy of their thumb bone. A flat, broad bone of the thumb similar to that of humans suggests that they had a pad of muscle with skin on it that helped them make tools. Shaped rocks have been found from this period of prehistory, which confirms that they had this ability. The tools made by *Habilis* were used to slice meat off small animals they hunted and to shear carrion. Their appetite for meat perhaps explains why *Habilis* had a brain that was 30 per cent larger than that of *Australopithecus*. But meat is much more difficult to obtain than nuts, foliage or fruit, and early hominins like *Habilis* did not at first possess the skills or tools for hunting fleet-footed herbivores much larger than themselves. Nor did they have the size or

Homo habilis *are perhaps the earliest members of the genus* Homo. *They had a slightly larger braincase and smaller face and teeth than* Australopithecus, *but they retained some ape-like features, including long arms and a forward-protruding jaw. An adult male would have been no bigger in size and weight than the average six-year-old boy. Although* Australopithecus *had discovered the use of stone tools, it was* Habilis *who took them to the next level of sophistication.*

the physical strength to steal from other predators and scavenge, and their diet would still have been predominantly one of fruit and tubers, with only an occasional small animal as a bonanza.

Around 2 million years ago, two distinct species of *Homo* existed side by side in Africa—*H. habilis* and *H. erectus* ('upright man'; '*Erectus*', from here on). *Erectus* was about 1.8 metres (nearly 6 feet) tall, with a build

By the time Homo ergaster *(or the working man, on the left) and* Erectus *emerged 1.9 million years ago, the last* Australopithecus *(right) were perhaps struggling to survive. It is also possible that they never came face-to-face. In just 1 million years, the ape-like* Australopithecus *had given way to a less hirsute, more vocal, sophisticated toolmaker and hunter, the genus* Homo. Ergaster *is believed to have diverged from the lineage of* Homo habilis *between 1.9 and 1.8 million years ago in Africa before disappearing from fossil records around 1.4 to 1.3 million years ago.*

similar to that of modern humans, but with a brain only about two-thirds the size of ours. *Erectus* had developed a taste for meat and hunted in groups or often stole prey from other carnivores. Like other large carnivores, *Erectus* followed their prey and dispersed successfully across the tropics in pursuit of game. In difficult times, *Erectus* would almost certainly have scavenged. Although carrion is not palatable, the bones of large mammals contain bone marrow which does not rot quickly and is otherwise eaten only by hyenas. Marrow contains fat and protein and is the only part of an adult animal that produces red blood cells. *Erectus* developed a fondness for marrow but to smash very large bones, especially those of elephants, hippos, rhinos and other mammals, he needed sharper and stronger tools. Feeding on raw and dead meat also introduced intestinal worms like tapeworm into our ancestors and these persistent parasites continue to live within us even to this day.

Gradually, sometime between 2.2 and 1.4 million years ago, as some parts of Africa became drier and hotter, the hair on the bodies of *Erectus* began to thin down over the next few generations. As their skin became less hirsute, they developed more sweat glands which helped them cool down more efficiently on a hot day. Some palæoanthropologists believe that, unlike many carnivores, the less hirsute ape no longer needed to pant with its tongue out. It is estimated that by about 1.2 million years ago *Erectus* had lost nearly all their body hair except in places where we modern humans still have hair. Because there was no need to extend the tongue out of their mouth, *Erectus* were able to relax the jaw and muscles around their vocal chords and use them better for making sounds. The vocal chords became stronger and perhaps even versatile at producing a variety of sounds including clicks, roars, grunts, whistles and imitations of animal calls. This improved communication between members of a group. The gradual reduction of hair also meant that larger breasts or mammary glands evolved from smaller teats. Mammary glands developed when sweat and sebaceous glands began to concentrate around the teats and formed a knot-like structure under the skin to form nodules around which fat began to deposit gradually. Sebaceous glands produce an essential oil (sebum) that reduces infections of the skin and prevents it from cracking. In human females, breasts are made of glandular tissues encased in fat, with the entire complex suspended by ligaments. It is this fat that gives human breasts their hemispherical shape.

In most grazing mammals, cattle and even whales, breasts are located in a region around the groin, but in primates (and in some large mammals that can stand on two feet, such as bears) the teats are on the chest. The breasts had moved to the upper chest in primates as they began climbing trees and walking upright and some scientists believe this could have happened as early as 11 to 7 million years ago. Smaller mammals typically have many pairs of mammary glands which correspond to the size of their litters, but apes have always had only a single pair of breasts, from the time of our earliest ancestors to the present.

Descendants of bipedal apes also evolved new means of childbirth. Modern monkeys and apes give birth on trees, squatting on their hind legs or crouching on all fours. During childbirth, their babies emerge face up, i.e. with their faces looking up towards their mothers. Monkey infants are occasionally strong enough to wriggle out on their own, using their hands to pull themselves out. When an infant monkey squeezes out of the birth canal, the mother's hand reaches out to guide her baby towards her breasts for its first feed, but before this she wipes the mucous and amniotic fluid from its face, nose and mouth to facilitate breathing and feeding.

Human babies are birthed differently and trace this modification to early bipedal hominids. The modification of the pelvis that enables us to walk upright necessitates that a human infant comes out of the birth canal face down. The arrangement of bones and the size of her hips suggest that *Erectus* was probably among our first ancestors to give birth like modern humans do. *Erectus*'s knees were angled in and her thigh bone fitted into her hips towards the side, rather than in front. This helped lower the centre of gravity and rearrange the pelvis. In anatomical terms, the back of the head of an infant rubs against the pubic bone of the mother's pelvis, so that it faces in the opposite direction to its mother—*occiput anterior*, as a gynaecologist would put it. For this reason, it is difficult for a bipedal mother who is delivering a baby to reach out and guide the baby like a tree-dwelling monkey does. Any attempt at pulling out the baby runs the risk of snapping its delicate neck and spine or twisting the baby's head. This prevents a mother from being able to clean the fluids that block the nose and mouth of the baby or cutting the umbilical cord on her own. It is more than probable that childbirth made *Homo* mothers more 'social' as they assisted each other during this difficult period. Cooperation would have established a closer bond among females, and mothers and children began to live in tightly knit groups. But it was not just during birthing that an entire team of caregivers was available to help the mother and infant. Human infants are helpless and need special care for longer than any other mammal and caregiving in humans is always a long drawn-out process, usually involving a number of individuals.

Our early ancestors developed uniquely long childhoods as the culmination of a remarkable chain of events. We are reasonably sure that early apes had short childhoods and brief lifespans. They also had wide hips which suggests that the brain of the foetus developed while it was still inside the womb. This is only to be expected of an animal that needed to acquire skills quickly in order to become independent sooner. In our early bipedal ancestors (such as *Erectus*, and perhaps another species called *Homo ergaster* too, which evolved just before *Erectus*), the human pelvis could no longer accommodate a well-developed infant's skull and from this point on—sometime after 1.8 million years ago—human infants were born at a very early stage of their brain development. This is especially noticeable relative to other primates. But unlike *Erectus*, who attained adolescence early, our later human ancestors underwent a long time of dependency as infants and required a large investment of support from parents, caregivers, midwives and helpers (a process called 'alloparental care'). The evolution of the family played a crucial role in creating a stable, secure environment for our early human ancestors. Along with the adaptation of communal birthing in females, humans became more sophisticated at hunting in cooperative groups and more proficient in the use of tools and fire, which further increased cohesion among small bands or groups of hominids.

It is not widely known that chimpanzees hunt and eat fair quantities of meat and their rate of success at obtaining meat increases dramatically whenever they gang up and cooperate in a hunt. The nature of giving birth, tending infants and hunting—all make it advantageous for chimps to form social coalitions. Within the chimpanzee social unit, however, there is the imperative of maintaining order. Uneven-aged social units of chimpanzees are characterized by constant struggle because males compete for females and for social rank. It is perfectly plausible that our early ancestors faced similar social pressures, but it is just as clear that the story of hominid evolution is also a story of the progressive development of social skills. Bands of ancestral humans would have hunted, foraged, scavenged and migrated to new areas when food ran out, and this could not have been without power struggles and setting boundaries for territories and apportioning responsibilities to members of the group. Gradually, early humans developed another key ability

Homo erectus *were agile and their feet were arched for walking and running. They could fix their eyes on a distant target. Better running and tool-making turned* Erectus *from scavenger to hunter. They lived in a hunter-gatherer society, hunted in coordinated groups and perhaps had begun to develop the earliest protolanguage.*

that enabled them to communicate better with each other: they evolved vocal chords and, in all likelihood, indulged in mimicry and imitation. This helped human ancestors evolve means to communicate with one another which gradually enabled them to communicate more complex ideas. This also helped them acquire new skills, refine existing abilities and pass them on to their children in a structured way. Learning how to pull out termites using a saliva-laced straw is a commendable feat for a chimp, but it does not compare with what it takes to shape stone or wood into tools, to dress up in animal pelts, or to learn how to use fire to cook.

Some anthropologists believe that the 'wasteful' eating habits of large carnivores like the leopard, the sabre-toothed cat and the bear dog enabled early human ancestors to scavenge on their leftovers and this whetted their appetite for meat. Our early ancestors caught small prey, as modern chimpanzees do, and there is good evidence from at least 2.6 million years ago which shows that they butchered large carcasses. It is possible that men alone went to hunt game while women collected tubers, fleshy stems and fruits. Studies on modern hunter-gatherer communities suggest that families derive more calories from what their womenfolk gather from forests than the meat that menfolk hunt. This may have been true millions of years ago, too.

Our ancestors were prey to fierce predators for millions of years. For *Erectus* and successive hominids, leopards, other large cats and hyenas would have posed dangers that could only be mitigated when they learned to become predators themselves. Leopards in these times were 25 per cent larger than they are today (about the size of an Asiatic lion) and there would have been frequent bloody encounters. *Erectus* lived in huddles in the open savannah and occasionally sheltered under overhanging rocks or in caves, but these were just the kinds of spaces sought by the big cats too. Spears could ward off leopards and the giant hyena (*Pachycrocuta* or 'thick hyena') but the stealth and cunning of these animals would often outwit and overpower *Erectus*. All of this changed dramatically when *Erectus* discovered fire.

We can only speculate about how early man tamed fire. One conjectural scenario goes something like this: a bolt of lightning caused a conflagration in a patch of dry grass or dry tinder. As the fire spread and animals ran in blind panic for safety, a band of *Erectus* might have subdued their instinctive fear of fire to examine it with curiosity. It could have taken a long time before *Erectus* learned how to tame fire. They would have noticed that small sparks flew when they knapped flints to shape spearheads and concluded that if they could concentrate these sparks over dry wood and leaves, fire could be kindled. Once *Erectus* learnt to create and control fire, it was only a short step before they began to use it as a weapon. The ability to start and manipulate fire also helped *Erectus* travel long distances away from their home range and venture out into new places without fear.

Around 1.2 million years ago, Earth's climate saw short bursts of intense cooling followed by longer periods of warm, dry spells. In winter, the temperature dipped to less than 10°C in the northern hemisphere in places that were not covered with ice. The East African lakes dried up and were filled again in short cycles, thereby creating brief times of plenty followed by long-drawn-out famines. The only way to tide over the vagaries of climate was to include a lot of fat and proteins in their diet, and *Erectus* may have come to depend on meat around this time. Armed with the knowledge to make fire, *Erectus* soon learnt the art of cooking meat (perhaps roasting it, more than any other method). Cooking fuelled the next big step of human evolution and transformed early hominins to near-human. Cooking breaks down complex fats and proteins into simple ones that can easily be absorbed in the body, reducing digestion time in humans. Uncooked meat mostly passes out of the human digestive system without much absorption. Cooking also reduces the energy needed to digest and improves the retention of energy from the food itself. This allowed human metabolism to devote more energy to other bodily functions as eating cooked food saves a quarter of the energy expended in eating raw food.

Easier access to calories also had a more surprising effect: the human body was redesigned. The ribcage became compact and the size of the diaphragm, stomach and intestines was reduced. Additionally, in females this meant that they had a larger and more durable uterus and a rich and spongy placenta, which ensured greater reproductive success. The teeth of *Erectus* (and of our later ancestors, *Sapiens*) decreased in size (too bad for dentists and toothpaste makers!). The small gut-big brain theory suggests that with cooked food—especially meats—less energy was needed to maintain the body, and more energy could be diverted for growing a larger brain. It takes a disproportionately large amount of energy to power the human brain. Although the brain is only 2.5 per cent of our body weight, it consumes 20 per cent of our total energy. In comparison, other living primates use only 7 to 10 per cent of their energy to power their brains. Brains are an expensive investment, but they paid dividends many times over for our species. This prehistoric instinct explains our craving for energy-rich fatty and sugary food, that our brains have been programmed for, a weakness exploited by the food industry. Our brains respond more to fats (think of the inviting aroma of ghee on hot rice or melting butter on warm toast) and have developed stimuli through our sense of smell, taste and sight to seek such food. With easy access to food that was once so difficult to obtain, fast food rewards our brain without our bodies having to make much effort.

Better diets also improved the chances of living longer. In the females of most animal species, the quality of life wanes after their reproductive service is over. Not so for us. Our mothers and grandmothers and great-grandmothers live well beyond their reproductive years, and this is partly because of our better diets, but also because we value our elders. Humans, like elephants, see elders of the community as a repository of knowledge, as those who transfer skills, assist young mothers during childbirth and help in rearing offspring. The length of our life is therefore shaped not only by our biological, but by our social evolution as well.

Climate change, driven by the rising Himalaya and the growing influence of the Indian monsoon, tinkered with the evolution of the ancestors of *Homo*. The intermittent drying and greening of East Africa and the rest of Eurasia, Africa and Australia affected the life of all land-living animals in these parts. The only species of apes and human ancestors that survived were those who were better problem-solvers and more adaptable to survive climatic hardships. Climate instability pushed early humans to adapt or perish, and together, three triggers caused the apes to evolve into the earliest ancestors of humans (*Homo*)—the ability to walk on two legs; a constantly growing brain; and intelligence, which led them to make tools, tame fire and cook meat. Human ancestors were durable creatures produced from climate change. Around 1.8 million years ago, a series of hardships pushed *Erectus* (and perhaps other hominin species as well) to undertake an epic journey of 11,000 kilometres from Africa, across Arabia to reach the Levant, and then into Europe and India, from where they would travel further to South East Asia and China. Although few fossils have been discovered from these times, our human ancestors left behind tools during their journey that, like Hansel's breadcrumbs, have helped us chart their course.

14

CITIUS, ALTIUS, FORTIUS

Around 2 million years ago, a sea of grass extended all the way from West Africa to China, with only a few groves of trees in between like islands. A variety of herbivores and their hunters, like big cats and hyenas, lived off these grasslands. There were virtually no large barriers in this sea of grass and the herbivores found here were common across this entire region. We know this from rodent fossils. Small mammals like rodents do not usually migrate over long distances, and their fossils from this time unearthed in West Africa, Arabia, the Siwalik Hills of India and in China are all similar. Large rivers (the Nile, Tigris-Euphrates and Indus) had emerged from their respective mountain ranges but many of these rivers were still young and seasonal and not the perpetual rivers of today. However, with every passing season, they were growing in might. The Indian monsoons were still gaining strength and were erratic but they had become a regular feature around 2 million years or so ago.

The strait connecting the Red Sea to the Gulf of Aden—the Bab-al-Mandab—divides Yemen (on the Arabian Peninsula, above) from Djibouti and Eritrea and north of Somalia (in the Horn of Africa, below). At their narrowest the straits are about 30 kilometres wide and between these lie desolate rocky islands like the Perim Island, seen as one of the brown specks on the right. The strait in Arabic roughly translate to the 'Gates of Grief' which allude to the dangers associated with navigating this stretch. According to an Arab legend, several hundred sailors drowned after a violent earthquake rocked the sea many centuries ago. All along the coast and islands, palæoanthropologists have found stone tools which have helped them chart out the possible paths taken by early humans, out of Africa.

Whenever the monsoons arrived, they transformed the tough savannah stalks into sweet, succulent grass, and in a matter of days the dry ochre fields turned into lavish grasslands.

It was around this time that our ancestors, *Homo erectus*, began losing their ape-like character and developed a body and physical abilities much like our own. *Erectus* had a larger braincase that housed a brain 20 per cent larger than their predecessors, and they had developed a distinct forehead, a feature conspicuously missing in apes. Anthropologists have discovered the butchered bones of deer, pigs, other mammals and tortoises from this time, with the charring marks of fire that are seen when we cook meat, which meant that they used fire for their cooking as well.

It is exciting and tantalizing to try and puzzle out what *Erectus* may have been like. Let me disappoint you by saying upfront that very few fossils of *Erectus* (or any of our early human ancestors) have been found so far on the Indian subcontinent, and no *complete* skeleton has been found anywhere in the world. Outside of Africa, some fossil sites and remains have been found in Eurasia—a partial skullcap in India, and several bones in a site called Dmanisi in Georgia, China and Java—and not all palæoanthropologists agree that these can definitively be attributed to *Erectus*. The precise dating of the fossils also poses problems. Several dating techniques have been developed but each one has limitations, leaving plenty of scope for scholars to bicker and contest each other's findings.

In the absence of incontrovertible fossil evidence, anthropologists use stone tools to study the path and behaviour of *Erectus*. They study the shape of the tools, the style and the number of facets that were shaped as indicators of the age of a site. Finding stone tools like cleavers, hand-axes, choppers and flakes is relatively simple. The earliest stone tools made by *Homo* from around 3.3 (perhaps more realistically about 2.8) to 1.6 million years ago were discovered in Olduvai Gorge in Tanzania and these distinctive tools are called 'Oldowan'. A toolmaker would use one rock to chisel and shape the edges of another rock. One end of the tool retained its original margin and only one edge of the stone was worked. These distinctive Oldowan tools are a major milestone in human evolutionary history: they are, quite literally, the earliest preserved evidence of material culture that can be attributed to our ancestors. While the last *Australopithecines* developed the ability to make stone tools, it was probably *Habilis* who first manufactured Oldowan tools. Oldowan technology is typified by 'choppers', which are basically stone cores with flakes removed from part of the surface, creating a sharpened edge for cutting, chopping and scraping. Microscopic surface analysis of the flakes struck from the cores has shown that some of these flakes were used as tools for cutting plants and butchering animals.

A second stone-tool technology emerged between 1.8 and 1.7 million years ago in Africa. This was called the 'Acheulean' (from St Acheul, a town in northern France where the first of these tools was discovered). Acheulean tools are the most well-known stone tool-making tradition of *Homo erectus* and others after them. Hand-axes have been discovered from southern Africa to northern and western Europe to the Indian subcontinent. The earliest known Acheulean artefacts from Africa have been dated to 1.76 million years ago; those found in India are only slightly younger. In Europe, the earliest Acheulean tools appear just after 800,000 years ago, as *Erectus* moved northwards, out of Africa and the Levant. Acheulean technology is best characterized by different kinds of stone hand-axes. Most common are teardrop-shaped hand-axes that are usually 12–24 centimetres long and flaked over at least part of the surface of each side of a margin (bifacial). There is considerable variation in the size of the tools and quality of workmanship. Hand-axes were multipurpose tools used for many different tasks. Studies of surface -wear patterns reveal the uses of the hand-axe which included butchering and skinning game, digging soil and cutting wood or other plant materials. Acheulean tools have also been found alongside animal bones. These tools were more effective than Oldowan but crafting and wielding them required greater strength and precision. The toolmakers shaped their stone cores into heavy-bodied axes and cleavers that were usually bifacial. Most Acheulean tools were heavy-duty butchering tools, like hand-axes, cleavers and blades, and these were discarded after they lost their edge and new ones were made. The style with which stone tools were made tells us of the presence of *Erectus* at any point of time. Acheulean stone tools have been found in parts of Africa, the Levant, western and central Europe, Arabia and India.

Around the time that Acheulean technology was emerging, the more robust *Australopithecus* species and sabre-toothed cats were becoming extinct in southern and eastern Africa and although it is tempting to correlate this to the widespread use of the new hand-axes and stone technology, there is little evidence to prove this conclusively. By 1 million years ago, all species of early *Homo* had either left Africa or perished, except for *Erectus* who persisted in Africa from 1.9 million until 70,000 years ago. Despite the tools, and despite being much bigger than any other hominins of their time, *Erectus* remained largely a scavenger and gatherer who hunted only occasionally.

Erectus sites have been known since the late 1880s to archæologists and new sites are discovered quite regularly. Chances are there is a site reasonably close to you no matter where you live in India. If you are in Chandigarh, Chennai, Allahabad, Bhopal, Jabalpur or Pahalgam, you would actually be spoilt for choice given the number of sites near you that you can visit. All you need to spot an Acheulean stone tool on the ground is a keen eye, and once you acquire the knack of spotting one, you are likely to find many more. But first you must know what you are looking for. A word of caution, though—not all Acheulean tools were made by *Erectus* and given the longevity of this tool technology (from 1.7 million years ago to about 60,000 years ago), other hominin species also must have made and used these tools. So it is not necessary that an Acheulean tool you find was fashioned by *Erectus*—it could have been made by another species of hominin as well.

It is simple to spot hand-axes if you know of a likely site. Hand-axes were shaped like teardrops and are about the size of an adult's palm. A cleaver is larger and rectangular. Blades and flakes were often chips and shards of rock that broke off while shaping hand-axes and cleavers. They are about 5-8 centimetres in size with one edge worked to make then sharp. They were used to separate skin from meat, to shear and size fur and to make spearheads. One of the best ways to spot blades and flakes is to look for shards of rock that are distinctively different from surrounding soil and stones.

From the evidence of their tools, it is conjectured that *Erectus* left Africa following herds of migrating beasts. They followed a coastal route to Arabia and arrived at a place called the Potwar plateau before crossing over to Riwat in the Sohan Valley of Pakistan. Estimates about the time of arrival (but not departure) of *Erectus* are made through the trail of tools left by them. The tools *Erectus* carved after they reached the Siwalik Hills of Himachal Pradesh are slightly different from those found in Pakistan.

When *Erectus* arrived, the Siwaliks were still a sprawling grassland interspersed with trees. India was continuing to push and shove into Eurasia and the Karakoram, Hindu Kush and Himalaya had already become an imposing barrier. Around 1.2 million years ago, the mountain range west of Srinagar called the Pir Panjal was beginning to impede the free-ranging movement of herbivores between the subcontinent and Tibet. The Pir Panjal had risen by nearly 3000 metres over the past 50,000 years and soon other mountains were rising around it and the rivers that flowed off them were becoming unpredictable. It seems likely that the seekers moved south from here, to the lower hills around Jammu and the Punjab. The uplift of the Lower Himalaya and the Siwalik Hills intensified around 1 million years ago, but these were gentle rolling hills then (as they are now), and did not deter *Erectus*'s migration. The Karewas, the flat-topped plateau of Kashmir which was once a giant lake, also began to rise between the towering peaks of the Pir Panjal range and Siwalik Hills. The lake drained and created a large grassland that became home to giant herbivores like the hippo, rhino and giant deer which arrived from Africa and Central Europe. Here *Erectus* found an abundance of game. But the hills of north-western India were still not the preferred home for the first migrants because they were gradually getting colder and the grasses had started to wither and disappear. The herbivores began to leave, and *Erectus* followed them, travelling south, towards the plains of the Himalayan rivers and the centre of the Indian Peninsula.

It seems fairly clear that *Erectus* found sanctuary in the foothills of the Siwalik Hills. In modern Atbarapur, a village 25 kilometres north of Hoshiarpur, you can find a wealth of bifacial tools. After a decade of study, researchers believe that this is probably the largest collection of Acheulean artefacts in the Siwalik Hills. The tools are dated to be around 600,000 years old, after which this site seems to have been abandoned. The sites around Atbarapur show that cobbles were used to make small flakes, and large boulders were broken to obtain smaller rocks from which hand-axes and cleavers (typical Acheulean tools) were shaped. This strategy of toolmaking is observed in most other parts of India but in this site the technology is more refined and the stones better sharpened. Some anthropologists have compared the tools found here to those found in Africa and East Asia, and have come to the conclusion that Atbarapur stands at a junction where old toolmaking technology met the new and from where the latter spread to East Asia and Europe.

Other sites along rivers like the Jhelum which emerges from the Siwalik Hills (on the Pakistan side) show a fair number of stone tools and Pahalgam near Srinagar has choppers and hand-axes that are thought to be between 700,000 and 500,000 years old. These dates, however, are contentious and some anthropologists believe that these tools are even younger.

A few *Erectus* bands settled around the lower Siwalik Hills, and others travelled south and reached Didwana (Lat: 27°24'N, Long: 74°35'E) near Nagaur in Rajasthan. Like Atbarapur, Didwana shows a diversity of stone tools which are 200,000 years old. This region has a reddish-brown layer of soil which extends to about 3–3.5 metres below the surface, suggesting that it underwent a prolonged humid phase when there was a large lake here, surrounded by trees and tall grass, which was fed by streams. The lush vegetation supported diverse animals, and fossils of mammals and a large number of Acheulean tools have been found here too.

Gradually the climate changed and the soil became friable and sandy, and as the desert began to encroach around 80,000 years ago, drying the lake and leaving the grasses withered, *Erectus* must have decided to move on. A small group travelled west along the Luni and other seasonal rivers and entered Chittorgarh and the Mewar region. But since only a few scattered tools have been found in secluded sites, it is presumed that this region was abandoned quickly, and that the settlers moved further south and east. They travelled across the scrub forests and rocky hills to reach what is now the town of Maheshwar on the banks of the Narmada, famous today for its woven saris and temples. Bands of *Erectus* journeyed down the Chambal river and then along the Narmada Valley, going west towards Saurashtra in Gujarat and east towards the origin of the Narmada. A few settlers set off west along the coast of Gujarat and reached the vicinity of the modern coastal towns of Umrethi and Madhuban in Saurashtra, and then possibly continued down the west coast. Between 1910 and 1930, stone tools were found in Pali Hill, Backbay reclamation, Kandivali and in the fields that lay between Churchgate and Colaba in Mumbai, and in 1985, stone tools were discovered in the southern ridge forests of Delhi. It is also possible that from southern Rajasthan (the Mewar region) *Erectus* reached Saurashtra directly, travelling along small perennial rivers. From Saurashtra a few bands went south towards the Orsang Valley which lies near the mouth of the Narmada, close to the town of Bharuch. In addition, a few caves have been discovered along the Konkan coast of Maharashtra, which hold tremendous promise of finding tools and perhaps even bones.

The emigrants who travelled east along the Narmada found limestone outcrops that provided them with natural shelters, and within these *Erectus* found sanctuary. Early British explorers, naturalists and geologists frequently found the remains of large elephants, hippos and big cats in Central India. In 1859-60, William Theobald, a palæontologist with the GSI, Kolkata, went to central Narmada in search of extinct animals and he writes in his memoirs that he discovered an incomplete fossil of an early human, but this was somehow lost after it arrived at the GSI's office in Kolkata. Some other records and correspondence from that time talk of a similar discovery made near Gorakhpur, in eastern Uttar Pradesh. However, neither the Narmada nor the Gorakhpur fossil was ever deposited at the Kolkata museum, and no other evidence of the presence of early human fossils has been documented.

In October 1982, Arun Sonakia, a geologist from the GSI office in Nagpur, travelled to Hoshangabad in Madhya Pradesh to search for early mammalian fossils around Hathnora (Lat: 22°52'N, Long: 77°53'E), at a place where a bend in the river had eaten into the limestone to create nooks and outcrops. Late one afternoon, on the eighth day of his expedition, 24 October to be precise, he found a curved bone embedded in a rock that appeared to be a skull (specifically, the parietal bones, or the two bones which make for roof of the skullcap) of a monkey or an ape. The 'find' was protected at night and intensified excavation resumed early the next morning. Gentle tapping around the rock exposed part of the skull and its imprint on the rock. However, over the next hours, the glabella—the smooth area between the eyebrows and above the nose which makes up the most forward projecting point of the forehead—got damaged and the thin bones that were embedded in the rock could not be retrieved.

On 5 December 1982, Sonakia announced to his office that he had discovered South Asia's first archaic human in Hathnora but was prevented by bureaucratic formalities from being able to make his discovery public. It took another six months before the Government of India sent a press release to all newspapers stating that such a discovery had been made. The Indian press dubbed it the 'Narmada Man', but later analyses done in the US revealed that it was actually a woman, between 27 and 32 years old. Dating the bones of bovines found alongside it suggested that these fossils were about 236,000 years old. No sooner had the news been made public than an army of anthropologists from India and abroad combed the area around this village, turning over rock and soil with trowels and spades and unearthing tools and artefacts from earlier and later periods, but no more hominid (a term covering *all* members of the genus *Homo*) bones turned up. The few bovine

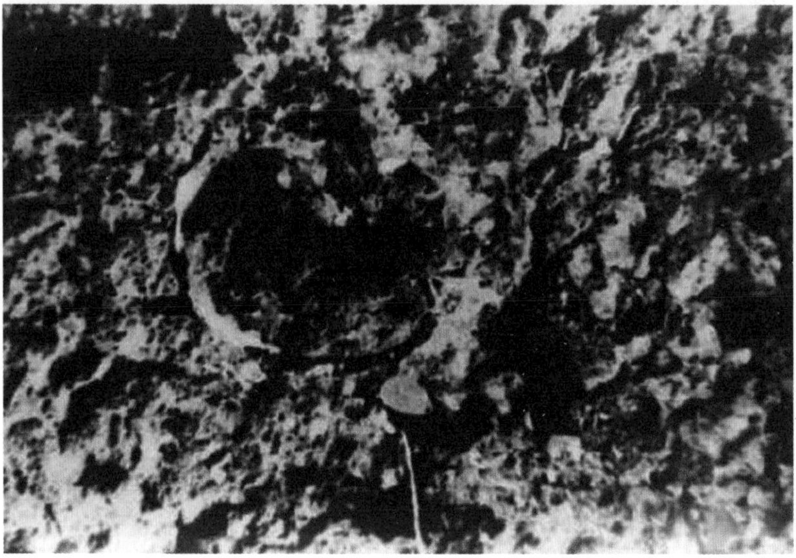

Missed it, almost! This bean-shaped skullcap was embedded in the Surajkund rock formation and took over six hours to extract.

bones that were found alongside suggested the age of Sonakia's find but these fossils could well have been transported by the river, mixing up old and recent fossils and tools to confound later researchers!

Just below the formation where the hominid fossil was found, stone tools were discovered and further exploration unearthed the bones of boars, small horses, hippopotami and an almost complete fossil of an extinct elephant (*Stegodon*). Although the stone tools that were found here have never been dated, experts believe that they could be between 800,000 to about 10,000 years old, a range that covers virtually the middle and end of India's stone age. Several sites around Hathnora and along the Narmada have shown the presence of a wide variety of hand tools. The Narmada Woman is now thought to be neither *Erectus* nor the later *Homo sapiens*, but an intermediate to both (termed 'archaic *Sapiens*') because she possesses features of both species. Some palæoanthropologists haven't ruled out the possibility of it being another *Homo* species, *H. heidelbergensis*. And although the age of Narmada Woman remains debatable, most experts estimate it to be between 2 to 0.7 million years old.

Just 3 kilometres south of Hathnora, across the Narmada river, is Dhansi, where 'Oldowan-like tools' (or perhaps some flakes resulting from biface production) have been found, raising questions about how

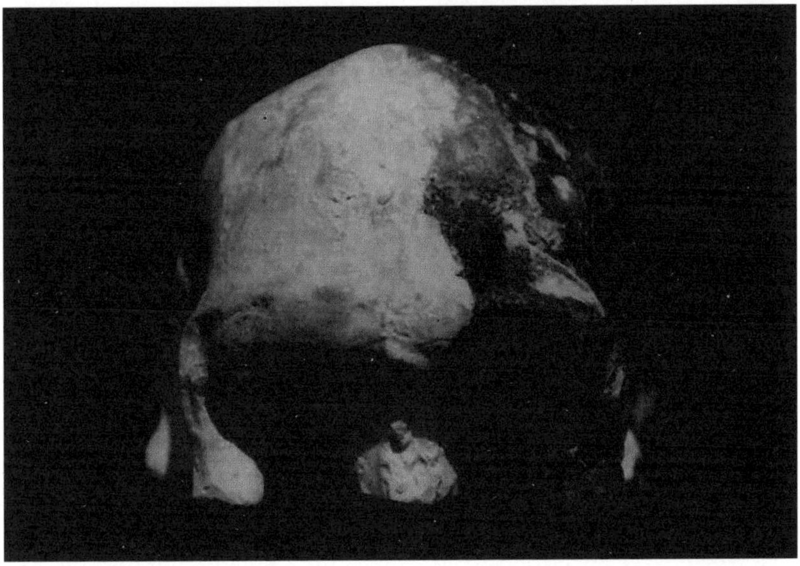

After several months, the skullcap was restored, with a little help from American and French academics. The red-brown part was excavated from the rocks and the white portion is plaster. Notice the pronounced glabella, the bone where the eyebrow rests.

they got there. Further downstream, Netankheri has yielded tantalizing new fossils that need to be dated and analysed before any conclusions can be drawn. One thing is certain: the entire Narmada belt—from near Vadodara in the west to Piparia at the base of Pachmarhi in Madhya Pradesh, and further east to the Son river—remains largely unexplored and offers immense possibilities of finding more bones and artefacts. There may yet be many more surprises in store for us!

After having reached the Narmada, *Erectus* perhaps travelled south to the open forests of Maharashtra and then along peninsular rivers—the Krishna, Bhima and Godavari—reaching central Andhra Pradesh and northern Karnataka, where several sites have been found with abundant tools littered around. If you start going east from Pune, you will find some good sites that yield stone tools. Morgaon (named after the Indian deity Moreshwar or Ganesh) is famous for its temples dedicated to the elephant-headed god. Along rounded rock faces and irregular trees and between temple sites are scattered rocks and amidst these you can find stone tools. When the first palæoanthropologists from Deccan College, Pune, came here in the mid-1980s, the devout locals were amused to see them collect these stones. But the hand-axes that were collected were distinctive and appeared more evolved. These tools were found below a bed of white volcanic ash and alongside lay the remains of extinct mammals. At Nevasa (Lat: 19°56'N, Long: 74°93'E), another site in Maharashtra, about 60 kilometres south of Ahmednagar, the evolution of these tools becomes clear. The Nevasa tools were shaped differently which shows that the toolmakers were adjusting their technique to the character of locally available rocks which were soft and shattered easily when pounded by a heavier rock. But these rocks were light and easy to shape and are found in large quantities in the surrounding areas. Nevasa tools are dated to be between 390,000 and 350,000 years old.

Change in technology happens very slowly, and it needs the keen eye of a seasoned field archæologist to spot these subtle transformations. Over time these refinements become part of general practice and the tools become distinctive to a local 'culture'. Archæologists use the term 'culture' when they find an assemblage of tools or artefacts of a particular style in a single layer. Such tools need to exist for a long period to show that they were used by many generations and were different from those of the previous and the following period. There is no clear evidence of culture in *Erectus* other than through tools but clearly the tools from Nevasa are different from those found in Hathnora and Didwana. Tools, like those found in Nevasa, had become thinner and had a sharper edge, and were called 'flake blades' (and from this time all blades which resemble these are called the 'Nevasan culture').

We see the use of Nevasan culture patchily beyond Maharashtra. Tools found in some immediately adjoining areas, like the ones south-east of Nevasa in north Karnataka, are different. These sites exist in the black soil region of the Deccan and hold within them another exciting set of stone tools. About 300 kilometres west from Hyderabad is the district of Gulbarga where two rivers called Hunsgi (also a town, Lat: 16°27'N, Long: 76°31'E) and Baichbal (also a village by this name, Lat: 16°33'N, Long: 76°33'E) drain into the Krishna river and form a continuous valley in Yadgir district of Karnataka. The Hunsgi-Baichbal Valley extends over about 500 square kilometres and this holds another rich record of tools with ages ranging from 700,000 to 500,000 belonging to *Erectus* years ago and about 60,000 years (in Isampur and the rest of the region, probably belonging to the later *Homo* called the *Sapiens*). Isampur (Lat: 16°27'N, Long: 76°30'E) is also in Yadgir district, and is about 16 kilometres west of Baichbal. At last count, there were 220 Acheulean sites in the Hunsgi-Baichbal Valley.

Interestingly, in Isampur, scientists found the site littered not just with tools but also with toolmaking debris. They realized that they had discovered the first Acheulean quarry and workshop or 'factory'. This discovery offers an opportunity to study the toolmaking process and provides an understanding of how *Erectus* used local stones to shape their tools. However, many new sites are threatened and over the last forty years, since the first expedition was conducted, several sites have been 'developed' and lost. The construction of a dam at Narayanpur on the Krishna river, roads, several irrigation canals that have been dug and concretized since 1982 and the intensification of agriculture along the canals has flattened gentle contours and disturbed many sites. Many settlements and grave sites (from the time of *Homo sapiens* and the earliest agriculturalists) have been permanently erased.

The last leg of *Erectus*'s travel probably led further south through north and central Andhra Pradesh. South of Hunsgi and Baichbal in Yadgir district of Karnataka and dead centre of the highway that connects Bangalore to Hyderabad is the district of Kurnool. Kurnool is famous for its caves, many of which are still to be discovered. One

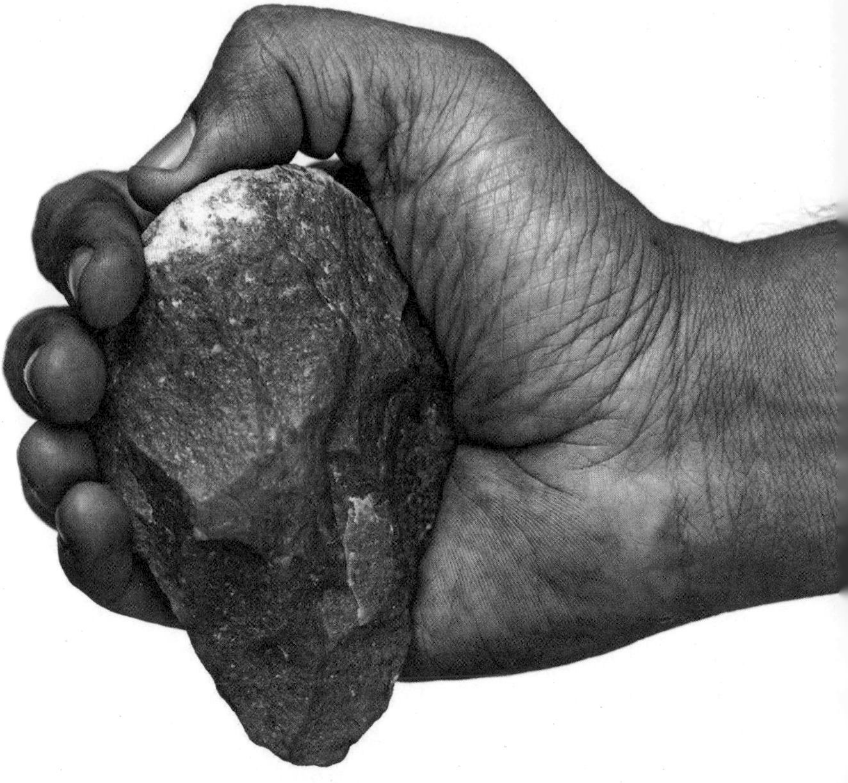

The earliest known Acheulean artefacts from Africa have been dated to 1.76 million years ago and the oldest Acheulean sites in India are only slightly younger. The hand-axe (above) from Attirampakkam has been worked on both sides (or bifacial) on a quartzite rock and tools from this site are believed to be nearly 1.5 million years old. This new time record revealed that bifacial technological advances in the region had actually occurred earlier than previously believed. In Europe, the earliest Acheulean tools appear just after 800,000 years ago, as Erectus *moved north out of Africa. Flaked stone tools were made by hitting a piece of stone, called a core, with a 'hammerstone', often a pebble. The best types of stone to make tools were hard and brittle and rich in silica, like quartzite, chert or flint.* Erectus *and* Sapiens *both quarried for rocks from outcrops or collected them from stream beds. If these rocks were not found locally, they were prized by their makers and often carried over long distances. Rocks like quartzite were used to make better weapons because these hard silica-based rocks were durable and could resist strong impact, and yet they could be shaped to provide a sharp edge. The use of silica-rich rocks to make stone tools perhaps led to the first silicon revolution!*

limestone cave system of interest is Billasurgam (*Billa* and *surgam* mean 'cave' in Telugu) where tools made of bone were found, along with a variety of bones of mammals. This is the first site where both stone and bone tools were found together. But scientists are divided on who the makers of these tools were.

Onwards we follow *Erectus* into central Tamil Nadu and travel along the eastern coast towards the peninsula's southernmost point, Rameswaram. Along the way, a few kilometres from the sea, it is relatively easy to find Acheulean stone tools. The types of tools found in central Tamil Nadu are similar to those found in Hunsgi in Karnataka (early archæologists had called these 'Madrasian industry' but now they are referred to as Indian Acheulean). And we are in for a big surprise as we come to the end of our journey. For over a century, archæologists who had been studying sites around the coast of Tamil Nadu had found a variety of stone tools. One particular site, Attirampakkam (Lat: 13°13'50"N, Long: 79°53'20"E) in Thiruvallur district, the northernmost district of Tamil Nadu, had been studied since 1863, but when an intensive exploration was undertaken for over twelve years (from 1999 to 2011) by researchers from the Sharma Centre for Heritage Education in Chennai it was found that the tools there were 1.5 million years old (though most archæologists find 1.1 million years old more acceptable). When this finding was published in the journal *Science* in March 2011, it made archæologists sit up; the new dates meant that scholars needed to rethink the timing of when *Erectus* evolved and when they left Africa. Frankly, any attempt at retracing the steps of *Erectus* is largely speculative and nobody really knows for certain if these indeed were the routes taken by *Erectus*. There is consensus among most palæoanthropologists, however, that around 1.5 million years ago *Erectus* was present in East Africa, India, the Levant and, perhaps earlier, about 1.8 million years ago, in Georgia where they remained until the 'arrival' of other human species.

How was it possible for a species to emerge in Africa and reach so quickly across continents? Based on what we know thus far, there are four competing scenarios, all equally plausible:

Scenario one: *Erectus* migrated from Africa to Asia soon after their species evolved. Their dispersal would have to have been very rapid and they would have spread latitudinally, crossing Arabia to reach the Levant, then marching on to South Asia and then to Indonesia. It is also possible that the ancestors of *Erectus* left Africa almost as soon as they evolved and reached Asia, and both the African and Asian cousins then evolved separately and invented Acheulean tools. In such an event, palæoanthropologists suggest that the Asian ancestors be called *H. erectus*, and the African ones *H. ergaster*.

Scenario two: It is possible that a very early African ancestor may have evolved in places other than Africa. In some places in Asia, Acheulean tools have been found dispersed soon after they appeared in Africa. So it is possible that there were two populations of *Homo erectus*, one in Africa and the other in South and East Asia that evolved separately but almost simultaneously. It was probably the African population that invented Acheulean tools around 1.76 million years ago, and there may be fossils and even older tools in Africa that are yet to be discovered.

Scenario three: It has been contended that *Erectus* in Africa, the Levant and Asia, all emerged from a common ancestor and evolved separately into related but distinct species. This seems a little improbable because, when geographically isolated, a population will most likely diverge into distinct species, and the technology of toolmaking would evince clear differences.

Scenario four: The more likely scenario is that scientists have got some dates and some dating techniques used for these tools mixed up!

It is also possible for two scenarios to exist even simultaneously, so again there can be no certainty about how and when *Erectus* reached and spread across India unless we get better dates for Acheulean tools and, if possible, find an *Erectus* fossil that is older than the African *Erectus*.

The ape *and* human story (not ape *to* human, as it is generally represented) is a work in progress and is highly controversial. Palæoanthropologists remain divided about discoveries and the timing of their arrivals and departures, and migration patterns. Facts and evidence need to be pieced together more precisely and reconciled so that a clear picture emerges.

This, of course, is my interpretation of what might have happened, based on published research by scientists who have worked extensively in this area. Unless we learn how to travel back in time or make a series of path-breaking fossil discoveries, we may never be able to say with absolute certainty that we have got the *Erectus* story absolutely right in every detail. Some facts, however, remain unequivocal. *Homo erectus* is by far the most 'successful' of all human species. Their career began about 2 million years ago in Africa and perhaps ended around 70,000 years or so ago across Africa and Eurasia. They lived in every possible land habitat that was interconnected. Their Olympian spirit took them faster, higher and further than any other species before them. *Erectus* was undoubtedly intelligent and had adapted to outsmart many vicious predators and massive herbivores.

In the final chapter, we begin our journey around 195,000 years ago in East Africa where we will retrace the steps that *Erectus* took from East Africa to India, but this time we will travel with the last species of *Homo* to evolve—*Homo sapiens.*

15

THE PROMISED LAND

Erectus walked like us, looked like us, behaved like us and ate like, well, at least *some* of us, but he wasn't *quite* us. After the emergence of *Erectus* another 1.8 million years passed before our species, *Homo sapiens* (Latin for 'man who knows', or 'the wise man'), appeared in East Africa, that is roughly 190,000 years ago, give or take 20,000 years.

As we have seen, between 1.6 and 1 million years ago *Erectus* spread across Africa and Eurasia, including China, Java and probably India, with a few other species of *Homo* living alongside. With different *Homo* species having to share the same space and food resources, there were bound to be competitive encounters. We do not know how *Sapiens* began their domination over their contemporary hominids (a term covering *all* members of the genus *Homo*) but palæoanthropologists surmise that *Sapiens* assimilated many adaptations of their ancestors (some of whom existed alongside them) and improved upon them. These include the making of better tools and the smarter use of fire, which helped them survive and outlast others. *Sapiens* also prevailed over *Erectus* because

The Gulf of Zula on the Red Sea in Eritrea (left) is flanked by mountain range and volcanic ridges from the Great Rift. It is believed that some prehistoric hunter-gatherer groups of the early Homo sapiens *may have trekked along the western coast of the Red Sea northwards to the Sinai or southwards over newly created land bridges across Bab-al-Mandab, and travelled along the coastline in Arabia before they entered India. The mangrove-covered coasts along the gulf provided a variety of shellfish which may have supplied modest food to the migrants.*

their diet was more cosmopolitan and they developed stronger bonds among families which led to greater cooperation and helped make stronger communities. But perhaps the most important of all was *Sapiens*'s invention of the earliest languages, using click sounds and signals.

Around 200,000 years ago, the shores of lakes in the rift valleys of Ethiopia were covered with lank, luxuriant grasses with hosts of feasting grazers. These were times of plenty, but as the population of *Sapiens* (along with other *Homo* species) grew, competition for food and shelter intensified. To make matters worse, seasonal rains started to become erratic and began a process of drying up East Africa. Soon, bands of *Sapiens* began to leave East Africa in search of new lands. Some moved south and inched onwards till they reached Southern Africa around 145,000 years ago. A few travelled west to reach Central and West Africa where they settled in the tropical forests. Over time, as fewer *Sapiens* migrated out of the rift valley and reached other parts of Africa, these communities became isolated and did not advance further technologically, and there was little diffusion of new genes within these populations. A few bands ventured north across a Saharan landscape that was verdant, and reached the western Mediterranean coast. Over the next 12,000 years (around 130,000 years ago), the climate suddenly cooled and this caused forests to disappear and grasslands to dry up, turning the Sahara into a sprawling, inhospitable desert. Earth was entering a phase of alternate cool and mildly warm periods. The Sahara dried up and the population of the coast dwellers dwindled and eventually perished. By around 110,000 years, *Sapiens* had reached the far corners of Africa but survived only in places where the climate was favourable. There was still intense competition around the lake regions of East Africa and the changing climate made matters worse for those who stayed back.

Life could not have been easy for the early *Sapiens*, our direct ancestors. Without natural weapons like claws or canines they were more or less defenceless against powerful predators. Their eyes, like those of other apes, were adapted to see only during the day, making them vulnerable at night. Large predators such as leopards, hyenas and dogs are nocturnal and *Sapiens* would have been easy targets as night fell. And although *Erectus* and *Sapiens* made better stone tools and spears, these tools and even fire were not effective enough to ward off stealthy predators who could ambush a sleeping huddle.

Evolution, however, bestowed *Sapiens* with certain assets, some of which we possess to this day. *Sapiens* (and, to a lesser degree, *Erectus*) were blessed with muscular, rotund buttocks (*Maximus gluteus*) and a good heel, attached to the ends of long leg bones. All this, together with extendible muscles, made walking and running long distances more efficient. Although no hominin species has ever been able to match the explosive burst of speed of cats over short distances or has the innate ability to outrun wild dogs, the legs of *Erectus* and *Sapiens* were built for endurance and they could run for long periods to wear down their prey. They could also climb and run across difficult terrain. In addition, the absence of hair and the presence of sweat glands helped them stay cool on hot afternoons more efficiently than their prey. We can imagine how a band of *Sapiens* perhaps ran in relay to tire out a large animal like an antelope, before killing it.

There was one more attribute which contributed to the hunting efficiency of our ancestors: their more evolved brains developed a language based on 'clicks'. Click languages, unlike the languages we speak today, consist of lower frequency sounds that are inaudible to most herbivores. Wild dogs and some cats with acute hearing can hear low-frequency clicks but not a herd of grazers like buffaloes or antelopes. Click languages also helped hunting parties of *Sapiens* to communicate with one another while orchestrating an ambush and attack.

How do we know all this? A few hunter-gatherer communities that live in isolation, like the Kung San people (pejoratively called 'Bushmen') of the Kalahari Desert in South Africa and some pygmy tribes of western Africa have not changed their way of life and *still* use clicks (some readers may recall seeing them in the hilarious 1980 South African movie called *The Gods Must Be Crazy*). We also know that the Kung San people are prolific runners (watch the film!), have close family and community bonds and are incredible trackers of game, which makes them expert hunters. So *Sapiens* perhaps left East Africa with these endowments which helped them survive in new and unknown lands. Gradually, as they improved their hunting methods and developed more sophisticated languages, most of the click languages began to die out.

During the last Ice Age, which started about 80,000 years ago and ended 11,000 years ago, ice sheets grew and sea levels dropped, exposing large areas of coastal reef and continental margins and separating the islands by shallow seas. As the climate began to change again, *Sapiens* made an attempt to leave Africa about 75,000 years ago, give or take 2000 years or so. Just around the time when *Sapiens* were leaving Africa, something catastrophic happened. Across the Indian Ocean, in the middle of North Sumatra is the volcanic island of Toba, 6600 kilometres east of the Horn of Africa and about 300 kilometres west of Kuala Lumpur as the crow flies. About 74,000 years ago, Toba was like any of the other 18,000 idyllic tropical islands that rise out of the sea in the Indonesian archipelago. But this region had been restless for over 2 million years as the underlying plates constantly jostled and caused small volcanoes to rise up from below the sea. Then Toba exploded in a cataclysmic outburst that was among the most violent volcanic eruptions ever known. The Toba eruption released 2800 cubic kilometres of material, which may only be a small fraction of the magma ejected in the Deccan episodes, but it did so in a *single* explosive burst. The violence of the eruption was such that a third of the volcano's top blew off. The Toba explosion created a great disequilibrium in global climate. Today it exists like a placid lake, the largest in South East Asia, the second-deepest and the largest volcanic lake in the world, at an elevation of more than 900 metres above sea level and is surrounded by the sea.

Sapiens were still at the margins of Africa when the Toba explosion happened but those *Erectus* who were in India (Eurasia and China as well) were directly in harm's way. For all life in peninsular India, the effect of Toba was catastrophic. Toba's ash blanketed eastern and central India which lay nearly 2000 kilometres west of the volcano. A strong westward wind carried the grey-white volcanic ash high into the sky and smothered the forests of India. Remnants of this layer of ash can still be seen where a cross section of soil is exposed in western Tamil Nadu, parts of central Andhra Pradesh and Telangana, near Baripada in central Odisha and along the banks of rivers like the Narmada and Son in Madhya Pradesh. The pattern of ash dispersion suggests that the eruption occurred during the early summer monsoon season. An eruption of this magnitude would have enveloped the atmosphere with ash and dust, making it hot and dark. The impact would have been devastating for all forms of life close to the volcano but its effects would have extended a lot further. Until the volcanic ash settled, the air would have been laced with particles of tiny quartz-like crystals which would have caused a burning sensation and suffocation in the windpipe and lungs leading to bleeding and eventually death among most land-living animals. Much of the population of *Erectus* in India would have been wiped out. But stone tools suspected to have been made by *Erectus* have been found in the layers of soil below and above the volcanic ash fallout in Andhra Pradesh, suggesting that some of them at least did survive the cataclysm, though in all likelihood their numbers would have been greatly decimated. Sulphate released from the Toba explosion was deposited in Greenland ice cores and its analysis shows that the eruption caused an intense six-year-long volcanic winter, following which the cooling subsided, but only just, and it led to the onset of an 1800-year period of the coldest temperatures of the last 125,000 years. This volcanic winter froze plants and animals even in some parts of the tropics, and those who had possibly been too far away to have come under the cover of the volcanic ash were also affected. The forests of south-eastern and central India (Andhra Pradesh and eastern Tamil Nadu) were smothered with ash, and trees died a slow death in the sudden burst of heat followed by the rapid cooling that came after. It would have taken several hundred, perhaps even a thousand, years for some plants to reclaim the ashen land. Some studies suggest that this volcanic event was an important reason why even today there are no luxuriant forests in places (like eastern Andhra Pradesh and north eastern Tamil Nadu) that were worst affected by Toba; one study has even suggested that Toba is an important reason, perhaps the *only* reason, for the absence of large mammals like tigers, elephants and even langurs in these specific areas.

For the last remaining *Erectus*, it would have been difficult to cope with life after Toba. As the forests recovered, many mammals returned within a few hundred years after the eruption. But among the victims of the climate change that followed Toba was *Erectus* whose populations dwindled and never recovered. The last *Erectus* probably died around 70,000 years ago, by which time *Sapiens* had taken their first steps that would propel them further than *Erectus* in claiming new territory.

Post-Toba, the climate cooled and North and East Africa began to dry up. The volcanic winter had triggered a sharp decline in temperatures, causing the poles to freeze intensely. This caused immense distress to all life and *Sapiens* began to move out of Africa in the hope of finding a new Eden. Staying back was not an option.

There were three possible routes that *Sapiens* could have taken: the first batch of emigrants travelled north across the Sahara in search of a better home but the changing climate had desiccated the remaining lakes and rivulets, making any life here difficult. The second sortie was conducted across the Sinai towards the Levant. This group made better progress than its predecessors but here too the desert had begun to encroach and their success was short-lived. A third set of refugees used the tried-and-tested coastal route that *Erectus* had taken before them (though, of course, they would not have known this). The earliest remains of *Sapiens* to be found anywhere outside of Africa are from a cave system in Israel. These are thought to be around 100,000 years old. Anthropologists have found no evidence of *Sapiens*'s migration out of this area and it is likely that these remains belong to a relict band from the East African population that ventured out but got no further. It is believed that a band of *Sapiens* arrived on the shores of the Red Sea on a thin peninsula called Ras Siyyan in Djibouti, on the tip of East Africa. This is the easternmost point of the Horn of Africa and it overlooks the Bab-al-Mandab strait (from Arabic, 'Gate of Grief') which today is a 30-kilometre gap between Africa and Arabia. But at the time *Sapiens* arrived here, spreading polar ice had caused sea levels to plummet and the strait at this time was less than 11 kilometres wide, dotted with several small islands and surrounded by shallow water. This probably made island hopping possible for *Sapiens* who travelled further along the southern Arabian coast, cutting through the Yemen highlands, the Dhofar region and Oman Mountains, to reach the Strait of Hormuz. About the time when they reached the Persian Gulf, the Indian monsoon, which took rain-bearing clouds to the coastal highlands of the Arabian Peninsula, weakened or failed completely, forcing the *Sapiens* to continue their journey westward in search of a haven. The monsoons in Arabia failed repeatedly for about 200 years and gradually the grasslands of the

Arabian Peninsula began to turn into a desert and its rivers and stream began to dry up. Stone tools have been found in *wadis* (river valleys) all along the margins of the Arabian Peninsula like a confetti trail left behind by these pioneers but Arabia was just a transit and not a home for the *Sapiens*.

We don't know quite how long after, but this band of *Sapiens* has left further evidence to show that it ventured onwards and travelled along the highlands of Arabia, the coast of the Arabian Sea and crossed the neck of land in the Gulf of Oman. A 40-kilometre stretch of water—the Strait of Hormuz—separates the Khasab jutland in Oman from the ports of Bandar-e-Abbas and Sirik in Iran. But with the Ice Age at its zenith, the seas would have receded and the strait would have been reduced to being only about 18 kilometres wide. Anthropologists hypothesize that the islands dotting this shallow sea were like stepping stones for migrating bands of *Sapiens*. After the pioneers had crossed the Gulf of Hormuz, the climate warmed again, which melted the ice sheets and caused sea levels to rise once again. This prevented any further dispersal of *Sapiens* out of Africa. Therefore, it seems that it was just a handful of emigrants who left Africa and went on to populate the rest of the world!

This small band travelled further along the coast, as evidenced by the characteristic *Sapiens* tools that anthropologists in Iran and Pakistan have found all along this trail which comes to an end in front of a large abandoned alluvial fan of the Indus. The mouth of the mighty Indus would have posed an intractable challenge for early *Sapiens*. The trail of stone tools suggests that they turned inland till they reached Potwar (or Potohar) in Rawalpindi district of the Punjab region of Pakistan. To the east and west of Potwar flow the Jhelum and Indus. These rivers flow in a leisurely fashion, like most mature tropical rivers and, at the height of summer, their water recedes and dries up at several places. For the first *Sapiens*, it would have been easy to wade across these rivers and journey beyond into India.

Anthropologists have found the oldest stone tools made by *Sapiens* along the banks of the Soan river (or Sohan, and this stone-tool tradition is called 'Soanian'), a distributary of the Indus not far from the city of Rawalpindi which lies on the Potwar plateau. When the first *Sapiens* arrived on the banks of the Indus, the mightiest Himalayan river at that time, its far bank would have been scarcely visible, much like the Ganga and Brahmaputra today at their widest. The waters of the Indus would have been filled with fish, turtles and crabs. Hippopotami would have lazed in its slow-moving waters, and several varieties of deer and horses, eleven species of elephants and four of rhinoceros would have roamed the land, and waterfowl of immense varieties would have skimmed over the river. The sandy riverbanks would have been lined with tall grasses and, from between these, ostriches and long-necked antelopes like *Sivatherium* would have watched out for new arrivals. The grasslands would have led to rows of short trees and these would have graduated into thick groves. By about 65,000 years ago (give or take 5000 years), the effect of the volcanic winter was over, and migratory animals and birds from Western Europe, Siberia and Tibet returned, seeking the perennial waters of the Indus.

From Potwar, some members of the band of *Sapiens* ventured further into the subcontinent, travelling through central Rajasthan (Nagaur and its environs) into Gujarat along the river valleys of central and western

The Potwar (or Potohar) plateau in north-eastern Pakistan is an extension of the Siwaliks. It is bound on the east by the Jhelum river, by the Indus in the west, and low hills in the north and the south. By the time Sapiens *arrived on the banks of the Indus, its might had begun to wane, and they would have encountered a wide, deep and slow-moving river. The Potwar plateau is characterized by white sands which are derived from the 50-million-year old limestone which formed when the Tethys began to close. Stone tools crafted by* Erectus*, and 70,000- to 50,000-year-old tools made by* Sapiens *have been found in these valleys and hills.*

India. When they reached the gentle rolling hills of Tamil Nadu, one band settled there and others moved onwards. Their descendants still live among us and we know this from several genetic studies that have compared the genetic structure of the oldest inhabitants in Africa, Asia, Australia and those who live in India. The oldest population of *Sapiens* that lives among us is the Kurumba who arrived close to 65,000 to 60,000 years ago and inhabited a tract along the border of north-eastern Kerala and south-western Tamil Nadu. A few families migrated further east and, over time, they developed small mutations to become distinct populations such as the Yannadis, Chenchu and Irula tribes. A small band continued its journey eastwards and reached close to Puducherry and Chennai and from there they travelled north along the east coast to reach Bengal. We know this because in a small pocket of Bengal lives a community called the 'Rajbanshis' (or the Rajshahis) who are genetically very similar to the tribes of Tamil Nadu.

The odyssey of nearly 1200 kilometres east from Potwar down into southern India and then to Bengal can be traced by following the Acheulean hand tools left behind by these travellers. The tools tell us how long the journey was but apart from the skullcap found in Hathnora in the Narmada Valley and just one other bone (a collar bone or clavicle, also found in Hathnora) no other bones or skeletons of early humans have been found in India. Perhaps we should ask our

grandparents where their grandparents are buried, but we will need to persevere for a thousand generations or more to reach the time when *Sapiens* first arrived in India!

Other than stone tools, our DNA holds the information that tells us how *Sapiens* colonized the world, and with more ingenious methods in genetics, scientists have estimated how many Out-of-Africa migrants it took to populate the world. The DNA in all of us who live outside Africa is virtually identical, whether we are Nordic, Slav, Chinese, Inuit, Malay, Tamil, Kashmiri or anyone else. The difference in our genes is no more than 0.01 per cent. Yet it is this subtle distinction that makes us who we are. The variations we see in our features, in our skin colour, for example, or curly hair or slanted eyes are due to a few heritable mutations that took place in specific periods of time. Most of these mutations are relatively recent but they are traits preferred by some populations and therefore passed on to successive generations.

On the other hand, there is more genetic diversity in a small region of Africa than the entire world put together! This tremendous genetic diversity *within* Africa is perhaps because *Sapiens* migrated several times within the continent itself, and several bands became isolated and were later joined by others, creating a diverse gene pool.

Outside of Africa, the human gene pool is more or less homogenous and this similarity in all our genetic make-up suggests that it took just a handful of *Sapiens* to create all the human races that exist today. Geneticists estimate that between a hundred and, at most, less than a thousand *Sapiens* in all would have left Africa. If the number of *Sapiens* bands to leave Africa had been larger, we could have expected far greater diversity in the DNA of those who live outside of Africa. This suggests that there was only one episode in which a successful exodus took place, following which—for reasons that can only be guessed at—it was not possible to leave Africa again. There seems to have been a huge amount of luck involved in every step of this journey of Man. The few migratory bands that left Africa were totally at the whim and mercy of the climate, and wherever the sea had receded and land bridges had surfaced, it allowed the migrants to venture on and beyond. Is it the case that *Sapiens* took advantage of a colossal coincidence of climate change in order to be able leave Africa? That the climate changed at *just* the right time to allow

Sapiens to leave Africa and they were able to expand their populations geographically from this single founder population? At this point in time, given what we know, I think it is safe to presume that *Sapiens* expanded geographically from a single founder population that left Africa hopping across land bridges in a shallow sea.

The genes of individuals from different communities also come into play and help us piece together our story further. Genetic analyses show that the inhabitants of the Andaman Islands are closest to the early *Sapiens* that arrived in India. The volcanoes arose to form the Narcodum and Barren Islands (the two weakly active or near-dormant volcanoes that exist in India) in the Andaman Sea and these lie along the same arc as the Toba. The mainland closest to the Andaman and Nicobar Islands is Tamil Nadu, to their west—Chennai is about 1200 kilometres west as the gull flies—that lies along the same latitude (around 13°N), while Bengal to their north is about the same distance. So how might the early *Sapiens* have managed to reach the Andaman Islands? Anthropological studies suggest that some hunter-gatherer tribes acquired the skills to craft a seaworthy dugout boat from timber. They used these boats to fish and hunt in the open sea and estuaries just as the tribes in the Andamans do today. It is possible that a few *Sapiens* were swept away from around Tamil Nadu by the Bay of Bengal gyre, a weak to moderately powerful cyclonic swirl created by the Indian monsoon, and this is perhaps how a viable population of castaways reached the Andamans and continues to inhabit it. South of the Andamans are the Nicobar Islands and their residents are called 'Nicobarese'. Studies have found that the Nicobarese are genetically closer to South East Asian groups than to the Andamanese. Genetic studies also show that the Nicobarese descended from early agriculturalists and colonized these islands relatively recently, around 8000 to 10,000 years ago.

Onwards from Bengal, a few pioneers travelled along the coast and reached Thailand and pushed further to reach Malaysia around 60,000 years ago where they evolved into a distinct group called the *Orang asli* (Malay for 'original people'). A handful carried on, crossing the Sunda Strait, and a few others ventured east into the Philippines when Australia and East Asia were still connected through islands and sandbanks, and the sea was still shallow. Geneticists and anthropologists agree that

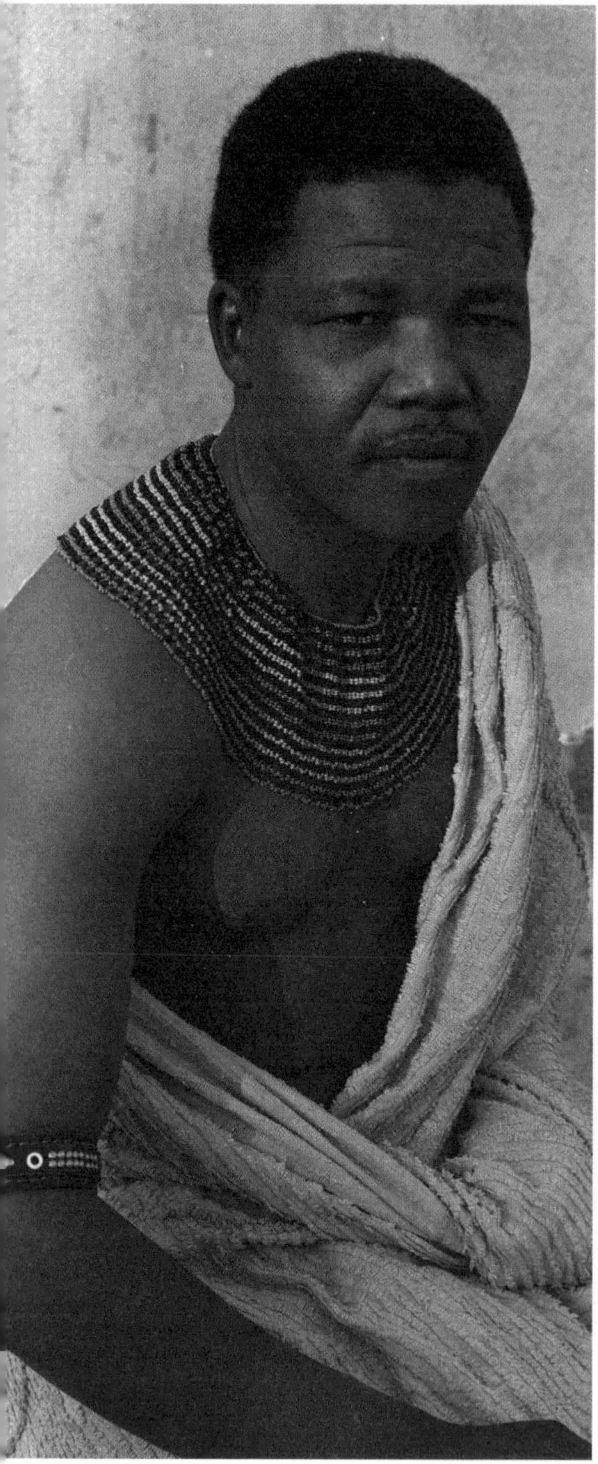

The indigenous hunter-gatherer people of southern Africa, the Kung San people (top left of this page) are among the earliest descendants of the Sapiens. They colonized parts of southern, central and western Africa and were the progenitors of all tribes and races of Africa, like this Xhosa statesman from South Africa (left, opposite page). The Namibia-South Africa border, the homeland of the Kung San people, was the starting point for a southwest-to-northeast migratory route that carried people through Africa and across the Red Sea into Eurasia. The nearest 'Out-of-Africa' descendants of the earliest migrants are the Jarawas of the Andaman Islands (like this elder, centre left on this page) and they were the first population to colonize India between 70,000 and 65,000 years ago, reaching the Andamans a little later. The Australian aboriginal people like this musician (opposite page, top right) are genetically similar to the Jarawa and are believed to have reached the southern tip of Australia between 55,000 and 50,000 years ago. As migrants travelled out of India, they settled along the way, quite like this Orang Asli hunter-gatherer in Malaysia (above) or Papuan highlander selling dolls (this page, left), moving onwards into Australia, along the east coast of China and Oceania. Colonization within India took place sporadically over the next 20,000 years or so, which established distinct communities like the Chenchus of Andhra Pradesh (centre, opposite page) and the Rajbanshis on the borders of West Bengal and Bangladesh (bottom, opposite page).

isolated populations of so-called Negritos like the Jarawas and Onge in the Andamans and in South East Asia have many features that connect them with the San people and Pygmies of Africa, including their short stature, dark skin, tightly curled hair and epicanthic folds (the skin fold of the upper eyelid that covers the inner corner of the eye). The population of intrepid *Sapiens* that left along the coast of eastern India reached New Guinea and then travelled onwards to Australia and the Philippines. Mamanwa (a Negrito group from the Philippines), the highland hunter-gatherers of New Guinea and the aborigines of Australia arose from this single population and it is estimated that these groups split from each other somewhere between 42,000 to 36,000 years ago. This is also approximately when changes in tool technology and food processing appear in Australia's archæological record, suggesting that these may be related to the migration from India. Could it therefore be possible that the Andamanese were not swept away accidently from Tamil Nadu (or Bengal) but were members of the band that was purposefully attempting to cross the Sunda Strait to end up here, instead? The journey of human evolution endures within these isolated indigenous groups like a genetic trail of breadcrumbs left behind by the pioneers.

It was also the time when the dingo appeared in Australia's archæological record. It was the pioneers who reached Australia between 50,000 and 44,000 years ago, who probably brought the dingo along with them because this canid has its ancestry in India and South East Asia. Genetic studies of dingos show that it coincides with the time of arrival of *Sapiens* in Australia. After early *Sapiens* arrived in Australia, sea levels rose again and effectively blocked further opportunities of crossing over. The Andamanese and Australian aborigines were both isolated now, since they had reached land's end. This restricted any possibility of intermingling with other *Sapiens*. Such isolated communities remain closely related to each other genetically, as they are also closely related to Africans. Yet they live many thousands of miles from each other.

Sapiens in the southern tip of India, around Rameswaram, would have reached the end of the mainland. But there are large rocks along the shallow sea whose tip extends from Rameswaram and tapers further into the sea, before gradually fading under the water. Rocks and islands rise in the sea, some continuous and others a little far apart. From space, it appears like a highway broken by water. This umbilicus has linked India with Sri Lanka since 500,000 years ago and served as a virtual highway for mammals like elephants, monkeys and leopards to cross over to Sri Lanka before *Sapiens* reached here about 50,000 years ago. In Sri Lanka it is called 'Adam's Bridge' and in India it is called 'Ram Setu'. Around 7000 years ago, parts of this became submerged in the Palk Strait but it was still relatively easy to cross this 18-kilometre-long land bridge. *Sapiens* reached the shores of Sri Lanka and went along the western coast. Some others also ventured inland to the central highlands but fossil records suggest that the coastal route along the west of the island was taken first. Since the 1930s, exciting discoveries have been made in the southern tip of the island in a place called Fa Hien Cave. This large cave complex, named after the famous Chinese monk who visited it in the fourth century AD, towers 140 metres above the surrounding forests and provides a panoramic view of the sea. Here a complete skeleton of a *Sapiens* from around 40,000 to 36,000 years ago has been unearthed, and around it ornaments made of beads and weapons made of animal bone have been excavated as well. This is, probably, as far as *Sapiens* could go within the subcontinent.

While one band entered India and took a coastal route, a few generations later another band left from close to Potwar and the Siwalik Hills, ventured north and then turned west to reach Central Asia around 62,000 years ago, travelling further to arrive in Eastern Europe about 50,000 years ago. Here they encountered their cousins far removed, the Neanderthals (*Homo neanderthalensis*, named after the Neander Valley in Germany), and from them they acquired a set of genes. The crossing over of genes from Neanderthals to *Sapiens* perhaps made modern humans more robust but it also gave them diseases and ailments that manifest in us till this day. How the Neanderthals had arrived in Europe remains a mystery but some archæologists believe that they were among the early *Sapiens* that left Israel and entered Europe. They survived the intense episodes of the Ice Age and stayed back because several mammals, especially herbivores like the antelopes, deer, elk and moose, mammoths and woolly rhinos, adapted to these changing times and provided enough food to these enterprising human ancestors. The Neanderthals ranged from Western Europe through Turkey into

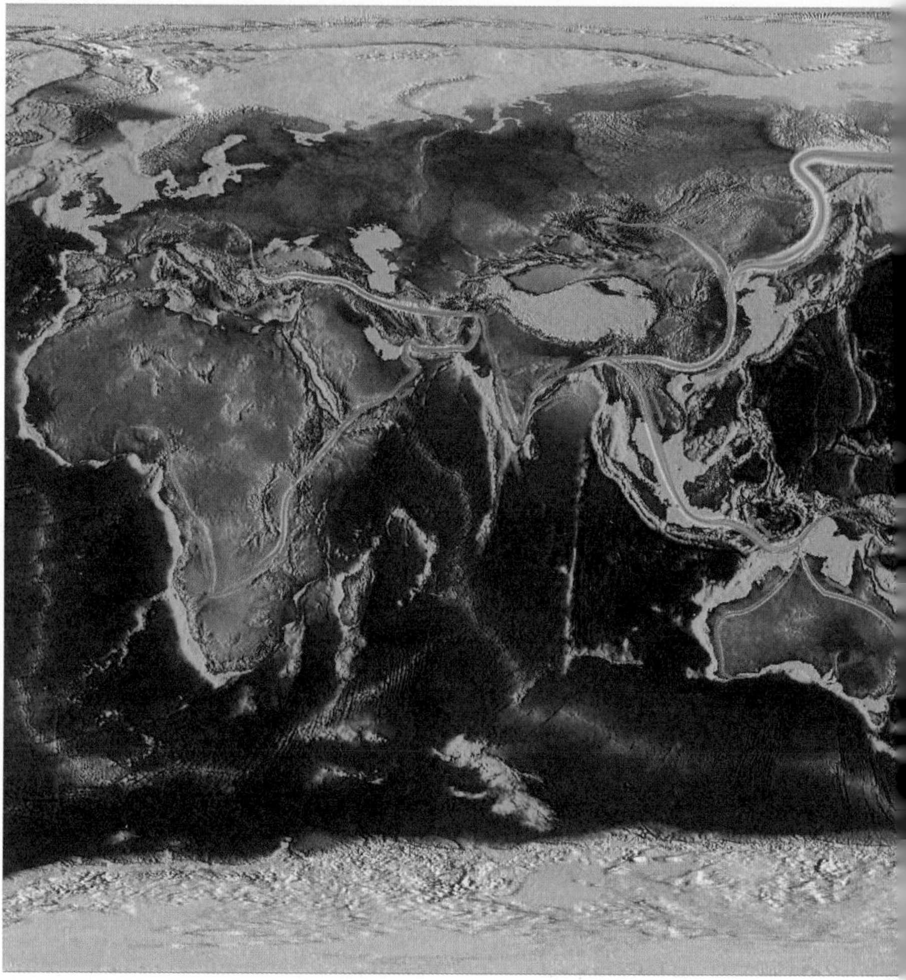

western Afghanistan. They were big-bodied and adapted to tolerate the cold climate. They had lived in Europe for nearly a quarter of a million years but the increasing fluctuation in climate and severe chill between 74,000 and 58,000 years ago drastically reduced their numbers. When the *Sapiens* who reached Europe from South Asia via Central Asia, encountered the Neanderthals, they had some very ugly skirmishes. This also created opportunities for chance mating which led to the intermixing of genes. Some geneticists estimate that at least two to three episodes of exchange of DNA took place during the intermixing among *Erectus, Neanderthalensis* and *Sapiens*. In 2012 a comprehensive genetic study found that Europeans and some Asian populations have

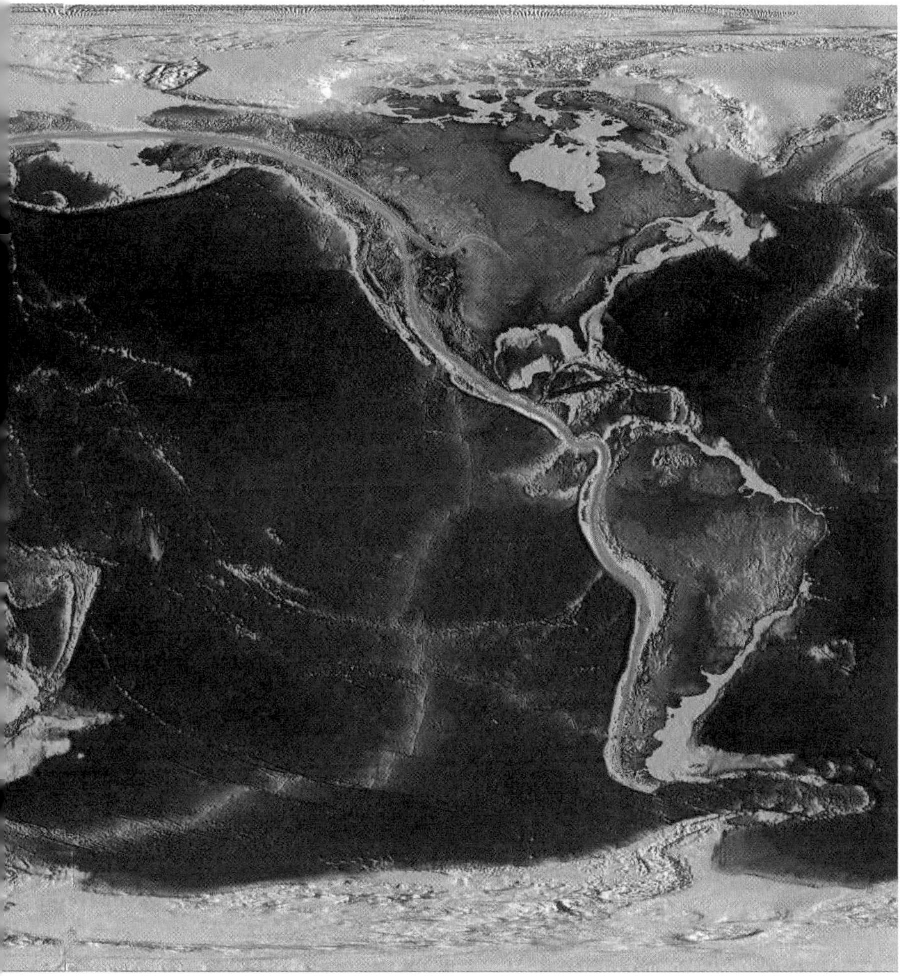

Early Homo sapiens *initially travelled within Africa, first south of the Rift Valley and then along to the east, reaching as far as the southern tip of Africa by about 200,000 years ago.* Sapiens *could have taken two possible routes out of Africa. The most accepted among the two is the southern route in which they travelled from Eritrea across the Bab-al-Mandab Strait into Yemen and around the Arabian Peninsula. As sea levels dropped between 74,000 and 71,000 years ago, rocky islands from the shallow straits which separate Arabia from Africa and Iran were exposed. These shorelines were teeming with marine life which provided food the human exodus. The northern routes would have taken our ancestors from their home in the East African Rift Valley across to the Sahara, then through the Sinai Peninsula into the Levant. Although the Sahara was alternating from wet and dry periods which could have supported the human migration, little evidence from between 80,000 to 50,000 years ago has been found in this corridor. In September 2016, a team from Oxford University and Saudi Tourism unearthed close to the northern Saudi town of Tayma what is possibly a 90,000-year-old human finger bone which—if confirmed—puts the date of when humans left Africa at least 30,000 years earlier than previously thought.*

roughly 2 to 5 per cent Neanderthal DNA. Neanderthals and *Sapiens* are thought to have coexisted for thousands of years and the interbreeding and mingling of these ancient genes helped early *Sapiens* and their descendants (modern Europeans) to adapt better to colder climates. Recent studies show that some Indians may also carry these genes (predominantly those in the north and north-east of the Indian subcontinent). So in a sense, some of our extinct ancestors and their cousins continue to survive within us. The Neanderthals died out and perhaps were even eliminated by *Sapiens*. The relict *Erectus* populations too probably met a similar fate. Scientists think it is improbable that some natural catastrophe or disease could have targeted and affected *Erectus* and *Neanderthalensis* but spared *Sapiens*; and some scientists have even suggested that *Sapiens ate* the other species. But it is most likely that the branches of other *Homo* species were pruned by *Sapiens*. Could this have been the first ethnic cleansing by our kind?

The idea that all non-African humans are descended from a single group of individuals contradicts previous theories that all the different modern races evolved separately from an earlier human ancestor known as *Homo erectus* in different parts of the world. Some anthropologists in China touted the idea that the people of their country had evolved directly from a lineage of *Erectus* that arrived there 1.6 million years ago and not from African *Sapiens*. One prominent Chinese anthropologist proclaimed that they were 'the earth's most ancient original inhabitants'. The political agenda in anthropological studies were disproved later by genetic studies, which showed that the Chinese too were descended from the *Sapiens* who had left from Africa.

With the elimination of all their immediate ancestors and cousins, *Sapiens* were left alone and became a family unto themselves. Like a few isolated and quirky species who are the only member of their family (such as the New Zealand horned lizard—tuatara—or the gingko trees and other such survivors), such single species face the threat of extinction because they have a restricted gene pool. Any 'shock' to this gene pool can cause the death of its members and eventually extinction. Can our genetic 'uniqueness' threaten our existence? Isolated tribes (like some Andaman and Nicobar islanders, Melanesian islanders, Native American Indians, among others) perished when they contracted infectious diseases (like measles, smallpox, STDs) from more recent human contact, against which they possessed no immunity. But are we truly so different? Going by the classification of all living things, we have been designated as a family unto ourselves, but the reality is not quite so straightforward. If we were to draw our family tree starting from the earliest primates, our closest relatives are chimpanzees, followed by gorillas, gibbons, baboons and macaques and so on . . . Human DNA shares nearly 99.97 per cent similarity with chimpanzee DNA. The chimpanzee (*Pan troglodytes*; *Pan* after the mischievous Greek god of the forest; *troglodytes*, cave-dweller) is not too far away in the ape-man family tree, just a few cousins removed from *Sapiens*. But other apes and primates are not too far removed either (geneticists call this 'genetic distance'). Genetic difference between individual humans today is minuscule—about 0.1 per cent, on average. Studies show that the chimpanzee genome differs from that of the human genome by a very small fraction. The bonobo (*Pan paniscus*) is a close cousin of the chimpanzee and differs from humans to the same degree. The DNA difference with gorillas is about 1.6 per cent. Most importantly, chimpanzees, bonobos and humans all show this same amount of difference from gorillas. A difference of 3.1 per cent distinguishes us and the African apes from the Asian great ape, the orangutan. So humans and others apes (like chimpanzees and gorillas) diverged from an orangutan-like ancestor and became equidistant in genetic terms from them. That makes gorillas, chimpanzees and humans contemporaries. How do monkeys compare? All of the great apes and humans differ from rhesus monkeys by about 7 per cent in their DNA. All this can be understood thus—around 6 million years ago, a single ancestor diverged and gave rise to different species of ancestors one of whom became the ancestor of humans while the other led to the evolution of the chimpanzee. This common ancestor of the chimpanzee and *Homo* is therefore a shared great, great (and many, many more 'greats') grandmother. So technically, *Homo sapiens* (and all other *Homo* species together) and chimpanzees belong to the *same* family. Even the specific name—*troglodytes*, or cave dweller—needs revision since we know that chimpanzees don't live in caves! For many reasons, none of them

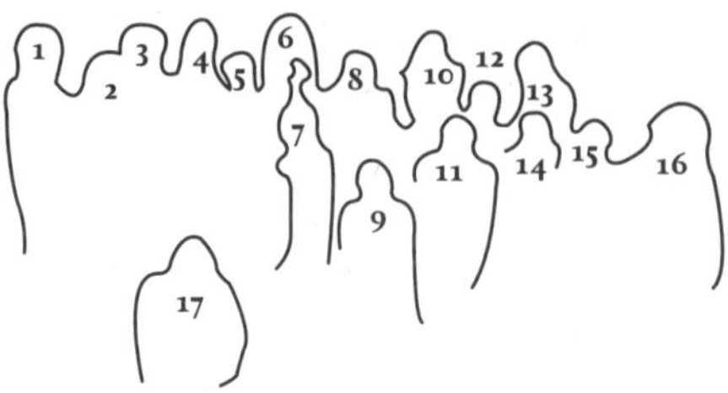

A four million year family portrait.
From left to right 1: Homo ergaster *from Nariokotome, Kenya, 1.55 million years before present; 2 and 7: First* Homo erectus *from Dmanisi, Georgia, about 1.7 million years before present; 3: A male Neanderthal from La Chapelle-aux-Saints, France, 50,000 years before present; 4: A female Neanderthal from Saint Cesaire, France, 36,000 years before present; 5: A Neanderthal boy from Roc-de-Marsal, France, 40,000 years before present; 6 and 13: Male Neanderthals from La Ferrassie, France,*

50,000 years before present; 8: A male Australopithecus afarensis *from Hadar, Ethiopia, 3.1 million years before present; 9:* Homo floreseinsis *from Flores, Indonesia, 18,000 years before present; 10:* Homo sapiens *from Cro-Magnon, France, 30,000 before present; 11: A female* Australopithecus afarensis *from Hadar, Ethiopia, 3.4 million years before present; 12:* Paranthropus bosei *from Olduvai, Tanzania, 2.5 million years before present; 14:* Sahelanthropus bosei *from Toros-Menalla, Chad, 7 million years before present; 15: A male* Australopithecus afarensis *from Sterkfontin, South Africa, 2.7 million years before present; 16: A Neanderthal woman from Krapina, Croatia, 36,000 years before present; and 17:* Homo habilis *from Koobi Fora, Kenya, 3 million years before present.*

scientific we persist with this flawed nomenclature, the intent of which is probably to distance ourselves as much as possible from animals closely related to us.

While *Sapiens* were spreading throughout the world, they often faced life-threatening conditions and many of them died in new environments. This caused populations to shrink, and it would take several generations before they recovered their numbers. This happened over and over again. When populations shrink, the gene pool (or the number of diverse individuals) shrinks too, and this results in fewer genetically similar adults who, over time, produce children with similar genetic make-up. This erodes genetic diversity, and scientists call these 'genetic bottlenecks'. Lack of diversity within a population is deleterious in the long run, as adults become less fertile and produce fewer children. And only a few of those who are born survive successive generations. The *Sapiens* populations passed through several such bottlenecks each time they faced disease and death.

Around 20,000 years ago, a small band of hunters from the plains of Central Asia headed into the icy void of the East Asian Arctic during a time when great ice sheets had reached the midlatitudes to create a large frozen wasteland. Sea levels, during this time, had dropped by more than 100 metres and this period was called the 'Last Glacial Maximum'. As the freezing intensified, the sea level dropped even further and for one last time, it exposed a land bridge that connected Asia to the Americas. The hunters who crossed this icy wasteland took the final stretch of the last frontier on land, the Americas. Somewhere between 18,000 and 15,000 years ago this band entered North America through the Bering Strait, and within a thousand years or so *Sapiens* reached the tip of South America. The dispersal of *Sapiens* occurred over a relatively short period of time (even by human standards) and as they travelled, they left behind new stone technologies. Unlike *Erectus*, *Sapiens* were more cosmopolitan and took to living outside of the tropics relatively quickly, about 65,000 years ago. And they started to live in higher latitudes (more than 50°) about 42,000 years ago.

With the arrival of *Sapiens* on every continent, the 4.6-billion-year-long story of the evolution of life reached its finale. What followed from here on is what scientists call 'prehistory', the period of the first human settlements till the emergence of the earliest civilizations. From the time of the formation of the first civilizations until the present day is a period that we call 'history'. After the arrival of *Homo sapiens* on the banks of the Indus, it took roughly 50,000 years before the first civilization arose along its banks, and from here begins human 'history'.

To put things into perspective: in the life of our 46-year-old Earth Woman, *Sapiens* emerged less than four hours ago. The life of the Earth Woman (or natural history) shows us that the existence of any species, or for that matter of life itself, is because of an incredibly improbable set of events that has led eventually to making us who we are. In the story of evolution, life forms do not have a clear destiny or direction, and the extinction of (almost all) species is inevitable. *Homo sapiens* is the most recent entrant in the evolutionary scene (of a dominant species). Had it not been for just the right events that wiped out our competitors and predators at just the right time, none of us would be here—neither the ancestors of the apes, their mammalian ancestors, nor their mammal-reptilian ancestors and so on. There is no guarantee that the replay of the time sequence, of all the chance events and accidents, starting from what happened one afternoon in a shallow stranded pool nearly 3.5 billion years ago, will again give rise to life as we know it. Perhaps if it were possible to perform such an experiment we may have very different animals and plants inhabiting our world today. The last untouched hot springs and deep-sea vents provide hope that life may perhaps re-emerge even if all of it gets obliterated. But what are the chances that life will restart? More importantly are we willing to gamble and, even if we are, who will remain to claim the stakes?

ACKNOWLEDGEMENTS

It is intimidating to write a book which describes the history of a land mass, from the time when it did not exist to how it got its oldest rocks, and end with the arrival of our ancestors about 70,000 years or so ago on the banks of the Indus river and their spread through the region. Such an effort requires the foolhardiness to be potentially exposed to criticism from experts and people in the know. I cannot for a moment say that I have understood it all, or that I have asked or answered all pertinent questions. But what you've read is the story of how the Indian subcontinent probably looked over billions or millions of years and the life it held, culled from hundreds of scientific papers and thousands of conversations. To weft and weave the different strands of the sciences into a story has taken me more than twenty-two years, although the process of organizing my research and writing began only about four years ago, after some serious introspection. There is much self-doubt which assailed me when I submitted my final manuscript to the editors at Penguin-Random House: am I the person best qualified to write this book? There were after all, more eminently suitable experts from a range of fields who understand natural history (or at least a part of it) better than I did. These include, though are not limited to, geographers, geologists, palæontologists, geneticists, anthropologists, climate experts, and experts from several sub-disciplines of physical and life sciences. There are also polymaths, eccentrics, enthusiasts and deft writers who could grapple with the subject better than I.

Four years ago, I finally began putting fingers to the keypad, the result of which you hold in your hands. From the moment my first draft began to take shape, I badgered scientists and academics for clarifications, and many of them patiently answered what now seem to be outstandingly stupid questions. So to those of you who may be curious like me, this book saves you the embarrassment!

No work of non-fiction to my mind is ever really the work of one person, and this book truly could not have been written without the intellectual, practical and mentorial contributions of many people, some of whom I now count as among my closest friends. Two people are very special in the making of this book.

In 2010, many palæontologists and geologists who I had met were beginning to lose their patience with my complicated questions. They directed me to meet one person, who perhaps could answer them. This was Professor Ashok Sahni. I had met several of his students who are now faculty in palæontology or geology in universities and colleges across India, and I could see that I was heading towards the wellspring of wisdom. Since my first meeting with him on a summer afternoon in the veranda of his Lucknow residence, I have called Professor Sahni at odd hours with a barrage of questions, which he has answered with wit, clarity and deftness. Professor Sahni has been a patient listener and an epistolary raconteur who provided encouragement and sage advice at every step.

I met Pradip Krishen, quite accidentally, to discuss the degradation of dry deciduous forests across India. That discussion strayed into our areas of common interest—forests, trees, soils and rocks—and Pradip has since inspired, provoked and goaded me to think beyond the obvious, brought ideas and suggestions with an air of gentleness and care, and has always been willing to lend me his critical gaze, including in places where my writing or my fervour has faltered. This book may never have been begun without the enthusiasm of Professor Sahni and it certainly would not have been finished without the endless encouragement of Pradip.

My first drafts began to take shape organically and my wife, Vandana, became its insightful first editor, helping me unravel many knotty problems with my storylines. She has been my prism to the world, a soother of ruffled feathers, a patient listener to my early morning ranting, a fellow dreamer; I would have never come this far without her by my side. For their wide-eyed listening to the stories of my travels, for being patient with my occasional surliness and for always making me laugh, a special thanks to my two lovely children, Aria and Avie. They have taught me that nature and humanity is infinitely more beautiful and varied than anything that could ever be in a book.

As I began writing, a fellow science writer warned me of his experience of falling into 'research rapture', an over-indulgence to dig deeper and not knowing when to let go. While working on this book I succumbed, wholeheartedly, to this. I relied heavily on interviews with discoverers and students of the sciences, and the research they pointed me to. They narrated to me how they imagined life had emerged over millions of years and I have tried to recount that faithfully and correctly. Any mistakes that I may have made are entirely due to shortcomings in my understanding and not because of any fault in their description.

Professor Rajeev Patnaik of Panjab University was a ready-reckoner for all my small queries, and I bothered him endlessly with my questions. I thank him deeply for his patience and humour. Professor Patnaik introduced me to Professor Parth Chauhan, a brilliant palæoanthropologist who tempered my grand assumptions on how our ancestors came to the subcontinent.

Professor Ashok Sahni and Pradip saw this ship set sail from its initial, rather rickety shape to what you see before you, and to them I am very grateful. I am deeply appreciative of comments from Anil Jacob and Anita Buragohain on early drafts and for candid advice which helped me structure the book. Ray Marcelo read early drafts of the book and encouraged me, saying that I was on the right path.

Special thanks to Meru Gokhale, who shared my enthusiasm and signed me on. Fazal Rashid was the first person at Penguin Random House to help me navigate a tortuous negotiation. I owe big thanks to Ahlawat Gunjan, the wonderful designer, for being patient and coming up with delightful design ideas for the book, and Tanvi Nathyal for bringing them to life. Rajni George who patiently held it all together, and Shatarupa Ghoshal, the deft editor who understood the nuance of science and made the text accessible; and others at Penguin Random House who oversaw this book to completion are gratefully acknowledged.

Finally, I am grateful for the friendship and comfort of strangers who I met during my travels who often offered me food and shelter, conversation, warmth and hospitality in far-flung places of India. Some of these are ordinary people making extraordinary effort to conserve India's amazing natural heritage. A physics teacher in Dhar district of Madhya Pradesh, Vishal Verma has been meticulously cataloguing fossil sites in western and central Madhya Pradesh and saving them from being lost forever. Like him, a war veteran in Kutch, a schoolteacher in Jhalawar, Rajasthan, a quarry hand in Ariyalur, Tamil Nadu, and many others who conserve geological monuments and fossil sites with no resources in the face of administrative apathy, and bear the brunt of public and personal ridicule—to me they are true heroes.

This book would not have been possible without the generosity and patience of the following individuals: Arun Sonakia and Suresh Srivastava, both formerly with Geological Survey of India; D.K. Pandey, Rajasthan University; Ajoy Bhaumik, Indian School of Mines, Dhanbad; Abhijit Chakrabarty, Jogamaya Devi College, Kolkata; Hukum Singh and Mukund Sharma, Birbal Sahni Institute of Palaeobotany, Lucknow; Hema Achyuthan, Anna University, Chennai; Saswati Bandyopadhyay and T.S. Kutty, Indian Statistical Institute, Kolkata; K. Thangaraj, Centre for Cellular and Molecular Biology, Hyderabad; Jeff Wilson, University of Michigan, Ann Arbor; Kenneth D. Rose, Johns Hopkins University; and several scientists and staff at Geological Survey of India, Jaipur and Kolkata, and other government institutions who wish not to be named.

The book in your hand is testimony to where I have been for a large part of my adult life and I hope that this provides sufficient explanation to my larger family—the Lals, Singhs and Bhatnagars and close friends—who have been patient with my absence on many a social occasion.

My biggest gratitude is to you, dear reader, for packing your mental suitcases and joining me on this fabulous journey.

NOTES

General Reading
Many subjects were covered in this book. While it is difficult to assign a single source or set of references, I found the following most useful:

Geological maps
Geological maps are location specific and give the age of rock horizons based on studies of mineral deposits, fossils and other indicators. Between 1879 and 1978, the Geological Survey of India (GSI) produced manuals, memoirs and monographs on geology, most of which remain relevant. These are made accessible through open access repositories such as www.biodiversityheritagelibrary.org. Geological maps of India are prepared by the GSI; while fairly accurate, they are often difficult to come by. The GSI's regional offices sell state-wise geological maps. High-resolution scanned maps are available online but these violate laws of the land. User discretion is advised.

Reconstruction maps
For reconstructions of how the world would have looked millions of years ago I turned to Professor Christopher Scotese of University of Texas (www.scotese.org) who has pieced together current geological information to create accurate palæo-maps and he has generously shared these with researchers, writers and enthusiasts such as myself. Tragically, there are no palæo-cartographers in India and therefore there are few accurate India-specific maps from various periods.

Stratigraphic maps
A map depicting the distribution of rock formations by age is called a stratigraphic map, and often cross-sections from formations which have layers arranged based on the various ages are depicted. As a starting point, I recommend that readers use a simple stratigraphic chart to understand the various periods of Earth's history and how a particular formation holds rocks from these times. For this book I used the 2013 version of the Geological Timescale developed by the International Commission on Stratigraphy (www.stratigraphy.org). As new information on the age of fossils and rocks is determined, the timescales are recalibrated and revised. In the stratigraphy chart (the most recent was prepared in 2015), you will notice that the colourful stratigraphic columns give the time span of each geological time interval. Although Earth is over 4.2 billion years old, the columns show only parts that are dated using indicator fossils.

Theoretical underpinnings
On theories proposed and the current understanding on the birth and arrangements of continents, I recommend Naomi Oreskes, *The Rejection of Continental Drift: Theory and Method in American Earth Science* (Oxford: Oxford University Press, 1999). On the history of geology and natural history, I recommend John Gribbin, *Science: A History* (London: Penguin, 2010). Any natural historian attempting a wide-ranging survey of literature, seeking archival material and scouring for samples in repositories owes an enormous debt to the countless scholars and researchers in whose work he has sought inspiration, ideas, thoughts and interpretations and guidance. I have entirely relied on the work of others to piece arguments in this book, and although my personal bibliography comprises thousands of books, scientific papers, pamphlets and newspaper articles, I have tried summarizing these with literature which simply yet comprehensively presents all sides of these arguments. Where one school of thought dominates, I have tried to balance these with references from other researchers. The endnotes which follow, I hope, will serve as some acknowledgement of where my greatest debts lie, but such is the vitality and range of work of the pioneers of the twentieth

century that neither of these can, within the confines of this or any book, be comprehensive. I have only been able to include references which I believe are important and most helpful for anyone wishing to understand a particular theme or period of time in greater depth, and added a few others which either brought a certain joy to me or, I believe, may become significant in shaping the future course of research in that area. This brings me to another problematic issue—that of bias. I may have during the course of my interactions with experts leaned slightly towards one school of thought or belief over the other. I may also have been callous or rash enough to have altogether ignore a few, less dominant thoughts and ideas which appear in the margins of scientific discussion, in these notes and acknowledgements. If so, then to those whom I have ignored and to you, dear reader, I owe an apology.

1. Why on Earth

1: The Archean province comprises the oldest cratons and shields and for details on how these were formed I recommend the following papers: O.P. Pandey and P.K. Agrawal, 'Lateral Zonation around Archean Nucleus of the Dharwar Craton, India: Its Deformation, Segmentation and Subsequent Breakup', *Advances in Earth Sciences: Series-I*, 2009, *e-Journal Earth Science India*, Vol. I (II), pp. 46-57. R.S. Sharma, *Cratons and Fold Belts of India* (Lecture Notes in Earth Sciences, 127; Berlin: Springer-Verlag, 2009); and Joseph G. Meert and Manoj K. Pandit, 'The Archaean and Proterozoic history of Peninsular India: tectonic framework for Precambrian sedimentary basins in India', in *Precambrian Basins of India: Stratigraphic and Tectonic Context*, ed. R. Mazumder and P.G. Eriksson (Memoirs, 43; London: Geological Society, 2015), pp. 29-54. The creation of younger rocks from melting and remelting of Archean rocks is described in Jean Francois-Moyen et al., 'Multi-element geochemical modelling of crust-mantle interaction during late-Archaean crustal growth: the Closepet granite (South India)', *Precambrian Research* 112 (2001): 87-105. Also see a shorter summary by K.S. Valdiya, 'Tectonic resurgence of the Mysore plateau and surrounding regions in cratonic Southern India', *Current Science* 81, 8 (2001): 1068-1088 which has a discussion on the evolution of rocks in the Dharwar craton.

2: Over most of where the Dharwar craton lies, the granite and gneiss that forms the basement is not visible because it was covered by soils in a gigantic volcanic event that took place about 65 million years ago. A younger craton that lies east of the Dharwar craton is the Bastar craton, whose boundary is marked by a depression through which the Godavari river flows. Younger rocks, although still over 3 billion years old, are found in East India in the Singhbhum region of Jharkhand which lies further east of the Bastar craton; here the boundary is marked by the Mahanadi river. Slightly to its north-west lies the Vindhyan Supergroup (a supergroup is a collection of rocks which are of similar age and origin), comprising the Bundelkhand craton and the Aravallis. The geological history of India is comprehensively covered in a graduate textbook by K.S. Valdiya, *The Making of India: Geodynamic Evolution* (New Delhi: Macmillan, 2010).

3: For comparing the age of cratons and rocks from across the world, see G. Brent Dalrymple, *The Age of the Earth* (Stanford: Stanford University Press, 1991). This is an elegant account on the history and methods used to determine the age of Earth and other planetary bodies. It also describes in simple terms the history and principles of scientific methods used to determine the age of Earth, the Moon and other planetary bodies. The pursuit to determine Earth's age is eloquently presented in Phillip C. England et al., 'Kelvin, Perry and the Age of the Earth', *American Scientist* 95 (2007): pp. 342-49. For more recent advances in dating methods using isotopes of potassium, argon, uranium, lead and strontium, see Martin Rudwick, *Earth's Deep History: How it was discovered and why it matters* (Chicago: Chicago University Press, 2014).

4: The story of how Earth was formed is pieced through several interviews and readings. To understand the making of our planet and the Moon, I benefited from reading Frank D. Stacey and Paul M Davis, *Physics of the Earth* (New York: Cambridge University Press, 2008).

Two articles on black holes and their natural history that are easy to ready and make complex astronomical science accessible are: Dennis Overbye, 'Gravitational Waves Detected, Confirming Einstein's Theory', *New York Times*, 11 February 2016 (http://www.nytimes.com/2016/02/12/science/ligo-gravitational-waves-black-holes-einstein.html?_r=0); and Joanna Klein, 'An Earthling's Guide to Black Holes', *New York Times*, 8 June 2015 (http://www.nytimes.com/interactive/2015/06/08/science/space/guide-to-black-holes.html)..

5: To explain the enormity of geological time, I am borrowing the analogy of the Earth Woman from Nigel Calder, *The Restless Earth* (London: Penguin, 1972).

6: The idea of using microscopic zircon to measure the age of Earth probably first dawned on John Joly (1857–1933), an Irish physicist and professor of geology at Trinity College, Dublin. A pioneer in the development of radiotherapy in the treatment of cancer, Joly developed the process for extracting radium from ores, proposed (with Henry Horatio Dixon) the cohesion-adhesion theory for capillary uptake of water in plants, and developed the first colour photographic plate. Mark Harrison calculated the age of these 4.4-billion-year-old crystals from the Jack Hills in Western Australia by measuring the radioactive decay in the two isotopes of the element hafnium (Hf_{177} and Hf_{176}). The solar system is estimated to be 4.567 billion years measured after dating components of meteorites, while Earth took 95 million years to form and the collision of Theia and proto-Earth took place 4.470 billion years ago (± 32 million years). For more on this method, see: Seth A. Jacobson et al., 'Highly Siderophile Elements in Earth's Mantle as a Clock for the Moon-forming Impact', *Nature* 508 (2014): 84–87.

7: Charles H. Lineweaver and Marc Norman, 'The Bombardment History of the Moon and the Origin of Life on Earth' (paper presented at the eighth Australian Space Science Conference, Canberra, Australia, 29 Sept–2 Oct 2008); available online at http://www.mso.anu.edu.au/~charley/papers/LHBASSC2008.pdf (accessed 12 August 2014).

8: In modern living corals, daily rings are hard to see unless viewed through cut-sections after staining and under a microscope. The chemical changes that occur during the fossilization process, however, make it easy to see the rings in fossils. Corals secrete aragonite, a form of limestone characterized by a special crystalline structure, with different densities at different times of the year. Seasonal changes in water temperature are responsible for the different densities, and coral produce these in response to the changing temperature and salinity in sea water. In 1963, Dr John W. Wells, emeritus professor of geology at Cornell University, was the first to count the daily growth rings on fossil coral from around 400 million years ago, and estimated that there were between 410 to 400 days in a year (John W. Wells, 'Coral Growth and Geochronometry', *Nature* 197 (1963): 948–50) and the average day would have been only about 21.5 hours long. With the discovery of older coral fossils, scientists have found that the ratio of daily rings to annual rings in fossils had increased (S.J. Mazzullo, 'Length of the Year during the Silurian and Devonian Periods: New Values', *Geological Society of America Bulletin* 82 (1971): 1085. However, the length of year has been nearly constant for the past 2 million years or so. The Moon's gravitational pull reduces Earth's rotation so very slowly that the time taken for Earth to revolve around its axis (or a complete day) has increased. This slowdown caused by the action of Moon's gravity is called 'tidal acceleration', and is infinitesimally minute. Scientists estimate that nearly 1 million years ago, a day was about 20 seconds shorter than it is now. This phenomenon is called Munk's Enigma (Walter Munk, 'Twentieth Century Sea Level: An Enigma.' *Proceedings of the National Academy of Sciences of the United States of America* 99 [2002]: 6550–55) and hypothesizes how the anomalous rise and fall in sea levels over millions of years has contributed in retarding Earth's spin.

9: Marine and freshwater creatures like bivalves, corals, brachiopods, and stromatolites which grow rings one over the other, sandstones like tidalite (it is formed when fresh sediments deposit over older sediments with every tidal cycle) from the past are good indicators of the distance between Earth and the Moon. The first conclusive studies using shells measured the external growth increments in a species of bivalve called the Paleorotation bivalve (*Clinocardium nuttalli*) (Wells 1963, 948–50); C.T. Scrutton, 'Periodic growth features in fossil organisms and the length of the day and month' in *Tidal Friction and the Earth's Rotation*, ed. P. Brosche and J. Sundermann (Berlin: Springer-Verlag, 1978), pp. 154–96, and J.W. Evans, 'Tidal increments in the Cockle *Clinocardium nuttalli*', *Science* 176 (1972): 416-17; and J.W. Evans, 'Growth and micromorphology of two bivalves exhibiting non-daily growth lines', in *Growth Rhythms and the History of the Earth's Rotation*, ed. G.D. Rosenberg and S.K. Runcorn (New York: John Wiley, 1975), pp. 119–33. The study of annual rings in stromatolites from the Precambrian period is a good indicator of tidal periodicity. Despite the abundance of shelled fossils and stromatolites over a long period of time in India, I have not found scholarly work on this theme. However, the interested reader is advised to consult R. Lathe, 'Fast tidal cycling and the origin of life', *Icarus* 168 (2004): 18–22, and responses to this article in subsequent issues.

10: How water was formed on a very hot Earth has been a mystery. There are two schools of thought, the 'dry' formation theorists believe water came to Earth through meteors, while 'wet' theory adherents contend that it came from the rocks themselves. In August 2014, the European Space Agency's spacecraft *Rosetta* used Ion Spectrometer while orbiting the Comet 67P/C-G to analyse the chemical fingerprint of gases in the comet's fuzzy envelope, and confirmed that it contained water. See: K. Altwegg et al., '67P/Churyumov-Gerasimenko, a Jupiter Family Comet with a High D/H Ratio', *Science* 347 347, 3–6 (2015). It is also possible for both these events to have occurred. This is the 'middle path' position I have taken as far as this debate is concerned in this book. A recent paper has concluded 'water enriched in deuterium contributed significantly <20% of the water in the Moon. Therefore, our work places important constraints on the types of objects impacting the Moon ~4.5-4.3 billion years ago and on the origin of water in the inner Solar System', see: Jessica J. Barnes et al., 'An asteroidal origin for water in the Moon', *Nature Communications* 7 (2016): 11684, available online at http://dx.doi.org/10.1038/ncomms11684.

11: For how meteors have impacted early Earth see: S. Marchi et al., 'Widespread Mixing and Burial of Earth's Hadean Crust by Asteroid Impacts', *Nature* 511 (2014): 578–82. Metals that mix easily with iron (called siderophiles or iron-loving metals) in Earth's mantle, including gold, associate closely with iron in their liquid forms. Scientists believe that interplanetary objects brought these elements to Earth (and Moon and Mars) just after it formed its core—the process was caused by a small number of enormous, random impacts roughly 4.5 billion years ago, the largest one perhaps roughly the size of Pluto (or 3220 kilometres wide). The energy of this impact recast the core and made it largely of iron and iron-like elements. Metal-rich asteroids fell on Earth because it had a stronger gravitational force, missing the Moon in doing so. This explains why gold and other elements are found in such minute traces on Moon but in relatively greater abundance on Earth (see: W.F. Bottke et al., 'Stochastic Late Accretion to Earth, the Moon, and Mars', *Science* 330 [2010]: 1527–530). Willbold et al. estimate that a staggering 20 billion billion tonnes of meteorite matter, including gold and platinum, slammed into Earth during the 200-million-year-long shower, see: Matthias Willbold et al., 'The Tungsten Isotopic Composition of the Earth's Mantle before the Terminal Bombardment', *Nature* 477 (2011): 195–98.

12: The difference between asteroids and meteorites is described on the NASA webpage on the Near Earth Object Program, available as http://neo.jpl.nasa.gov/faq/ as viewed on 1 June 2015.

13: On the role of the Moon and its influence on the formation of proto-Earth, see Dalrymple (1991). A compilation of recent research on the Moon can viewed on *Nature*'s website, http://www.nature.com/nature/focus/moon-origin/ as viewed on 7 January 2014.

14: For the Dhala impact event, see: Jayanta Kumar Pati et al., 'First SHRIMP U-Pb and $_{40}Ar/_{39}Ar$ chronological results from impact melt breccia from the Paleoproterozoic Dhala impact structure, India', in *Large Meteorite Impacts and Planetary Evolution* (Special Paper No. 465), ed. U.W. Reimold and R.L. Gibson (Boulder: Geological Society of America, 2010), pp. 571-91. I am grateful to Prof. Pati of Allahabad University for his comments and review of the section on the Dhala meteor.

15: For a simple and accessible article on Lonar impact event, see: Adam C. Maloof et al., 'Geology of Lonar Crater, India', *Geological Society of America Bulletin* 122 (2010); 109-26. On how the Lonar crater formed and its impact in the locality, see: Shiloh Osae et al. 'Target rocks, impact glasses, and melt rocks from the Lonar impact crater, India: Petrography and Geochemistry', *Meteoritics & Planetary Science* 40 (2005): 1473-92.

16: The hypothesis that the Dal Lake is a possible impact crater was presented in a working paper by N. Iqbal et al., 'A Probable Meteor Impact Crater in Kashmir Valley (India)', at the Inter-University Centre for Astronomy and Astrophysics, Pune, India (2008). It is available at http://www.iucaa.ernet.in/-library/0708.pdf. The discovery of the crater in Kutch is described in R.V. Karanth, P.S. Thakker and M.S. Gadhavi, 'A preliminary report on the possible impact crater of Kachchh', *Current Science* 91 (2006): 877-79. The paper hints that the meteor which fell about 4000 years ago may have been responsible for wiping out a Harappan Age town.

17: I am grateful to M. Sivaram, former director of the Archaeological Survey of India in Chennai, for pointing me to the Brahmagiri megaliths. Few pre-Indus valley civilization sites receive protection in India and many early human sites like cave paintings and rock structures remain neglected. A special issue on megalith structures of India was published by the Archaeological Survey of India see A. Ghosh, ed., *Indian Archaeology 1960-61: A review* (New Delhi: Archaeological Survey of India, 1996). On the Brahmagiri structures, see: N. Kameswara Rao, 'Astronomical orientations of the megalithic stone circles of Brahmagiri', *Bulletin of the Astronomical Society of India* 21 (1993): 67-77.

18: The simple proof which shows interference by cosmic debris with radio frequencies was demonstrated to me by astronomers-in-residence at the Kodaikanal Observatory in May 2010.

2. Breath of Life

1: Life is a hard-to-define phenomenon. One of the first, most eloquent essays to be written on the subject was by Erwin Schrödinger, *What is life? The Physical Aspect of the Living Cell* (Cambridge: Cambridge University Press, 1943), p.17. A better resolution to the philosophical questions raised by Schrödinger is offered by David Deamer, *First Life: Discovering the Connections between Stars, Cells, and How Life Began* (Berkeley: University of California Press, 2012). For a useful review on the most recent developments and controversies on the definition, origins and evolution of life, see Peter Ward and Joe Kirschvik, *A New History of Life—The radical discoveries about the origins and evolution of life on Earth* (New York and London: Bloomsbury, 2014).

2: In 2010, biochemist Douglas Theobald of Brandeis University, Massachusetts, conducted a mathematical analysis to test the most likely ways in which twenty-three proteins in a wide array of different life forms had evolved. His results showed that it is very likely that all life had indeed emerged from one common ancestor. Further, Theobald's study found that life originated multiple times and on several separate occasions. See his webpage (http://theobald.brandeis.edu/) for details.

3: For those inclined to venture further on various hypotheses on how early microbial life forms evolved and diverged, see: Peter P. Sheridan et al., 'Estimated Minimal Divergence Times of the Major Bacterial and Archaeal Phyla', *Geomicrobiology Journal* 20 (2003): 1-14. Another interesting hypothesis on emergence of first life is presented by Armen Y. Mulkidjanian et al., 'Origin of First Cells at Terrestrial, Anoxic Geothermal Fields', *Proceedings of the National Academy of Sciences of the United States of America* 109 (2012): E821–30. The paper suggests that it was possible for cells to form in a zone where vapour emanated from geothermal systems fired by magma.

4: A readable account of how a proton pump could have jumpstarted life is Nick Lane, 'Why Are Cells Powered by Proton Gradients?', *Nature Education* 3 (2010): 18.

5: It is widely agreed among palæobiologists that the first living cells appeared around 3.8 billion years ago. The time has been estimated based on genetic analyses and the date assigned to the earliest microfossils found in the oldest sedimentary rocks so far remains debatable. The two main contenders for the oldest fossils are *Isuasphaera* from Greenland (see, Manfred Schidlowski, 'The Beginnings of Life on Earth: Evidence from the Geological Record', in *The Chemistry of Life's Origins*, ed. J.M. Greenberg, C.X. Mendoza-Gómez and V. Pirronello [Dordrecht: Kluwer Academic Publishers, 1993], pp. 389-414), which is dated to about 3.8 billion years ago and a slightly more complex macro-fossil of *Primaevifilum amoenum* from Australia which is estimated to be from about 3.5 billion years ago (see J. William Schopf, 'Fossil Evidence of Archaean Life', *Philosophical Transactions of the Royal Society B* 361 [2006]: 869–85). In September 2016, Australian scientists published findings of what they claim to be a fossilized microbial mat-like stromatolite from Greenland which is from about 3.7 billion years ago, or nearly 300 million years older than what until then was known to be the oldest fossil. They found reddish peaks and filamentous structures that are perhaps made by microbes in a shallow ocean, thus claiming that these could be the earliest known evidence of life on Earth (see: A. Nutman et. al., 'Rapid emergence of life shown by discovery of 3,700-million-year-old microbial structures', *Nature* 536 [2016]: 7618).

6: To know how life becomes comatose and awakens, see a lecture by Mark Roth in his TED talk: www.ted.com/talks/mark_roth_suspended_animation. Roth's papers are available at: http://labs.fhcrc.org/roth/.

7: The LUCA operating system is called the central dogma—the self-sustaining passage of information from DNA to RNA to protein.

8: When and how most elements especially oxygen appeared on earth is a mystery. A 2017 paper has reported traces of oxygen in 3.27-billion-year-old komatiite lavas from the Kaapvaal Craton in Southern Africa, suggesting that oxygen may have appeared on earth earlier than previously believed. (Benjamin L. Byerly, Keena Kareem, Huiming Bao & Gary R. Byerly, Early Earth mantle heterogeneity revealed by light oxygen isotopes of Archaean komatiites, Nature Geoscience 10, 871–875 (2017), 30 October 2017). The rise of oxygen in air is attributed to the emergence and proliferation of aerobes especially *cyanobacteria* that began to add oxygen to the air is well documented. See: E.G. Nisbet and N.H. Sleep, 'The Habitat and Nature of Early Life', *Nature* 409 (2001): 1083–91. For a more recent debate on the role of minerals in evolution, see: J.E. Johnson et al., 'Manganese-oxidizing Photosynthesis before the Rise of Cyanobacteria', *Proceedings of the National Academy of Sciences* 110 (2013): 11238–43. A recent study has estimated that Earth is home to upward of 1 trillion microbial species (Kenneth J. Locey and Jay T. Lennon, 'Scaling laws predict global microbial diversity', *Proceedings of the National Academy of Sciences* 113 [2016]: 201521291, available online at http://www.pnas.org/content/early/2016/04/26/1521291113.full.pdf).

9: There is a tremendous interest in the Great Oxidation Event since its physical proof was presented just two decades ago. For a comprehensive review of the discovery, see: H.D. Holland, 'The Oxygenation of the Atmosphere and Oceans', *Philosophical Transactions of the*

Royal Society B 361 (2006): 903–15, or the 'barren billion'. The period roughly between 1.8 and 0.8 billion years ago has been referred to as the 'boring billion'. At this time, life made very little progress because most land was clumped together and there was little glacial activity which could trigger the advancement of new life. For the role of oxygen and its impact on evolution, see: Grant M. Young, 'Precambrian Supercontinents, Glaciations, Atmospheric Oxygenation, Metazoan Evolution and an Impact That May Have Changed the Second Half of Earth History', *Geoscience Frontiers* 4 (2013): 247–61. Others have termed the period between 1.6 to 1.0 billion years ago as the 'dullest time in Earth's history' (see: Roger Buick et al., 'Stable Isotopic Compositions of Carbonates from the Mesoproterozoic Bangemall Group, Northwestern Australia', *Chemical Geology* 123 [1995]: 153–71). Scientists attribute the lack of any significant environmental, biological or geological events to the advance innovations in life forms, which caused the tempo of evolution of life to slow down.

10: For a good account of the events that caused changes in the chemistry and the environment that led to the Great Oxygenation Event, see: D.C. Catling, 'The Atmosphere—A History', in *Treatise on Geochemistry* 13 (2nd ed.), ed. D. Canfield, J. Farquhar, J.F. Kasting (Philadelphia: Elsevier, 2013). Also see more recent evidence of the Great Oxygenation Event from India in Joydip Mukhopadhyay et al., 'Oxygenation of the Archean Atmosphere: New Paleosol Constraints from Eastern India', *Geology* 42 (2014): 923–26; I thank the co-author of this paper, Quentin Crowley, for his comments. Scientists have calibrated the presence of certain minerals to determine oxygen levels and environmental conditions which existed at any given time. For example, pyrite or iron sulphide is formed at oxygen concentrations less than 0.1 per cent (at present atmospheric level, 21 per cent), Uraninite, an oxide of uranium, forms at one-tenth of levels at which pyrites form, and siderite, a carbonate of iron forms when oxygen levels are at a tenth of uraninite (a hundredth of pyrite). When geologists find these minerals in rocks they infer the conditions and environments in which they were formed.

11: The supercontinent Rodinia was named in early 1990s and several land and ocean configurations have been proposed, which remain largely unresolved (see: M.A.S. McMenamin and D.L.S. McMenamin, *The Emergence of Animals—The 550 Cambrian Breakthrough* (New York: Columbia University Press, 1990).

12: The earliest multicellular life forms to be discovered in India remain controversial (see: Stefan Bengtson et al., 'The Controversial "Cambrian" Fossils of the Vindhyan Are Real but More than a Billion Years Older', *Proceedings of the National Academy of Sciences* 106 [2009]: 7729–34) and have been dated to be around 1.65 billion years old. These are from the Tirohan dolomite found in Jankikund river near Chitrakoot in the Vindhyan range. Currently, one of the oldest and most unequivocal records of multicellular life from India (other than stromatolites) is that of algal mats from around 1.5 to 1.4 billion years ago or perhaps older (See: Mukund Sharma, 'Late Palaeoproterozoic [Statherian] Carbonaceous films from the olive shale [Koldaha shale], Semri group, Vindhyan Supergroup, India', *Journal of the Palaeontological Society of India* 51 [2006]: 27–35). For a handy guide on the subject, see: S. Kumar and Mukund Sharma, 'Field Guide to Vindhyan Basin, Son Valley Area, Central India' (published by the Palaeontological Society of India, Lucknow, 2010, for field workshop) which covers the geology of the Vindhyas and early life forms including microbial mats, *Grypania spiralis* and stromatolites sites that are found here. Also see a compilation of algal mats that are more than 1 billion years old in P.K. Bose et al., 'Mat-related features from sandstones of the Vindhyan Supergroup in central India', in *Atlas of Microbial Mat Features Preserved within the Cladistic Rock Record*, ed., J. Schieber et al. (Philadelphia: Elsevier, 2007), pp. 181–88.

13: How the Great Oxidation Event caused the rise of multicellular life is discussed in B.E. Schirrmeister, 'Evolution of Multicellularity Coincided with Increased Diversification of Cyanobacteria and the Great Oxidation Event', *Proceedings of the National Academy of Sciences* 110 (2013): 1791–96; and D.E. Canfield et al., 'Late-Neoproterozoic Deep-Ocean Oxygenation and the Rise of Animal Life', *Science* 315 (2007): 92–95. The paper by Canfield et al. shows the importance of iron sources from deep-sea sediments which turned the near barren, oxygen-starved depths to oxygen-rich depths leading to the emergence of diverse new life forms after glaciation episodes around 580 million years ago. This led to the emergence of animals that make up Ediacaran life and in the presence of sustained oxygen levels for the next 25 million years, animals developed the ability to move. A summary of the four major events that caused the rise of oxygen and the emergence of multicellular and complex life is elegantly presented in: Heinrich D. Holland, 'The oxygenation of the atmosphere and oceans', *Philosophical Transactions of the Royal Society B* 361(2006): 903–15. The paper argues that while shallow oceans were oxygenated, oxygen levels in the depth of the oceans fluctuated which were key for the emergence of life.

14: For discussions on the rise of multicellular organisms, development of body plans and symmetry, see: S. Conway Morris, 'Early Metazoan Evolution: Reconciling Paleontology and Molecular Biology', *American Zoologist* 38 (1998): 867–77; Douglas H. Erwin, 'The Origin of Bodyplans', *American Zoologist* 39 (1999): 617–29; and Mark Q. Martindale et al., 'The Radiata and the Evolutionary Origins of the Bilaterian Body Plan', *Molecular Phylogenetics and Evolution* 24 (2002): 358–65. Taxonomists are studying molecular data (mostly 18S rRNA sequences) of diverse life forms to piece together the time of when bilateral symmetry emerged (see C. Nielsen Animal Evolution: Interrelationships of the Living Phyla. Second Edition. Oxford University Press, Oxford, 2001.).

15: An overview of the rise of metazoans in the Kroll, Bhander, Semri and Chhattisgarh rocks and fossils and an attempt to reconcile the age of arrival of metazoans can be understood in a comprehensive paper by Abhijit Basu, 'Ediacaran Fossils in Meso- and Paleoproterozoic Rocks in Peninsular India Extend Darwin', *Journal of the Geological Society of India* 73 (2009): 528–36.

16: For a compilation on the different types of stromatolites from 3.5 to 2.5 billion years ago, see J. William Schopf et al., 'Evidence of Archean life: Stromatolites and microfossils', *Precambrian Research* 158 (2007): 141–55; and a catalogue of fossil stromatolite from India see S. Kumar, 'Stromatolite and the Indian biostratigraphy: A review', *Journal of the Palaeontological Society of India*, 23 & 24 (1980): 166-83. An overview of the diverse types of stromatolites from about 2.8 billion years ago from the Vanivilas Formation of Chitradurga, and 2.7 billion years ago from Kalche and Joldhal from the Dharwar Supergroup are described in H.J. Hofmann, 'Archean stromatolites as microbial archives', in *Microbial Sediments*, ed. R.E. Riding and S.M. Awramik (Berlin: Springer, 2000) pp. 315–27, while other sites in Chitradurga are also dated between 2.7 and 2.6 billion years old (R. Srinivasan et al., 'Archaean Stromatolites from the Chitradurga Schist Belt, Dharwar Craton, South India', *Precambrian Research* 43 [1989]: 239–50). Another interesting review on the importance of stromatolites is Tanja Bosak et al., 'The Meaning of Stromatolites', *Annual Review of Earth and Planetary Sciences* 41 (2013): 21–44.

17: An undated report of the GSI, 'Preliminary reporting of algal structure in the Singhbhum Group of rocks from Chandil area, Saraikela Kharsawan district, Jharkhand' (see: *www.portal.gsi.gov.in/gsiDoc/pub/cs_stromatolites_jh.pdf*) has added new forms of stromatolite that are being discovered in India.

18: For more on the role of hydrothermal vents and underwater volcanoes in the creation of banded iron formations (BIF) deposits formed west of Jamshedpur and north-west of Keonjhar, see: Subrata Roy and A.S. Venkatesh, 'Mineralogy and Geochemistry of Banded Iron Formation and Iron Ores from Eastern India with Implications on Their Genesis', *Journal of Earth System Science* 118 (2009): 619–41.

19: In 1987, Caltech biomagnetist and palæomagnetist Joe Kirschvink gave undergraduate Dawn Sumner a rock sample for analysis which was collected by palæontologist Bruce

Runnegar from Flinders Ranges in South Australia. Over the next fifteen years or so, fifteen such sites were found where the earliest Snowball Events were discovered (D.Y. Sumner et al., 'Soft-sediment paleo-magnetic fold tests of late Precambrian glaciogenic sediments', *Eos* 68 [1987]: 1251). For a comprehensive analysis and description of the Snowball Earth event, see: John J.W. Rogers, 'A History of Continents in the past Three Billion Years', *The Journal of Geology* 104 (1996): 91-107. It provides a comprehensive account of India in relation to other ancient land masses and describes how land masses and climate influenced early evolution. For evidence of the Snowball Earth event in India and glacial events, see: A.V. Sankaran, 'Neoproterozoic "snowball earth" and the "cap" carbonate controversy', *Current Science* 84 (2003): 871-73. The description of signs and deposits from the first global glaciation event are from the Son and Ken valleys in the Lower Vindhyas of Madhya Pradesh, which are believed to have occurred about 2.1 million years ago. (see: Anil Kumar, 'Rb-Sr Ages of Proterozoic Kimberlites of India: Evidence for Contemporaneous Emplacement', *Precambrian Research* 62 [1993]: 227-37).

20: Evidence of the emergence of soft-bodied multicellular animal life that arose around 620 to 540 million years ago (called the Ediacaran Period after a mountain range in Australia) can be found in the Krol-Tal belt of lesser Himalayas; the Upper Vindhyas of Madhya Pradesh; the Bhima and Badami river basins of Karnataka and the Kurnool basin of Andhra Pradesh. In the past decade remarkable Ediacaran fossils have been discovered around Jodhpur and Nagaur (see: S. Kumar et al., 'Five-armed body fossil from the Ediacaran Jodhpur Sandstone, Marwar Supergroup, western Rajasthan, India: a possible precursor of phylum *Echinodermata*', *Current Science*, 102 [2012]: 24-26). The origin of hard-bodied animals began with accretion of calcium which led to the emergence of shelled creatures (like *Cloudina* and *Namacalathus*) who exploded on the scene around 548 million years ago, laying the foundation of reefs in shallow seas. A good review of the origins of these creatures is in: Rachel Wood and A Curtis, 'Extensive Metazoan Reefs from the Ediacaran Nama Group, Namibia: The Rise of Benthic Suspension Feeding', *Geobiology* 13 (2014): 112-22. One group of minute marine creatures that evolved in large abundance were acritarchs. Among the oldest acritarchs were leiosphaerids which are found in abundance 1.8 billion year old rocks, and some palæontologists contend that a common ancestor of all acritarchs had arisen around 3.2 billion. Around 1 billion years ago some began acritarchs began to develop spines possibly to protect themselves from predation. The confirmed time by when hard-shelled marine animals became abundant was between 550 and 530 million years ago. See. Jonathan B. Antcliffe, Richard H. T. Callow and Martin D. Brasier, Giving the early fossil record of sponges a squeeze, Biological Reviews, April 2014.

21: The evidence of glaciation (with good photographs of geologic formations) in India from Talchir is available in Chandan Chakraborty and Sanjoy Kumar Ghosh, 'Pattern of sedimentation during the Late Paleozoic, Gondwanaland glaciation: An example from the Talchir Formation, Satpura Gondwana basin, central India', *Journal of Earth System Science* 117 (2008): 499-519.

22: The wonders of the chemical and physical properties of water have been written about simply in Kathy Wollard, *How Come? Every Kid's Science Questions Explained* (London: Workman Publishing, 2015). For a more scientific account on the behaviour of water, see: Anders Nilsson and Lars G.M. Pettersson. 'The Structural Origin of Anomalous Properties of Liquid Water', *Nature Communications* 6 (2015): 8998. The original paper (E.B. Mpemba and D.G. Osborne, 'The Mpemba effect', *Physics Education* 14 [1979]: 410-12), makes for very special reading!

23: The story of the co-evolution of minerals and life is elegantly presented in Robert Hanzen, 'Mineral Evolution', *American Mineralogist* 93 (2008): 1693-720.

24: The processes and condition under which fossils are preserved in rock and sediments is presented in David E. Fastovsky and David B. Weishampel, *Dinosaurs: A Concise Natural History* (New York: Cambridge University Press, 2009).

25: The most accessible account of Cambrian explosion is available in Stephen Jay Gould, *Wonderful Life—The Burgess Shale and the Nature of History* (New York: W.W. Norton, 1990). For a more global review which includes India, see Andrey Yu Zhuravlev and Robert Riding, ed., *Ecology of the Cambrian Radiation* (New York: Columbia University Press, 2001).

26: The 1.2 billion years old rocks in Canada that hold evidence of fossils of the oldest known microorganism called *Bangiomorpha* known to reproduce sexually, and look remarkably similar to living red algae (see Butterfield, N.J., 2000. *Bangiomorpha pubescens* n. gen., n. sp.: implications for the evolution of sex, multicellularity, and the Mesoproterozoic/ Neoproterozoic radiation of eukaryotes. Paleobiology, 26, 386-404.). Lynn Margulis provides a delightful and intellectually teasing account on the origin of sex (Lynn Margulis, 'Sex, Death and Kefir', *Scientific American* 271 [1994]: 96).

27: For evidence of the Ediacaran fossils and rock imprints see McKenzie, N.R., Hughes, N.C., Myrow, P.M., Xiao, S., Sharma, M., 2011. Correlation of Precambrian-Cambrian sedimentary successions across northern India and the utility of isotopic signatures of Himalayan lithotectonic zones. Earth and Planetary Science Letters 312, 471-483.

28: The period from 565 to 530 million years ago was critical for the advancement of metazoan body plans and the *Hox* gene played a pivotal role in this. For a summary, see: James W. Valentine et al., 'Developmental Evolution of Metazoan Bodyplans: The Fossil Evidence', *Developmental Biology* 173 (1996): 373–81; and Jeffrey H. Schwartz, 'Homeobox Genes, Fossils, and the Origin of Species', *The Anatomical Record* 257 (1999): 15–31.

29: The idea that 'nature is a tinkerer' was first suggested in François Jacob, 'Evolution and Tinkering', *Science* 196 (1977): 1161-166. For a wonderful book on how design of body plans evolved and the importance of symmetry, see: J. Scott Turner, *The Tinkerer's Accomplice: How Design Emerges from Life Itself* (Cambridge, MA: Harvard University Press, 2010), 32.

30: *Prototaxites* were perhaps the largest terrestrial organisms of their times (see: Francis M. Hueber, 'Rotted Wood-alga-fungus: The History and Life of Prototaxites Dawson 1859', *Review of Palaeobotany and Palynology* 116 (2001): 123–58.). Recent discoveries in India of land fungi like the *Aspidella* and a possible fungus analogue (called *Nimbia*) have confirmed that diverse fungi lived on land during the Ediacara (see: R. Shanker et al., 'Ediacaran biota from the Jarashi [Middle Krol] and Mahi [Lower Krol] Formations, Krol Group, Lesser Himalaya, India', *Journal of the Geological Society of India* 63 [2004]: 649–54]. However, this is not to say that modern fungi cannot be gigantic. One of the largest inhabitants of the United States, and perhaps the world, is a single honey fungus (*Armillaria gallica*) which covers an area of 600 hectare (slightly larger than the size of the island of Gibraltar), weighs more than 100 tonnes and is several thousand years old. There are other exceptional species of *Armillaria* and other fungi as well. Some edible puff balls which grow in temperate forests attain a few feet in girth.

31: Traces and trackways of worms and arthropods have been discovered in the Tal Formation in the Lower Himalayas. *Climactichnites* is a worm from the Early Cambrian Tal Formation of India and this has been recorded by observing thread-like tracks (about 4 millimetres long) that are made on siltstones (see: C. De et al., 'Ichnostratigraphic and paleoenvironmental significance of trace fossils from Tal Formation of Nigali Dhar Syncline, Sirmur District, Himachal Pradesh, India', *Journal of the Geological Society of India* 66 [1994]: 77-90). From Tal, six arthropod and two other worm tracks have been identified, (see: M. Tiwari and S.K. Parcha, 'Early Cambrian trace fossils from the Tal Formation of the Mussoorie Syncline, India', *Current Science* 90 [2006]: 113–

19) and for more recent discoveries, see: Shivani Pandey and Suraj Kumar Parcha, 'Ichnofossil from the Cambrian succession of Parahio Valley, Spiti Basin, India: Their stratigraphic and paleoenvironmental significance', *Geophysical Research Abstracts* 15 (2013): 2013-232.

32: The finest evidence of trackways by early arthropods in India was discovered and studied by Erich Draganits of University of Vienna, Austria. See: Erich Draganits et al., 'A Gondwanan Coastal Arthropod Ichnofauna from the Muth Formation (Lower Devonian, Northern India): Paleoenvironment and Tracemaker Behavior', *Palaios* 16 (2001): 126-47.; and Erich Draganits and N. Noffke, 'Siliciclastic Stromatolites and Other Microbially Induced Sedimentary Structures in an Early Devonian Barrier-Island Environment (Muth Formation, NW Himalayas)', *Journal of Sedimentary Research* 74 (2004): 191-202. Another study reports of trackways made by primitive crabs and other crustaceans in the Talchir formation (see Abhijit Chakraborty and H.N. Bhattacharya, 'Early Permian Xiphosurid Trackways from India', *Journal of the Geological Society of India* 80 (2012): 129-35.).

33: An extremely enjoyable account of life under the primitive sea as seen from the eyes of its most enigmatic inhabitant can be found in Richard A. Fortey, *Trilobite: Eyewitness to Evolution* (New York: Knopf, 2000). A shorter but immensely readable piece is, Fortey, 'The Lifestyles of the Trilobites', *American Scientist* 92 (2005), pp 446-453. For trilobite discoveries from Rajasthan, see: S. Kumar and S.K. Pandey, 'Discovery of trilobite trace fossils from the Nagaur Sandstone, the Marwar Supergroup, Dulmera area, Bikaner district, Rajasthan', *Current Science* 94 (2008): 1081-85; and for the Tethyan Himalayan region, see: Shanchi Peng et al., 'Cambrian trilobites from the Parahio and Zanskar valleys, Indian Himalaya', *Paleontological Society Memoirs (supplement to the Journal of Paleontology), 71: 1-95*

34: The only record of velvet worm from India was reported in Stanley Kemp, 'Preliminary note on a new genus of Onychophora from the N. E. Frontier of India', *Records of the Indian Museum* 9 (1913): 241-42.

35: For a good overview on earthworms and a description of *Drawida grandis* see Clive A. Edwards and P.J. Bohlen, *Biology and Ecology of Earthworms* (London: Chapman & Hall, 1996).

36: For a history of the evolution of plants from the plants to the present, see: Stephen F. Greb et al., 'Evolution and importance of wetlands in earth history', *Geological Society of America Special Papers* 399 (2006): 1-40. Also see: Paul Kenrick and Peter R. Crane, 'The origin and early evolution of plants on land', *Nature* 389 (1997): 33-39. An extremely accessible guide on the evolution of plants is B.S. Venkatachala et al., *Plant Fossils—A Link with the Past* (New Delhi: Publications and Information Directorate, CSIR, 1992). Another interesting account from the physiological perspective of how plants survived on water and land and in varying environments can be found in Patricia G. Gensel, 'The Earliest Land Plants', *Annual Review of Ecology, Evolution, and Systematics* 39 (2008): 459-77.

37: On the origins of cycads, I found a delightful review by Robert Buckley, 'A Brief Review of the fossil cycads', http://www.plantapalm.com/vce/evolution/fossils.pdf.
The newly discovered cycads in the Mandya district of Karnataka are described in Rita Singh and P. Radha, 'A New Species of Cycas (Cycadaceae) from Karnataka, India', *Botanical Journal of the Linnean Society* 158 (2008): 430-35. The relatively recent re-emergence of cycads is described in N.S. Nagalingum et al., 'Recent Synchronous Radiation of a Living Fossil', *Science* 334 (2011): 796-99. On the history of sago and the cycads and palms that yield it, see: George Watt, *A dictionary of the economic products of India: Sabadilla to Silica* (Cambridge Library Collection, digital edition, 2014).

38: Matthew B. Vrazo and Simon J. Brady, 'Testing the "mass-moult-mate" Hypothesis of Eurypterid Palaeoecology', *Palaeogeography, Palaeoclimatology, Palaeoecology* 311 (2011): 63-73, imagines the mating behaviour of early crustaceans.

Notes

3. Fins, Flippers and Feet

1: *Branchiostoma* (which means gill-mouth) was earlier called *amphioxus* (from Greek, 'pointed on both sides') and is considered a contemporary of the earliest vertebrates. For how *Branchiostoma* perceives light, see: Camilo Ferrer et al., 'Dissecting the Determinants of Light Sensitivity in Amphioxus Microvillar Photoreceptors: Possible Evolutionary Implications for Melanopsin Signaling', *Journal of Neuroscience* 32 (2012): 17977–87.

2: Perhaps the most enjoyable and popular account of early Cambrian radiation and the origin of the earliest invertebrates is Stephen Jay Gould, *Wonderful Life: The Burgess Shale and the Nature of History* (New York: W.W. Norton & Company, 1989). The book builds on the pioneering work by Gould and his contemporaries, Simon Morris Conway and Derek Briggs.

3: For a perspective on the processes of how land masses broke up between 770 to 250 million years ago, see Z.X. Li et al., 'Assembly, Configuration, and Break-up History of Rodinia: A Synthesis', *Precambrian Research* 160 (2008): 179–210. However, I favour the palæo-construction of Prof. Christopher Scotese (www.scotese.com).

4: Some of the oldest plant fossils in India are ferns and fern-like plants. See: A.K. Pal and W.G. Chaloner, 'A Lower Carboniferous Lepidodendropsis flora in Kashmir', *Nature* 281 (1979): 295–97.

5: Two books on early vertebrate life on land that present the background on how changing times shaped the body plans of vertebrates and prepared them for life on land are: Michel Laurin, *How Vertebrates Left the Water* (Berkeley: University of California Press, 2010); and Robert Carroll, *The Rise of Amphibians: 365 Million Years of Evolution* (Baltimore: Johns Hopkins University Press, 2009), p. 392. The latter is an excellent overview on tetrapod evolution and life of the amphibians on land. In addition, papers of Jennifer A. Clack at the University of Cambridge proved useful, especially 'Getting a Leg Up on Land', *Scientific American* 293 (2005): 100–7. A thought-provoking article by Gregory J. Retallack, 'Woodland Hypothesis for Devonian Tetrapod Evolution', *The Journal of Geology* 119 (2011): 235–58 presents how limbs and necks evolved for scavenging and hunting in shallow margins of woodlands and oxbow lakes.

6: I was told about flying fish by Dr M.R. Nair, former director of the Central Institute of Fisheries Technology, Kochi. He directed me to G. Pajot, 'Capture of flying fish: Findings of Bay of Bengal Project (BOBP) trials and studies', *Bay of Bengal News* 41 (1991): 4. Also see: S.D. Kamble et al., *Record of three species of flying fish from Mumbai waters*, available at http://eprints.cmfri.org.in/3347/1/8.pdf, as viewed on 14 January 2015.

7: The earliest recorded landfall in India by a vertebrate was at the Mamal formation and occurred in the *Gangamopteris* Beds of Kashmir. The fossils of fish (like *Amblypterus* and *Gardinerichthys*) and amphibians (*Kashmirosaurus*) are reported in S. Woodward, 'Permo-Carboniferous plants and vertebrates from Kashmir. II. Fishes and labyrinthodonts', *Memoirs of the Geological Survey of India—Palaeontologica Indica* 11 (1905): 10-11; and have been updated in A.P. Tewari, 'On a new species of Archegosaurus, from the Lower Gondwana of Kashmir', *Records of the Geological Survey of India* 89 (1962): 427–34. I especially benefitted from reading the analyses of the Permian *Glossopteris* flora of Talchir, Karharbari, Barakar and Lower Kamthi (see: Shreerup Goswami et al., 'Permian biodiversity of Mahanadi Master Basin, Orissa, India and their environmental countenance', *Acta Palaeobotanica* 46 [2006]: 101–18), and the coastal to deltaic environments which led to formation of Gondwanan coal beds of India, and the vegetation and landscape of the Mamal (P.C. Srivastava, 'Glimpses of Permian biodiversity in Mamal bed of Kashmir Himalaya: Floristic Analysis', in *Vistas in Palaeobotany and Plant Morphology: Evolutionary and Environmental Perspectives—Prof. D.D. Pant Memorial Volume*, ed. P.C. Srivastava [Lucknow: UP Offset, 2004], pp. 133–69).

8: The fossil records of the earliest insects discovered from India are those of cockroaches (*Blatta* spp.) and few other flying insects from around 258 to 245 million years ago (see, Thomas Schlüter, 'Fossil insects in Gondwana—localities and palaeodiversity trends', *Acta Zoologica Cracoviensia* 46 [2003]: 345–71; and H.M. Kapoor et al., 'On a fossil cockroach from the Mamal formation, Kashmir Himalayas', *Journal of the Palaeontology Society of India* 38 [1993]: 31–36).

9: There are several theories on how the Permian extinction took place. A detailed account is presented in Michael J. Benton, *When Life Nearly Died: The Greatest Mass Extinction of All Time* (London: Thames & Hudson, 2003). This immensely readable account focuses on the end-Permian extinction and its possible causes. The four most accepted explanations for the greatest extinction of life are: meteor (several possible sites have been proposed), magma (from the Siberian Traps), microbes (the most recent among all theories suggests that micro-organisms produced copious amounts of methane); and methane from hydrate reserves deep in oceans. For methane produced by microbes (called *Methanosarcina*) see, Daniel H. Rothman 'Methanogenic Burst in the End-Permian Carbon Cycle', *Proceedings of the National Academy of Sciences* 111 (2014): 5462–67. For a good review of why the Siberian volcanic event remains an important contender for Permian extinction, see-Andy Saunders and Marc Reichow, 'The Siberian Traps and the End-Permian Mass Extinction: A Critical Review', *Chinese Science Bulletin* 54 (2009): 20–37. It also compares Deccan and Siberian volcanic events and puts their implications in perspective. For the methane hydrate theory see Gregory Ryskin, 'Methane-driven oceanic eruptions and mass extinctions', *Geology* 31 (2003): 741–44. Possible scenarios using combinations of the above causes have been presented in Haijun Song et al., 'Anoxia/high temperature double whammy during the Permian-Triassic marine crisis and its aftermath', *Nature Scientific Reports* 4 (2014): 4132. In this paper the authors propose that because shallow waters became too warm and the depths became oxygen-depleted, life in the seas became severely restricted, leaving only a narrow refuge zone in between where life could survive. On the subject of how life recovered after the Permian aftermath I benefited from reading Paul B. Wignall, 'The End-Permian Crisis, Aftermath and Subsequent Recovery' (paper presented at the international symposium on 'The Origin and Evolution of Natural Diversity', Sapporo, Japan, 1–5 Oct 2007); available in *Proceedings* (2008): 43–48.

10: C_{12} is the naturally occurring isotope of carbon which decomposes from extensive chemical reaction, mostly released by rotting plants and decomposing animal matter, while C_{13} and C_{14} are short-lived, transitional forms of carbon. In nature, most carbon exists as C_{12} (98.9 per cent) and C_{13} (1.1 per cent) and a very tiny amount as C_{14}. Simply put, the ratio of C_{14}:C_{12} helps scientists determine the timing of an event (the process is called carbon dating) and tells them about the ecological and environmental conditions that prevailed at the time. For an introduction on carbon isotopes and methods used for carbon dating, see Tom Higham, Radiocarbon Laboratory, University of Waikato, New Zealand, at http://www.c14dating.com (as viewed on 11 April 2013).

11: For an overview on the Permian and Triassic provinces in India and the stratigraphy of peninsular India, the Greater Himalayas, Sikkim, Arunachal Pradesh, Lesser Himalayas and Tibet, see H. Wopfner and X.C. Jin, 'Pangea Megasequences of Tethyan Gondwana-margin Reflect Global Changes of Climate and Tectonism in Late Palaeozoic and Early Triassic Times—A Review', *Palaeoworld* 18 (2009): 169–92.

12: The method to estimate the vegetation it would require to make the Eastern Indian (or Permian) coal reserves was suggested by Santanu Dey, former director of the Indian School of Mines, Dhanbad. Smaller coal reserves in Maharashtra, Assam and other states, which constitute 15–20 per cent of India's coal reserves, are not included in the estimate.

13: In 1851, the British created the Coal Mission (which later became the Geological Survey of India) to identify large coal deposits and construct ports and inland waterways to send coal to factories in Britain. In 1860, 280,000 tonnes of coal was being extracted, and by 1868, 492,700 tonnes was produced. Since then, coal production in India has roughly doubled every twenty years. According to estimates made by the GSI and Coal India Limited, coal reserves in India to a depth of 1200 meters are estimated to hold 264.54 billion tonnes, and annually India mines for more than 565 million tonnes of coal from its mines (Coal Directory of India [2007–08], Government of India, Ministry of Coal, Coal Controller's Organisation, Kolkata). A brief history on the contribution of India's coal and other mineral wealth during the First and Second world wars, see Joel Mokyr, ed., *The Oxford Encyclopaedia of Economic History, Vol. 2* (Oxford: Oxford University Press, 2011).

14: Although *Glossopteris* emerged on the scene around 290 million years ago, they were dominated by *Gangamopteris*, and their time came during the Permian period (265 to 252 million years ago) and ended when *Dricoidium* species replaced them as the dominant flora from 252 to 201 million years ago. (See: Ashwini K. Srivastava and Rajni Tewari, 'Development of Glossopteris flora in Indian Gondwana sequence', *Geociencias* VI (2001): 42–49.) As a response *Glossopteris* species reduced in size in what is known as the Lilliput effect (Reshmi Chatterjee et al., 'Dwarfism and Lilliput effect: a study on the *Glossopteris* from the late Permian and early Triassic of India', *Current Science* 107 (2014): 1735–44.

15: Reptiles are central to the four classes of terrestrial vertebrates that live today. They evolved from amphibians and were the first true land vertebrates and gave rise to the ancestors of birds and mammals. The first reptiles appeared around 310 million years ago and have existed in some form or another since. Modern reptiles emerged round 70 million years ago, a mere fraction of the many diverse and bizarre varieties that existed before them. The number of species living today (about 6000 or so) is double the number of mammal species. Just as in the case of insects and fish, one type is more prevalent than the other. In the case of reptiles, it is lizards.

16: Discoveries of Triassic reptiles have largely been spearheaded since 1960s by the Indian Statistical Institute (ISI), Kolkata. I present the discoveries and descriptions in the order of possible emergence, as they appear in the chapter. For the first account of *Exaeretodon statisticae* from Maleri, which describes its relationship with similar species from South America, see Sankar Chatterjee, 'A New Cynodont Reptile from the Triassic of India', *Journal of Paleontology*, 56 (1982): 203–14; and Sankar Chatterjee, 'A primitive Parasuchid (phytosaur) reptile from the Upper Triassic Maleri formation of India', *Paleontology* 21 (1978): 83–127. For the first complete description of *Parasuchus hislopi* and an overview with illustrations, see Saswati Bandyopadhyay, 'Gondwana Vertebrates of India', *Proceedings of the Indian National Science Academy* 65 A (1999): 285–313. The Permian tetrapod assemblage from the Kundaram Formation, including specimens of *Endothiodon*, in the Pranhita-Godavari Valley has been described in T.S. Kutty et al., 'Permian reptilian fauna from India', *Nature* 237 (1972): 462–63. Also see S. Ray, 'Small Permian dicynodonts from India', *Palaeontological Research* 5 (2001): 177–91. A description of *Rechnisaurus* and *Wadiasaurus* is provided in Tapan Roy Chowdhury, 'Two new dicynodonts from the Yerrapalli formation of Central India', *Palaeontology* 13 (1970): 132–44. For more exciting fossil discoveries from the Permian and Triassic that are taking place in India, see D.P. Das and Abir Gupta, 'A New Cynodont Record from the Lower Triassic Panchet Formation, Damodar Valley', *Journal Geological Society of India* 79 (2012): 175–80. *Thrinaxadon* has been reported by P.P. Satsangi, 'The vertebrate faunas of the Permian and Lower Triassic sequence of India', in Mesozoic Gondwana Vertebrates; *Geological Survey of India Special Publication* 11 (1987): 165–78. The discovery of the early mammal-like reptile, *Deccanodon*, is reported in T.T. Nath and P.Y. Yadagiri, 'A new mammal-like reptile (*Cynodontia*) from the Upper Triassic Maleri Formation of Pranhita-Godavari Valley, Andhra Pradesh', *Journal of the Geological Society of India* 69 (2007): 57–60.

17: There is extensive literature on the commonalities and differences between the life forms of India and other Gondwana land masses. For a concise review of similarities and relationships between Australian and Indian Permian floral and faunal records, see H.G. Raggatt and Harold O. Fletcher, 'A contribution to the Permian-Upper Carboniferous problem and an analysis of the fauna of the Upper Palaeozoic (Permian) of North-West Basin, Western Australia', *Records of the Australian Museum* 20 (1937): 150–84; and Jose Bonaparte, 'Tetrapod faunas from South America and India: A paleobiogeographic interpretation', *Proceedings of the Indian National Science Academy* 65 A (1999): 427–37.

18: A paper which uses a list of related species of foraminifers, bryozoans, brachiopods, bivalves, gastropods and crinoids from the Vindhyas (Manendragarh and Umaria) and draws correlations to the marine faunas of Western Australia is J.M. Dickins and S.C. Shah, 'Correlation of the Permian marine sequence of India and Western Australia', in *Fourth International Gondwana Symposium, Papers, Vol. II*, ed. B. Laskar and C.S. Raja Rao (Delhi: Hindustan Publishing Corporation, 1979), pp. 387–408.

19: When the Department of Post and Telegraph, Government of India, came out with a commemorative stamp in 1951 for the GSI's centenary, it perhaps got the years of its constitution wrong. The origin of the Survey should be considered as being in 1856, when it was formally constituted, or 1846 when Hiram Williams was appointed India's first geological surveyor. This misconception perhaps arises because the authoritative manual, 'Geology of India' written by Thomas Oldham, who had taken over from Williams in 1851, was published in this year and it was after this that the GSI took root. The germ of the idea of a Survey of India perhaps originated in 1767 when George Rennell, a trained geographer was appointed the surveyor general of Bengal. He surveyed Bengal and its neighbouring lands and made a detailed account of the mineral wealth. The British East India Company was in the quest for quality coal and much of it was imported from England. In 1774, civil servant Suetonius Grant Heatly, then the commissioner of *Chutia* Nagpur (or Chota Nagpur, also see note 9 of Chapter 11) and Purnea, found coal seams in Ramgarh but these were of poor quality and the collusion of coal merchants and East India Company bureaucrats in London prevented any mining enterprise in India from taking off. But England required more and more coal to fuel industrialization in Calcutta, Bombay and Surat. The first domestic coal mine began extraction in 1859 from Raniganj in Bengal. This was a 133 *bigha* concession given by the Rani of Burdwan and the leased mine was named Raniganj by William Blanford in 1863. For the origins of the GSI and the history of coal mining in India, see E.R. Gee, 'History of the Development of the Coal Industry', *Proceedings of the Indian National Science Association* 6 (1940): 305–12; and Deepak Kumar, 'Economic Compulsion and the Geological Survey of India', *Indian Journal of History of Science* 17 (1982): 289–300.

20: S.L. Jain, '*Indocoelanthus robustus* n.gen, n.sp, (Coelocanthidae, Lower Jurassic), The first fossil Coelacanth from India', *Journal of Paleontology* 48 (1974): 49–62.

21: S. Ghosh and A.K. Pati, 'Circadian variation in phototactic behaviour of walking Indian catfish, *Clarias batrachu*,', *Biological Rhythm Research* 35 (2004): 367–75. The walking behaviour of *Polypterus* is described in Benjamin C. Wilhelm et al., '*Polypterus* and the evolution of fish pectoral musculature', *Journal of Anatomy* 226 (2015): 511–22. The discovery of the climbing behaviour of the waterfall-climbing cave fish (*Cryptotora thamicola*) is discussed in Brooke E. Flammang et al., 'Tetrapod-like pelvic girdle in a walking cavefish', *Nature Scientific Reports* 6 (2016): 23711.

22: The Permian period and the extinction that followed is an unsolved puzzle in palæontology. Many experts put their bets on the emission from Siberian volcanoes as the chief cause for the Permian extinction (see: Grzegorz Rackia and Paul B. Wignall, 'Late Permian double-phased mass extinction and volcanism: an oceanographic perspective', in *Understanding Late Devonian and Permian-Triassic biotic and climatic events: Towards an integrated approach*, ed. D.J.

Over et al. [Amsterdam: Elsevier, 2005]). Some scientists have analysed the current levels of extinction to those that occurred at the end of the Permian. One article has compared the fallout from Siberian volcanoes to understand contemporary loss to biodiversity and concludes that 'the end-Permian extinction may serve as an important ancient analog for twenty-first century oceans'. (see: Jonathan L. Payne and Matthew E. Clapham, 'End-Permian Mass Extinction in the Oceans: An Ancient Analog for the Twenty-First Century?' *Annual Review of Earth and Planetary Sciences* 40 [2012]: 89–111).

23: Two papers present an alternative explanation of how magma from below heats up large coal seams which may have led to the massive release of methane and other gases during the Permian. See: Gregory J. Retallack et al., 'Middle-Late Permian mass extinction on land', *Geological Society of America Bulletin* 118 (2006): 1398–1411; and Gregory J. Retallack and A. Hope Jahren, 'Methane Release from Igneous Intrusion of Coal during Late Permian Extinction Events', *The Journal of Geology* 116 (2008): 1–20.

24: Two papers that throw light on how life emerged after glaciations with specific reference to India are: Soumen Sarkar et al., 'Ichnology of a Late Palaeozoic Ice-marginal Shallow Marine Succession: Talchir Formation, Satpura Gondwana Basin, Central India', *Palaeogeography, Palaeoclimatology, Palaeoecology* 283 (2009): 28–45; and Biplab Bhattacharya, 'Sedimentary cycles related to the late Palaeozoic cold-warm climate change, Talchir Formation, Talchir Basin, India', *Earth Resources* 1 (2013): 12–25. Both these papers are easy to read and well-illustrated which help interpret the rocks and sediments in Talchir and related formations.

25: For the position of India and the extent of glacial cover with respect to the rest of the Gondwanan land masses during the Permian, see M.H. Stephenson et al., 'The Early Permian fossil record of Gondwana and its relationship to deglaciation: a review', *Geological Society Special Publication* (2007): 169–89.

26: '. . . we all but neglect the nature and origins of coal—one of the crust's most concentrated and extensive carbon reservoirs. In spite of a once-thriving research community, and coal's continuing economic importance and environmental implications, research on this fascinating substance has all but ceased.' See: Robert M. Hazen and Craig M. Schiffries, 'Why Deep Carbon?', *Reviews in Mineralogy & Geochemistry* 75 (2013): 1–6.

27: Diamonds are neither rare nor intrinsically valuable. Until the discovery of the Brazilian deposits in 1728, all the world's diamond supply came from India (see: Iran F. Machado and Silvia F. de M. Figueiro, '500 years of mining in Brazil: a brief review', *Resources Policy* 27 [2001]: 9–24). India's Golconda mines and, to a lesser extent, the region around the Panna mines in Madhya Pradesh were the only source of diamonds for the world (see: Aja Raden, *Stoned: Jewelry, obsession, and how desire shapes the world* [New York: Ecco/HarperCollins, 2015]). Diamonds lie embedded within kimberlite rocks called pipes in Panna and Golconda. These pipes were formed nearly three billion years ago deep within the mantle.

28: The Vindhyas have a rich variety of rocks and formations. These mountain ranges have contributed exquisite sandstone to build the Agra Fort, Fatehpur Sikri, Humayun's Tomb, Sanchi Stupa, Red Fort, Qutub Minar, and the four lions of the Ashoka Pillar—the national symbol—at Sarnath.

29: The 2-million-year gap in the fossil records (between 269 and 267 million years ago) of the Permian tetrapod globally is discussed in Spencer G. Lucas, 'A global hiatus in the Middle Permian tetrapod fossil record', *Stratigraphy* 1 (2004): 47–64.

4. Revival

1: In addition to the basics of how seafloors spread and land masses move (Oreskes 1999), the formations of the Gondwana and the position of India are described elegantly in Sankar Chatterjee et al.; 'The longest voyage: Tectonic, magmatic, and paleoclimatic evolution of the Indian plate during its northward flight from Gondwana to Asia', *Gondwana Research* 23 (2013): 238-67. For more details on how Gondwana was formed, see J.J. Veevers and R.C. Tewari, 'Gondwana Master Basin of Peninsular India Between Tethys and the Interior of the Gondwanaland Province of Pangea', *Geological Society of America Memoir* 187 (1995): 1-73.

2: The revival of life after the Great Dying took place under very difficult conditions. Scientists from the University of Leeds, the China University of Geosciences and the University of Erlangen-Nurnburg (Germany) analysed oxygen isotopes from the fossil records of survivors of the Permian extinction like snails, clams and conodont (or 'cone-toothed' marine creatures which resembled tiny eels) from South China and found that the oceans had been simmering and devoid of oxygen at that time. O_2 isotope studies can tell us about seawater temperature records and researchers found that the ocean's surface may have reached 40°C (see: Haijun Song et al., 'Anoxia/high temperature double whammy during the Permian-Triassic marine crisis and its aftermath', *Nature Scientific Reports* 4 [2014]: Article no. 4132, available online).

3: The terrestrial flora from these times and how *Dricoidium* took over from *Glossopteris* 252 to 201 million years ago is described in Srivastava and Tewari, 2001; and M.N. Bose and S.C. Srivastava, 'The genus Dicroidium from the Triassic of Nidpur, Madhya Pradesh, India', *Palaeobotanist* 19 [1971]: 41-51. For another good survey of fossils that tell us about the succession of plants after the Permian refer to M.N. Bose and J. Banerji, 'Some fragmentary plant remains from the Lower Triassic of Auranga Valley, district Palamau', *Palaeobotanist* 23 (1976): 139-44. A highly technical but eminently readable book on early gymnospermic plants is Karl Ulrich Kramer and P.S. Green, ed., *The Families and Genera of Vascular Plants Pteridophytes and Gymnosperms* 1 (Berlin: Springer-Verlag GmbH, 2013).

4: For a good review of the early discovery of fish, mammal-like reptiles and reptiles from the Triassic period (252 to 201 million years ago), see: Edwin H. Colbert, 'Relationships of the Triassic Maleri Fauna', *Journal of the Palaentological Society of India* 3 (1958): 68-81.
Scientists from ISI, Kolkata, have yielded a fascinating array of Triassic vertebrates (see: T.S. Kutty and D.P. Sengupta, 'The Late Triassic Formations of the Pranhita-Godavary Valley and their vertebrate faunal succession—a reappraisal', *Indian Journal of Earth Sciences* 16 [1989]: 189-206; and D.P. Sengupta, 'Triassic temnospondyls of the Pranhita-Godavari Basin, India', *Journal of Asian Earth Sciences* 21 [2003]: 655-62). I thank Sankar Chatterjee and T.S. Kutty for sharing valuable insights on their expeditions and discoveries. An exhaustive list of discoveries made of the Triassic period is mentioned in the endnotes of the previous chapter.

5: Papers which describe specific species discovered from the Pranhita-Godavary Valley:
Fossils of three species of Lystrosaurus have been discovered mostly from the Panchet Formation starting with H.C. Dasgupta's discovery in 1922 (H.C. Dasgupta, 'Notes on the Panchet Reptile,' *Annual Magazine of Natural History* 7 [1922]: 935-38) which was followed by discoveries of better-preserved fossils by C. Tripathi and collaborators in 1962-1963, and Pamela Robinson in 1967. A survey of the discoveries of *Lystrosaurus* and other amphibians can be found in S. Bandyopadhyay, 'Gondwana Vertebrate Faunas of India', *PINSA* 65A (1999): 285-313. For *Parasuchus hislopi*, see Sankar Chatterjee, 'A primitive Parasuchid (Phytosaur) reptile from the Upper Triassic Maleri formation of India', *Paleontology* 21 (1978): 83-127. For the first account of *Exaeretodon statisticae* from Maleri and its relationship with similar species from South America, see Sankar Chatterjee, 'A New Cynodont Reptile from the

Triassic of India', *Journal of Paleontology* 56 (1982): 203–14. For *Malerisaurus robinsonae*, I referred to Sankar Chatterjee, 'A primitive parasuchid (phytosaur) reptile from the Upper Triassic Maleri Formation of India', *Palaeontology* 21 (1978): 83–127.

6: For a review of the discoveries of early dinosaurs from the Pranhita-Godavari Basin, see Fernando E. Novas et al., 'Upper Maleri and Lower Dharmaram formations of Central India', *Earth and Environmental Science Transactions of the Royal Society of Edinburgh* 101 (2011): 333–49. The paper suggests that there was a greater diversity of dinosaurs during the Triassic than previously believed.

7: For similarities between early dinosaurs, especially *Eoraptor* and *Alwalkeria*, see K. Remes and O. Rauhut, 'The oldest Indian dinosaur *Alwalkeria maleriensis*. Chatterjee revised: a chimera including remains of a basal saurischian', in *Boletim de Resumos do II Congresso Latino-Americano de Paleontologia de Vertebrados* 12, ed. A.W.A. Kellner, D.D.R. Henriques and T. Rodrigues (Rio de Janeiro: Serie Livros Museu Nacional, 2005), p. 218.

8: Why did dinosaurs do better than large reptiles or amphibians? One possible explanation comes from their posture and body design. Early tetrapods on land faced a serious challenge to breathe during the oxygen crisis. The design of their shoulder and pelvic girdles, and their limbs was such that they had a sprawling gait much like that of most living amphibians and reptiles. This distortion of the chest prevents normal breathing. If the animal is walking, it may be able to breathe between steps, but sprawling vertebrates cannot run and breathe at the same time. Physiologists call this problem Carrier's Constraint, named after American zoologist David R. Carrier of the University of Utah, Salt Lake City. According to Carrier, an erect posture helps animals overcome this challenge, and for this reason some bipedal dinosaurs were quick and agile even when the air became suffocating and made them more efficient in chasing prey. The erect posture in humans is perhaps incapable of out-running many animals in short sprints but in terms of endurance they certainly can outdo most animals (see: D. Carrier, 'Born To Run—Humans can outrun nearly every other animal on the planet over long distances', *Discover Magazine* [2006]: 3; and David R. Carrier et al., 'The Energetic Paradox of Human Running and Hominid Evolution [and Comments and Reply],' *Current Anthropology* 25 [1984]: 483–95).

9: T.T. Nath et al., 'First record of armoured dinosaur from the Lower Jurassic Kota Formation, Pranhita-Godavari Valley, Andhra Pradesh', *Journal of the Geological Society of India* 59 (2002): 575–77. The discovery remains contentious. According to one review, it is probably a terrestrial crocodilian, as indicated by the pitted bone surface (see comments in Oliver W.M. Rauhut and Adriana Lopez-Arbarello, 'Archosaur evolution during the Jurassic: A Southern Perspective', *Revista de la Asociación Geológica Argentina* 63 [2008]: 557–85). This paper also presents a good review of the relationship between archosaurs, the stem group of crocodylomorphs, pterosaurs and dinosaurs from the Jurassic using fossil records from southern Gondwana. Also see F.E. Novas et al., 'New primitive dinosaurs for the Late Triassic of India', *Ameghiniana* 43 (2006): 48R.

10: I benefitted from the discussions on early mammals from India (*Gondwanadon, Tikitherium*, and *Nakulodon)* with professors Ashok Sahni, G.V.K. Prasad of Delhi University, and Emmanuel Gheerbrant of the Natural History Museum in Paris, France.

11: On how plants emerged and the structure of forests in the Gondwana, I benefitted from reading J.M. Anderson et al., 'Patterns of Gondwana plant colonisation and diversification,' *Journal of African Earth Sciences* 28 (1999): 145–67.

12: Two recent studies have found that the density of stomatal pores has reduced by 34 per cent in some tropical species as the carbon dioxide levels have increased in the last 150 years (see: E.I. Lammertsma et al., 'Global CO_2 rise leads to reduced maximum stomatal conductance in Florida vegetation', *Proceedings of the National Academy of Sciences* 108 [2011]: 4035-40; and H.J. de Boer et al., 'Climate forcing due to optimization of maximal leaf conductance in subtropical vegetation under rising CO_2' *Proceedings of the National Academy of Sciences* 108 [2011]: 4041-46).

13: For herbariums close to you where you can try and access leaf specimens from previous decades, see http://sweetgum.nybg.org/ih/herbarium_list.php.

5. The Making of a Dynasty

1: Other than reference 1 of Chapter 4, I also recommend Christopher Robert Scotese, 'Plate Tectonics Driving Mechanisms: Some Simple Rules that Explain Why the Plates Move the Way They Do', Technical Report, January 2015, available at https://www.researchgate.net/profile/Christopher_Scotese3/contributions, for more recent developments in plate tectonics. The idea of how continents moved was proposed in a paper which is now considered a veritable classic in the field: J. Tuzo Wilson 'A new class of faults and their bearing on continental drift', *Nature* 207 (1965): 343-47. For more details on the age and rates at which the continents have moved, see R. Dietmar Müller et al., 'Age, spreading rates, and spreading asymmetry of the world's ocean crust', *Geochemistry, Geophysics, Geosystems* 9 (2008): 36.

2: The basement rock of southern Florida is part of Gondwana (see: Paul A. Mueller et al., 'Precambrian zircons from the Florida basement: A Gondwanan connection', *Geological Society of America Bulletin* 22 [1994]: 119-22).

3: On oxygen levels and how it has impacted evolution and size of life forms, see Ian J. Glasspool and Andrew C. Scott, 'Phanerozoic concentrations of atmospheric oxygen reconstructed from sedimentary charcoal', *Nature Geoscience* 3 (2010): 627-30.

4: Until recently, scant vertebral discs and teeth remains long assumed to belong to an ichthyosaurus have been found in the Karai Formation of Cauvery Basin, the most recent discovery was 2011, about twenty kilometres west of Ariyalur in Tamil Nadu. These marine reptiles would have lived in shallow seas which were created as India, Australia and Antarctica began separating from one another around 137-132 million years ago (C.J. Underwood et al., 'Marine Vertebrates from the 'Middle' Cretaceous [Early Cenomanian] of South India', *Journal of Vertebrate Paleontology* 31 [2011]: 539-52). In January 2015, a team led by D.K. Pandey, then professor of palæontology at Rajasthan University, Japiur, found a couple of looser fossil bones on the surface and an in-situ rib. The site was later excavated and the skeleton was diagnosed by G.V.R. Prasad of Delhi University as an ichthyosaur (G.V.R. Prasad et al (2017) 'Discovery of the first ichthyosaur from the Jurassic of India: Implications for Gondwanan palaeobiogeography'. PLoS ONE 12(10): e0185851). The fossils found in the Karai Formation are now revised and reassigned to another marine reptile (*Platypterygius*) and are disccused in G.V.R. Prasad (2017). For an enjoyable account of life in the Jurassic seas I recommend the article by Ryosuke Motani, 'Rulers of the Jurassic Sea', *Scientific American* 283 (2001): 52-59.

5: I am grateful to Dr T.S. Kutty for his patience and the intimate details of the expeditions, discoveries and contributions of the scientists of ISI, Kolkata, since 1960, and the stratigraphy and formations of the Pranhita-Godavari Valley. I benefitted from reading his paper (T.S. Kutty et al., 'Basal Sauropodomorphs [Dinosauria: Saurischia] from the Lower Jurassic of India: Their Anatomy and Relationships', *Journal of Paleontology* 81 [2007]: 1218-40) which describes the structure and morphology of *Lamplughsaura dharmaramensis* and *Pradhania gracilis* which were discovered from the Upper Dharmaram Formation.

6: The first pterosaur fossil to be found in India was in a village near Sirolkhal near Shahbad in Kota district of Rajasthan. Some pterosaur experts believe that the fossil remains found were too scanty to attribute them to the flying reptile (see: V.S. Dubey and K. Narain, 'A note on the occurrence of Pterosauria in India', *Current Science* 15 [1946]: 287–88). The only flying reptile discovered in India was a pterosaur called *Campylognathoides*, whose fossil was found in Chandrapur district (called Chanda at the time of discovery) of Maharashtra (see: C.N. Rao and S.C. Shah, 'On the occurrence of Pterosaur from the Kota-Maleri beds of Chanda district, Maharashtra,' *Records of the Geological Survey of India* 92 [1963]: 315–18; and S.L. Jain, 'Jurassic Pterosaur from India', *Journal of the Geological Society of India* 15 [1974]: 330–35). I also benefitted from reading about the behaviour of *Campylognathus* in Mark P. Witton, '*Pterosaurs: Natural History, Evolution, Anatomy*' (New Jersey: Princeton University Press, 2013).

7: The discovery of *Barapasaurus* by Prof. Sohan Lal Jain and the team from ISI was considered a landmark and remains one of the few near-complete skeletal dinosaur discoveries made in India (see: Sohan Lal Jain et al., 'The sauropod dinosaur from the Lower Jurassic Kota Formation of India', *Proceedings of the Royal Society of London A* 188 [1975]: 221–28).

8: The first description of *Kotasaurus* was by its discoverers (P. Yadagiri, 'A new sauropod Kotasaurus yamanpalliensis from Lower Jurassic Kota Formation of India', *Journal of the Geological Society of India* 11 [1988]: 102–27; and P. Yadagiri, 'The osteology of Kotasaurus yamanpalliensis, a sauropod dinosaur from the Early Jurassic Kota Formation of India', *Journal of Vertebrate Paleontology* 21 [2001]: 242–52), but its relationship with other Gondwanan dinosaurs is described by D.D. Gillette, 'The geographic and phylogenetic position of sauropod dinosaurs from the Kota formation (Early Jurassic) of India', *Journal of Asian Earth Sciences* 21 (2003): 683–89.

9: Indian scientists have claimed that unlike the rest of the world, stegosaurs appeared early and died later in India. *Dravidosaurus*, the last-known stegosaurus from India, survived until the late Cretaceous. P. Yadagiri and K. Ayyasami, 'A new stegosaurian dinosaur from Upper Cretaceous sediments of south India', *Journal of the Geologicial Society of India* 20(1979): 521–30 described the *Dravidosaurus*, based on the eye-socket fragment discovered. This has been refuted by some other scientists.

10: The three major dinosaur lineages (theropods, sauropodomorphs and ornithischians) emerged in the supercontinent Pangæa during the Triassic (252 to 201 million years ago), when much of the land was hot and arid. See: Stephen L. Brusatte et al., 'The origin and early radiation of dinosaurs', *Earth Science Reviews* 101 (2010): 68–100. The authors argue that the rise of dinosaurs was due to opportunism which helped them edge out crocodiles and other reptiles.

11: The physiology, especially digestion of food of the giant herbivores, is described simply in Ashok Sahni, *Dinosaurs of India* (New Delhi: National Book Trust, 2001).

12: The geology of the Kutch region and its islands Pachcham, Khadir, Bela and Chorar is described in W.T. Blanford, 'On the Geology of a portion of Cutch', *Memoirs of the Geological Survey of India* 4 (1867): 17–38. Also see: N. Khonde et al., 'Environmental significance of raised Rann sediments along the margins of Khadir, Bhanjada and Kuar Bet islands in Great Rann of Kachchh, Western India', *Current Science* 101 (2011): 1429–34.

13: Tree ferns are 'molecular living fossils' that evolved around 240 million years ago and expanded their diversity from 200 to 165 million years, but they gradually declined after the emergence of the flowering plants, as did most gymnosperms which were relinquished to the margins. Only one group of ferns (the *leptosporangiates*) made an attempt by changing their strategy to survive. They evolved from being free-living to living on the crown and trunks of trees as epiphytes (an epiphyte is a plant that grows harmlessly on another plant and seeks moisture, nutrients, sunlight and protection, and sometimes feeds on organic debris that accumulates in trunk holes and bark). Although leptosporangiate ferns originated nearly 360 million years ago or about 200 million years before the rise of flowering plants, they adapted better than other ferns because they had evolved a photoreceptor that helped them orient their leaves towards sunlight, something that other ferns could not do. The radiation of new species of leptosporangiate ferns occurred in two bursts, once at the time when flowering plants were increasing their dominance (around 150 million years ago) and the second time when the major Cretaceous extinction (post Deccan-Cretaceous events) was followed by revival around 60 million years ago, the last one establishing them as the most abundant of their kind—nearly 80 per cent of the ~11,000 non-flowering vascular plant species (see: Eric Schuettpelz and Kathleen M. Pryer, 'Evidence for a Cenozoic radiation of ferns in an angiosperm-dominated canopy', *Proceedings of the National Academy of Sciences* 106 [2009]: 11200-05).

14: The earliest angiosperm traits such as small flowers, quick life cycle and the ability to colonize rapidly would have given these plants a competitive advantage in disturbed sites, thereby promoting their radiation. The major expansion of angiosperms started with their emergence 140 million years ago and then rapidly grew between 112 and 99 million years ago (the time of divergence has been refined in David Dilcher, 'Toward a new synthesis: Major evolutionary trends in the angiosperm fossil record', *Proceedings of the National Academy of Sciences* 97 [2000]: 7030-36) to about 340 taxa before declining around 89-85 million years ago and then rising to over 470 taxa by 60 million years ago (see: J.M. Beaulieu, 'Heterogeneous rates of molecular evolution and diversification could explain the Triassic age estimate for angiosperms', *Systematic Biology* 64 [2015]: 869-78). More recent evidence of the evolution in angiosperms comes from new fossils discovered in the Jehol Formation of China (see: Wei Wang et al., 'Accelerated evolution of early angiosperms: Evidence from ranunculalean phylogeny by integrating living and fossil data', *Journal of Systematics and Evolution* 54 [2014]: 336-41). An earlier study reports that the flowering plant crown group originated 217 million years ago (see: Stephen A. Smith et al., 'An uncorrelated relaxed clock analysis suggests an earlier origin for flowering plants', *Proceedings of the National Academy of Sciences* 107 [2010]: 5897-902).

Molecular studies have estimated the time of diversification of early flowering plants (*Mesangiospermae*) during the mid-to-late Jurassic and have correlated these to the corresponding evolution of pollinating insects which triggered the angiosperm radiation. See: Liping Zeng et al., 'Resolution of deep angiosperm phylogeny using conserved nuclear genes and estimates of early divergence times', *Nature Communications* 5 (2014): 4956. Also, N. Wikstrom et al., 'Evolution of the Angiosperms: Calibrating the Family Tree', *Proceedings of the Royal Society B: Biological Sciences* 268 (2001): 2211-20.

15: The earliest reports of footprints were from Anjar near Bhuj in Gujarat (see: Z. G. Ghevariya and C. Srikarni, 'Anjar Formation, its fossils and their bearing on the extinction of Dinosaurs', *IGCP Chandigarh Abstract book* [1990]: 106-09). There are few other reports of other dinosaur footprints chiefly from Bhuj area by Ghevariya and others. Footprints of bipedal, three-toed theropods and ornithopods (without claws) on compact calcareous sandstone and shale were found on the Katrol Formation in Tharauda, north-east of Ratanpur, Bhuj; and a trackway of quadruped adults, juveniles and young of a sauropod herd on hard, compact, current-bedded, burrowed quartzitic sandstone near Fatehgarh in Bhuj. The most recent discovery of footprints was made in Jaisalmer district (see: Grzegorz Pieńkowski et al., 'Dinosaur footprints from the Thaiat ridge and their palaeoenvironmental background, Jaisalmer Basin, Rajastan, India', *Volumina Jurassica* XIII [2015]: 17-26).

6. Beasts and Behemoths

1: The introduction to the landforms of India from 190 to 65 million years is described in K.S. Valdiya, *The Making of India* (2010).

Notes

2: P.S. Ranawat, 'Can Habur Limestone Curdle Milk?', *Current Science* 89 (2005): 729-30. The paper describes the geologic and chemical properties of habur limestone.

3: Jurassic fish fossils, foraminifera and marine molluscs discovered in India and the rest of Gondwana are a good way to tell which parts of India were covered by fresh or marine water, whether shallow or deep. For a good compilation of fossil records. see Adriana Lopez-Arbarello et al., 'Jurassic Fishes of Gondwana', *Revista de la Asociación Geológica Argentina* 63 (2008): 586-612. The other evidence that Narmada and Tapti were seaways comes from foraminifera records and sediments deposit studies. See: Gerta Keller *et al.*, 'Early Danian Planktic Foraminifera from Intertrappean Beds at Jhilmili, Chhindwara District, Madhya Pradesh, India', *The Journal of Foraminiferal Research*, 39 (2009): 40-55; and Sunil Bajpai, 'Biotic Perspective of the Deccan Volcanism and India-Asia Collision: Recent Advances', *Current Trends in Science* (2009): 505-516.

4: In intermittently warm Jurassic and Cretaceous seas, small shelled animals called coccolithophorids appear, and as they died began to create deposits of chalk during mid to late Cretaceous. Although Makrana and other marble from south of the Aravalli is older (about 1.2 billion years old), other deposits belong to the Cretaceous period. For a list of different types of marble found in Rajasthan, see, '50 Years of Marble Development in Rajasthan', Department of Mining and Geology, Government of Rajasthan, available at: "http://www.dmg-raj.org/docs/vol.%2021-4-b.doc"www.dmg-raj.org/docs/vol.%2021-4-b.doc.

5: The reasons for the rise and fall of oxygen levels and their impact on life are described in Peter Ward and Joe Kirschvik (2014).

6: On how climate spells influenced soils, especially the emergence of iron and aluminium rich laterites across Gondwana, particularly India, see, Y. Tardy, *Petrology of Laterites and Tropical Soils* (Rotterdam: A.A. Balkema, 1997) p. 419.

7: An excellent overview of the dinosaur families and their evolution is given by David E. Fastovsky and David B. Weishampel, *Dinosaurs* (2009), which also has excellent line drawings by John Sibbick. For Indian dinosaurs, see Sahni, *Dinosaurs of India* (2016).

8: Lameta is the most prolific formation to reveal Cretaceous fossils and the exposures and geologic details are provided in S. Kumar et al., *Lameta and Bagh Beds, Central India—Field Guide* (Lucknow: The Palaeontological Society of India, 1999).

9: A synthesis of titanosaur discoveries was presented by S.L. Jain and S. Bandyopadhyay, 'New Titanosaurid (Dinosauria: Sauropoda) from the Late Cretaceous of Central India', *Journal of Vertebrate Paleontology* 17 (1997): 114-36.

10: For details of the discovery and taxonomic position of *Indosaurus, Laevisuchus* and *Indosuchus*, see, Fernando E. Novas, Federico L. Agnolin and Saswati Bandyopadhya, 'Cretaceous Theropods from India: A Review of Specimens Described by Huene and Matley (1933)', *Revista del Museo Argentino de Ciencias Naturales* 6 (2004): 67-103. This paper presented a review of *Indosuchus, Indosaurus* and *Laevisuchus* and their possible relationship with other Gondwanan theropods. Earlier, F.V. Huene and C.A. Matley, 'Cretaceous Saurischia and Ornithischia of the Central Provinces of India', *Palaeontologia Indica* 21 (1933): 1-74, had described undesignated fossil remains that fit with these species.

11: R.S. Loyal, A. Khosla, and A. Sahni, 'Gondwanan Dinosaurs of India: Affinities and Palaeobiogeography', *Memoirs of the Queensland Museum* 39 (1996): 627-38.

12: Discoveries of eggs, especially those of titanosaurs, have been described simply in Sahni, *Dinosaurs of India* (2016). Eggs of small predatory dinosaurs vary from long and elongate to teardrop-shaped, while sauropod eggs (the most common ones found in India) are near spherical. Even the sizes of dinosaur eggs differ significantly. The smallest dinosaur eggs are just a few centimetres across (not in India but in Utah, USA, and China) and the largest reported ones, those of sauropods, are as large as 40 centimetres across. Recently an enormous 70-centimetre long, sausage-shaped egg from a giant oviraptor from China was discovered. Dinosaur eggshells were made of calcite with microscopic pores which allowed the developing embryo within to 'breathe', much like bird eggs today. But despite these enormous sizes, the largest dinosaur eggs were small compared to the mothers who laid them. The size of the egg is determined by the physiology of calcium and diffusion of gases. Had eggs been any larger, the eggshell would need to be thicker and harder. This would make it difficult for air to diffuse into and out of the shell. A thicker shell would also make it difficult for the feeble hatchling to break through it.

13: The rediscovery of *Rajasaurus narmadensis* can be found in Jeffrey A. Wilson et al., 'A New Abelisaurid (Dinosauria, Theropoda) from the Lameta Formation (Cretaceous, Maastrichtian) of India', *University of Michigan Papers* 31 (2003): 1-42. For a complete record of major dinosaur discoveries in India including the *Rajasaurus* and its relationship with its cousins (*Majungatholus* of Madagascar and *Carnotaurus* of South America), see Jeffrey A. Wilson et al., 'A New Abelisaurid (Dinosauria, Theropoda) from the Lameta Formation (Cretaceous, Maastrichtian) of India', *University of Michigan Papers* 31 (2003): 1-42. I am grateful to Dr Jeffery Wilson and Prof. Ashok Sahni for sharing their insights.

14: The giant snake *Sanajeh*, its discovery and relationships with other fossil discoveries have been presented in Jeffrey A. Wilson et al., 'Predation upon Hatchling Dinosaurs by a New Snake from the Late Cretaceous of India', *PLoS Biology* 8 (2010): 1-5. I am grateful to Prof. J.C. Rage, the foremost authority on extinct snakes at the Natural History Museum in Paris, France for providing me with inputs.

15: The discovery of *Bruhathkayosaurus* fossils were first presented in conference abstracts and presentations (P. Yadagiri and K. Ayyasami, 'A Carnosaurian Dinosaur from the Kallamedu Formation [Maestrichtian Horizon], Tamil Nadu', in Symposium on Three Decades of Development in Palaeontology and Stratigraphy in India, Vol. 1. Precambrian to Mesozoic, ed. M.V.A. Sastry et al. *Geological Society of India Special Publication* 11 [1989]: 523-528) but these were challenged, and the lack of fossils has made these claims contentious.

16: On the evolution of dinosaurs see, Max Langer et al., 'The Origin and Early Evolution of Dinosaurs', *Biological Reviews* 85 (2010): 55-110.; and Michael J. Benton et al., 'Models for the Rise of the Dinosaurs', *Current Biology* 24 (2014): R87-R95. For an excellent account of processes that cause dwarfing when animals are isolated on islands see, Michael J. Benton et al., 'Dinosaurs and the Island Rule: The Dwarfed Dinosaurs from Hateg Island', *Palaeogeography, Palaeoclimatology, Palaeoecology* 293 (2010): 438-454.

17: The dinosaur lineage which evolved into birds shrank in body size continuously for 20 million years, starting from the neotheropod 220 million years ago which began to shrink in size and became more agile, and after successive transitions took flight as the archaeopteryx, the first true flying reptile about 150 million years ago. The amazing flexibility of the dinosaur body plan, which adopted more shapes, styles and sizes than just about any other group of animals, also suggests that miniaturization was an essential prerequisite for flight. This adaptability and fast rate of evolutionary change allowed birds to survive the mass extinction event that killed off the dinosaurs 66 million years ago. Birds were better prepared for cataclysms: they could regulate their body temperature and stay warm in the post-impact winters (something like a nuclear winter in its effect); they had the ability to live in trees or underground; they could also glide or fly and escape from wildfires, and search for food. Because they were small in size they could survive on much less food than their dinosaurian cousins. See, Michael J. Benton, 'How Birds Became Birds', *Science* 345 (2014): 508–9; and for theories on the evolution of birds, see, Luis M. Chiappe, 'Downsized Dinosaurs: The Evolutionary Transition to Modern Birds', *Evo Edu Outreach* 2 (2009): 248–56. Interestingly some scientists believe that crocodiles are the closest relatives of the birds, and a few genetic studies show that crocs may be closer to birds than, say, lizards, tortoises and snakes, although taxonomists group them together and the birds on a different branch of the tree of life. Should this debate conclude that birds are indeed closer to crocodiles, the tree of life would have a major shake-up! (C. Sibley and J. Ahlquist, *Phylogeny and Classification: A Study in Molecular Evolution* [New Haven: Yale University Press, 1990].)

18: Vandana Prasad et al., 'Dinosaur Coprolites and the Early Evolution of Grasses and Grazers', *Science* 310 (2005): 1177. This breakthrough paper pushed back the dates of the origin of grasses (between 88 and 55 million years ago) and perhaps their place of origin (Greater India). A second paper (V. Prasad et al., 'Late Cretaceous *Origin* of the *Rice Tribe* Provides Evidence for Early Diversification in Poaceae', *Nature Communication* 2 [2011]: 480.) presents evidence that the family of rice is older than previously believed.

19: On the discovery of the plant-eating land crocodile (*Simosuchus*) which was found first in China (see: X.C. Wu and A. Sun, 'A Plant-eating Crocodyliform Reptile from the Cretaceous of China', *Nature* 376 [1995]: 678–80) and has been found in Tamil Nadu which finds a description in David W. Krause and Nathan J. Kley, '*Simosuchus clarki* (Crocodyliformes: Notosurchia) from the Late Cretaceous of Madagascar', *Society of Vertebrate Paleontology Memoir* 10: 1–236; and the supplement to *Journal of Vertebrate Paleontology* 30. The evolution of the family of gavial (or the gharials) is discussed in Christopher A. Brochu, 'Morphology, Fossils, Divergence Timing, and the Phylogenetic Relationships of Gavialis', *Systematic Biology*, 46 (1997): 479–522. That crocodiles can thrive on vegetarian diets is reported in S.G. Platt et al., 'Frugivory and Seed Dispersal by Crocodilians: An Overlooked Form of Saurochory?', *Journal of Zoology* (2013).

20: The Cretaceous were truly the most interesting times in evolution of life forms. For the first time, life on land became more diverse (only about 110 million years ago) compared to the seas. See, J. Vermeij Geerat and Richard K. Grosberg, 'The Great Divergence: When Did Diversity on Land Exceed That in the Sea?', *Integrative and Comparative Biology* 50: 675–82. For the plant perspective on the evolution of new families before and after Cretaceous period, see, J. John Sepkoski Jr, 'Biodiversity: Past, Present, and Future', *Journal of Paleontology*, 71 (1997): 533–39.

21: Molecular studies have estimated the time of diversification of angiosperms (*Mesangiospermae*) during the mid to late Jurassic and have correlated these to the corresponding evolution of pollinating insects which triggered the angiosperm radiation. See, Liping Zeng et al., 'Resolution of Deep Angiosperm Phylogeny Using Conserved Nuclear Genes and Estimates of Early Divergence Times', *Nature Communications* 5 (2014): 4956.

7. Isolation

1: For the early rejection of continental drift theory and an incisive account on why US scientists denied Alfred Wegener and others of their rightful place in the annals of geological sciences, see, N. Oreskes, *The Rejection of Continental Drift* (1999). On the brief history of Eduard Suess, an Austrian geologist (1831–1914) and the origins of the word Gondwana, see, John Gribbin, *Science* (2010). Simply put, magma generated at depth breaks through the crust and creates massive outpourings, and the place around the crust where this occurs is called a hotspot. Magma rises in these long columns (called mantle plumes) from the depths. As it rises, it faces resistance from a rigid lithosphere, but as the pressure continues to build from below, it causes the crust to lift up and the brittle rocks to rupture, causing an outflow of lava. A plume that starts to rise from below the seafloor has the potential to create a large plateau of flood basalt under the sea. It can also push up a thin chain of volcanic island forms above the tail of the plume along the path in which the plate has moved. The convection of magma that rises from deep below the continent has the power to tear apart a continent into smaller land masses and pack them off like flotsam in flood. New seas and oceans begin to open within a supercontinent, and the smaller land masses are then despatched at the speed of your growing fingernails. After the magma supply is drained, the tail narrows and the hotspot die out. Hotspots can take millions of years to rise up and sustain volcanic activity but a few may be short-lived and can die out within a few thousand or, perhaps, even a hundred years. Basalts made from magma plumes differ from where they have erupted because each basalt is exuded is partially derived from rocks that are taken deep into the mantle and re-melted and ejected again. It is the signature in these vestiges which help us to determine how and when India was connected to other Gondwanan land masses.

2: The tectonic evolution of the Indian plate, which started in the late Jurassic, about 167 million years ago, with the breakup of Gondwana, the timing of separation, the events that caused it and the relative position of land masses is eloquently presented in Chatterjee, Goswami and Scotese, 'The longest voyage' (2015).

3: The timing and separation of the East Coast of India with Antarctica is described in Ana D. Gibbons, 'The Breakup of East Gondwana: Assimilating Constraints from Cretaceous Ocean Basins around India into a Best-fit Tectonic Model', *Journal of Geophysical Research: Solid Earth*, 118 (2014): 808–22. The mid-ocean ridge that separated Greater India from Australia-Antarctica started from the north to south, began around ~136 Ma northwest of Australia, and reached the southern tip of India at ~126 Ma. Another paper (Arundhuti Ghatak and Asish R. Basu, 'Vestiges of the Kerguelen Plume in the Sylhet Traps, Northeastern India', *Earth and Planetary Science Letters* 308 [2011]: 52–64) put a range of dates to rock samples from the Sylhet and Rajmahal Traps and therefore the timing of the events. For the separation and the processes of seafloor spreading that led to the formation of the Indian Ocean, see, J.G. Watson et al., 'A Combined Rigid/Deformable Plate Tectonic Model for the Evolution of the Indian Ocean', (paper presented at the European Geosciences Union General Assembly, Vienna, Austria, 22–27 Apr 2012); published in *Geophysical Research Abstracts* 14.

4: An excellent description and illustration of the basalt columns of the Rajmahal Hills from around Sahibganj is available in R.D. Oldham, *A Manual of Geology of India* (Calcutta: Geological Survey of India, 1892), p. 175.

5: I am grateful to Prof. Ashok Sahni for sharing with me the interesting story of the Katthghar 'rice'.

6: For the origins of carnivorous plants, see, Adrian Slack and Jane Gate, *Carnivorous Plants* (Cambridge, MA: MIT Press, 2000). A short review of carnivorous plants in India has been done in K. Natarajan et al., 'A Note on the Identity of Carnivorous Plants of Karungalakudi, Tamil Nadu, India', *Ethnobotanical Leaflets* 12 (2008): 1073–77, and references therein.

7: The post-Kerguelen events and the orientation and migration of Greater India are described by Prakash Kumar et al., 'The Rapid Drift of the Indian Tectonic Plate', *Nature* 449 (2007): 894–97. The authors present the rates of separation of India with its neighbours caused by the Marion, Kerguelen and Reunion plumes and infer that the underside of the Indian plate was possibly melted, permitting faster motion. Another paper (K.R. Subrahmanya, 'Post-Gondwana Tectonics of the Indian Peninsula', *Current Science* 67 [1994]: 527–30) explains the tectonic process that caused the eastern and northern tilt of the Indian peninsula. The distance India has travelled has been calculated by several scientists (e.g., Scotese et al., 2015). The magma-driven process is described by Douwe J. J. van Hinsbergen, 'Acceleration and Deceleration of India-Asia Convergence since the Cretaceous: Roles of Mantle Plumes and

Continental Collision', *Journal of Geophysical Research* 116 (2011): B06101.

8: For the process of separation of Madagascar, India and Seychelles; the Marion event; and processes which led to the creation of St Mary's Island, see: T.H. Torsvik, 'Late Cretaceous India-Madagascar Fit and Timing of Break-up Related Magmatism', *Terra Nova* 12 (2000): 220-25; and S.K. Bhushan et al., 'Petrography and Geochemistry of St. Mary Islands, Near Malpe, Dakshina Kannada District, Karnataka', *Journal Geological Society of India* 76 (2010): 155-63. An interesting study links magmatic structures found in western Karnataka with the Marion event and relates it closely with rocks of St Mary's Islands, see, Anil Kumar et al., 'The Karnataka Late Cretaceous Dykes as Products of the Marion Hotspot at the Madagascar—India Breakup Event—Evidence from ^{40}Ar-^{39}Ar Geochronology and Geochemistry', *Geophysical Research Letters* 28 (2001): 2715-18.

9: A paper which presents the geographic evidence of how eastern Madagascar and the western margin of India were once part of a continuous land mass especially through the aerial views of the Palghat (Palakkad) and Ranotsara topographic gaps is, Yanni Gunnell and David Harbor, 'Structural Underprint and Tectonic Overprint in the Angavo (Madagascar) and Western Ghats (India)—Implications for Understanding Scarp Evolution at Passive Margins', *Journal Geological Society Of India* 71 (2008): 763-79.

10: Volcanic activity continued in Madagascar along its eastern and southern tip well after India split from it. Madagascar's largest volcano perhaps was Volcan de l'Androy which remained active from 86 to 78 million years ago. See, J.J. Mahoney *et al.*, 'Geochemistry of the Volcan de l'Androy Basalt—Rhyolite Complex, Madagascar Cretaceous Igneous Province', *Journal of Petrology* 49 (2008): 1069-96.

11: The Indian purple frog was described in S.D. Biju and Franky Bossuyt, 'New Frog Family from India Reveals an Ancient Biogeographical Link with the Seychelles', *Nature* 425 (2003). The relationship of frogs from the Indian Ocean region and possible modes of their dispersion are described in Franky Bossuyt and Michel C. Milinkovitch, 'Amphibians as Indicators of Early Tertiary "Out-of-India" Dispersal of Vertebrates', *Science* 292 (2001): 93-5; and Franky Bossuyt, 'Local Endemism within the Western Ghats-Sri Lanka Biodiversity Hotspot', *Science* 306 (2004): 479-81. The latter explains why similarities and uniqueness exist in the Western Ghats and Sri Lanka.

12: To see how corundum, rubies, sapphires and chrysoberyls are related to the Gondwanan Archean rocks found in India, Sri Lanka, Madagascar and Africa, see, Gaston Giuliani et al., 'Oxygen Isotope Systematics of Gem Corundum Deposits in Madagascar—Relevance for Their Geological Origin', *Miner Deposita* 42 (2007): 251-270; C.B. Dissanayake and Rohana Chandrajith, 'Sri Lanka-Madagascar Gondwana Linkage: Evidence for a Pan-African Mineral Belt', *Journal of Geology* 107 (1999): 223-35, and M. Santosh and A.S. Collins, 'Gemstone Mineralization in the Palghat-Cauvery Shear Zone System (Kakur-Kangayam Belt), Southern India', *Gondwana Research* 6 (2003): 911-18.

13: A study which charted the relationship between different cichlid species and calibrates this to determine the timing of the breakup of Gondwanan land masses is, Prosanta Chakrabarty, 'Cichlid Biogeography: Comment and Review,' *Fish and Fisheries* 5 (2004): 97-119.

14: Green chromide (*Etroplus suratensis*, or the pearlspot) is also popularly called *karimeen* in Malayalam. *Etroplus* species found in India are described in http://www.fishbase.org/home.htm (as viewed on 7 January 2011).

15: The exciting story of the discovery of the coelacanth and its history are expertly described in Samantha Weinberg, *Fish Caught in Time: The Search for the Coelacanth* (New York: Harper

Perennial, 2001). Fossil of coelacanth found in India as described in S.L. Jain, 'The first fossil Coelacanth from India' (1974).

16: Currently evolutionists are debating how life forms migrated during the various stages of breaking up between Gondwanan land masses. One of the first persons to put thought to the similarities between Indian and Madagascan life forms was India's leading palæobotanist of the time, Birbal Sahni (see: Sanjay Chandra, 'Birbal Sahni and India-Madagascar Fit', *Indian Journal of History of Science* 28 [1993]: 67–9) when he made early attempts to model how the two land masses may have split using plant fossils. A. Datta-Roy and K.P. Karanth, 'The Out-of-India Hypothesis: What Do Molecules Suggest?' *Journal of Biosciences* 34 (2009): 687–97, propose three possible modes of dispersion after studying plants, snails, fish, frogs, reptiles and flightless birds from India and Madagascar. Datta-Roy and Karanth propose that the Indian ark carried within it Gondwanan forms to Eurasia after it broke up with Madagascar and Antarctica. Although many Asian life forms may have reached via India, not all Gondwanan life forms dispersed out of India. Some life forms entered India instead of leaving India and then distributed over the rest of Asia. Datta-Roy and Karanth caution (as do other biogeographers) that there may be no one mode of dispersal and that the existence of species must be understood in the complex biogeographical context of India and its Gondwanan cousins. Amphibians like frogs are an excellent subject to study the timing and nature of dispersal (see: Franky Bossuyt et al., 'Phylogeny and Biogeography of a Cosmopolitan Frog Radiation: Late Cretaceous Diversification Resulted in Continent-scale Endemism in the Family', *Ranidae' Systematic Biology* 55 [2006]: 579–94) which suggests that ancestors of many contemporary frogs perhaps evolved in Madagascar-India and later in the Indian Ocean islands, and subsequently spread to Eurasia and Australia-New Guinea after India joined Eurasia. Studies on mammals (like rodents), however, suggest a different mode of dispersal. Rodents originated in Eurasia and entered Madagascar only after they reached India, and not from Africa as it was commonly believed. A second invasion of rodents from Madagascar into Africa took place, but there were no mammals or terrestrial vertebrates (except birds) that reached Madagascar from Africa. Rodents prove that other mammals could have also reached Africa via Madagascar from India—a reverse of what is seen in fish and frogs (see: Sharon A. Jansa et al., 'Molecular Phylogeny and Biogeography of the Native Rodents of Madagascar [Muridae: Nesomyinae]: A Test of the Single-origin Hypothesis', *Cladistics* 15 [1999]: 253–70). Currently 'out-of-India' migration is confirmed in ranid frogs, ratite birds, fish (arowanas, channids and cichlids). Among fish, this is particularly interesting. Tropical fish in Asia, Americas, Australia and Africa have complicated our understanding of models of dispersion and how they evolved in isolation. Knifefish found in western Africa and South and south-eastern Asia offer another contrasting perspective on dispersion and evolution of fish. Genetic research has confirmed that the Asian knifefish are of African origin. Researchers have proposed three possible models of how they dispersed and got isolated in West Africa and Southeast Asia. First, the fish got isolated in the early Cretaceous (157 million years ago) and had reached Eurasia, and after India docked with Southeast Asia, the knifefish found their way into South Asia (much as the frogs did). The second, more accepted model is that about 133 million years ago the knifefish were present in Africa, India and Madagascar and reached South East Asia when India carried these along with them, and later became extinct in India and Madagascar. The third, unlikely scenario proposed is that the knifefish diverged in Pangea before it broke up into Laurasia and Gondwana (178–157 million years ago) and two lineages were confined and evolved separately. However, if this was correct, then one would have discovered fossils from land masses connecting the two supercontinents (like southern and central Europe), but such fossils are missing. See: J.G. Inoue et al., 'The Historical Biogeography of the Freshwater Knifefishes Using Mitogenomic Approaches: A Mesozoic Origin of the Asian Notopterids (Actinopterygii: Osteoglossomorpha)', *Molecular Phylogenetics and Evolution* 51 (2009): 486–99.

17: Given the importance of India-Madagascar around 150 million years ago, in the evolution of frogs, limbless lizards and blindsnakes among other creatures, Nicholas Vidal and colleagues have coined the term 'Indigascar' for this land masss (Nicholas Vidal et al., 'Blindsnake Evolutionary Tree Reveals Long History on Gondwana', *Biology Letters* 6 (2010): 558-61.

18: On the mystery of the late origin of snakes see Nicolas Vidal and S. Blair Hedges, 'Molecular Evidence for a Terrestrial Origin of Snakes', *Proceedings: Biological Sciences*, 271 (2004): S226-S229.

8. Deccan's Inferno

1: For a wonderful overview of the spectacle of Ellora, see, Walter M. Spink, 'Ellora' in *Art of India: Prehistory to the Present*, ed. F. Asher (New Delhi: Encyclopedia Britannica, 2003). The size of Kailasa is nearly twice the size of the Parthenon of Athens according to Wikipedia. The Archaeological Survey of India measured the size as 276 feet in length, 154 feet in width and 107 feet in height (see: http://asi.nic.in/asi_monu_whs_ellora_brahmanical.asp). I benefitted immensely from reading Ariel Glucklich, *The Strides of Vishnu: Hindu Culture in Historical Perspective* (New York: Oxford University Press, 2007) on how Cave 16 was built.

2: On how the Satpuras and other peninsular mountains were raised, see Jyoti Singh et al. 'Structure and Evolution of Satpura Gondwana Basin over Central Indian Tectonic Zone: Inferences from Seismic and Gravity Data', *Journal of Indian Geophysical Union*, 19 (2015): 39-54.

3: The meteor impact (or the Yucatan or Chicxulub meteor) theory was proposed in 1980 by the father-son duo of Walter and Luis Alvarez. See: L.W. Alvarez et al., 'Extraterrestrial Cause for the Cretaceous-Tertiary Boundary', *Science* 208 (1980): 1095-1108. Deccan volcanism was for a short period an acceptable cause for the mass extinction of dinosaurs. Both the theories have been vociferously debated since the 1980s. Gerta Keller of Princeton University has advocated for the Deccan event as the reason for the death of the dinosaurs and critiqued the impact theory (G. Keller, 'KT Mass Extinction: Theories and Controversies, The Geological Society of London [2010]). For a point by point comparison of the two competing theories (Deccan volcanoes versus the Chicxulub event), see, Gerta Keller, 'Defining the Cretaceous-Tertiary Boundary: A Practical Guide and Return to First Principles' in *The End-Cretaceous Mass Extinction and the Chicxulub Impact in Texas* (Tulsa: Society for Sedimentary Geology, 2011; SEPM Special Publication 100), pp. 23-42. The debate on extinction of the dinosaurs has a notorious, ongoing history of contentious proposals backed by data, each equally convincing. It is evident that the population of dinosaurs was declining before the asteroid impact. See: Stephen L. Brusatte et al., 'The Extinction of the Dinosaurs', Biological Reviews (2014), available at: http://dx.doi.org/10.1111/brv.12128. This supports Keller's earlier claim that the second phase of volcanism had a catastrophic effect on life especially on climatic-sensitive planktic foraminifera (G. Keller, 'Deccan Volcanism Linked to the Cretaceous-Tertiary Boundary Mass Extinction: New Evidence from ONGC Wells in the Krishna-Godavari Basin', *Journal Geological Society Of India* 78 [2011]: 399-428). Because of the proximity of the intense Deccan event within Greater India, large land animals were at greatest risk here. See: G. Keller, 'Deccan Volcanism, the KT Mass Extinction and Dinosaurs', *Journal of Biosciences*, 34 (2009): 709-28. Also see: S. Chatterjee and D.K. Rudra, 'KT Events in India: Impact, Rifting, Volcanism and Dinosaur Extinction', *Memoirs of the Queensland Museum* 39 (1996): 489-532.

4: For types of lava flows and their forms I referred to the following papers: R.A. Duraiswami, 'Pulsed Inflation in the Hummocky Lava Flow Near Morgaon, Western Deccan Volcanic Province and Its Significance', *Current Science* 97 (2009): 313-16; Mamta Kashyap et al., 'Occurrence of Small Scale Inflated Pahoehoe Lava Flows in the Mandla Lobe of the Eastern Deccan Volcanic Province', *Current Science* 98 (2010): 72-6; and G. Lakshminarayana et al.,

'New Observations on Rajahmundry Traps of the Krishna-Godavari Basin', *Journal Geological Society of India* 75 (2010): 807–19.

5: For the timing of Deccan episodes and their impact on the biodiversity of the time, see, Ashu Khosla and Ashok Sahni, 'Biodiversity During the Deccan Volcanic Eruptive Episode', *Journal of Asian Earth Sciences* 21 (2003): 895–908.

6: Delicate frog fossils and crocodile eggshells can be found in the posh Malabar Hill and busy commercial district of Worli in Mumbai. See, G.K. Noble, 'The Fossil Frogs of the Intertrappean Beds of Bombay, India', *American Museum Novitates* 401 (1930): 1–13. Fossils of frogs (called toothed bufonid) have been found in India and are related to those found in Australia, proving that land bridges existed between the two land masses till about 60 million years ago. In the past two decades, many more excavations have revealed frogs from Malabar Hill, Chowpatty and even behind a garage in Santa Cruz—layers of shale and sandstone between volcanic lava flows have preserved delicate fossils of varieties of frogs. See: Swarn Deep Singh et al., 'Reptilia from the Intertrappean Beds of Bombay (India)', *Veröffentlichungen aus dem Fuhlrott-Museum* 4 (1998): 307–20.

7: A variety of crocodilians lurked in India, and this suggests that they survived the repeated episodes of extinctions. See, R.S. Rana, 'Dyrosaurid Crocodile (Mesosuchia) from the Infratrappean Beds of Vikarabad, Hyderabad District, Andhra Pradesh', *Current Science* 56 (1987): 532–34. An interesting paper which presents evidence that alligators as those found in the Americas were discovered in India, and suggest possible Laurentian life forms could have existed in India as recently as 65 million years ago is R.S. Rana, 'Alligatorine Teeth from the Deccan Intertrappean Beds Near Rangapur, Andhra Pradesh, India: Further Evidence of Laurasiatic Elements', *Current Science* 59 (1990): 49–51. For turtles and the Deccan episode, see, G.V.R. Prasad and H. Cappetta, 'Late Cretaceous Selachian from India and the Age of the Deccan Traps', *Palaeontology* 36 (1993): 231–248; the mammals that survived the Deccan episodes are discussed in: G.V.R. Prasad and A. Sahni, 'First Cretaceous Mammal from India', *Nature* 332 (1988): 638–40; and G.V.R. Prasad et al., 'Systematic Palaeontology (Vertebrate palaeontology): First Mammal Evidence from the Late Cretaceous of India for Biotic Dispersal between India and Africa at the KT Transition', *Comptes Rendus Palevol* 9 (2010): 63–71.

8: A paper published in June 2016 has challenged the view that mammals diversified after the extinction of the dinosaurs. The traditional view has been that there were a handful of early, small, tree-dwelling or burrowing, largely insect-eating animals. Over the past two decades more early mammals have been discovered including small hoofed mammals and some with diverse array of teeth, suggesting that mammals had begun diversifying earlier than previously expected. The mass extinction events that killed the dinosaurs also impacted mammals but while far greater numbers of the latter survived, virtually no dinosaurs did. D.M. Grossnickle and E. Newham, 'Therian Mammals Experience an Ecomorphological Radiation During the Late Cretaceous and Selective Extinction at the K-Pg Boundary', *Proceedings of the Royal Society B* 283 (2016).

9: The last ammonites are described in W.J. Kennedy and R.A. Anderson, 'Heteromorph Ammonites from the Upper Maastrichtian of Pondicherry, South India', *Palaeontology*, 35 (1992): 693–731.

10: Like the turtles and mammals, fish too benefitted from mass extinctions in the seas. See, Elizabeth C. Sibert and Richard D. Norris, 'New Age of Fishes Initiated by the Cretaceous-Paleogene Mass Extinction', *Proceedings of the National Academy of Sciences* 112 (2015): 8537–42.

11: The other volcanic rock (a'a) is not preferred for sculpting because the stones embedded within it make it difficult to shape or carve upon it. These stones form pockets of air which expand and cool, causing the rocks to crack and in time such rocks disintegrate and crumble if heavy rocks rest above it. Interestingly, the city of Pune straddles over both a'a and pahoehoe flows. The spectacular Maratha forts typically use both types of volcanic rock—the heavier dark grey pahoehoe basalt which makes for a stronger foundation and the lighter brownish grey a'a rocks to make walls, balustrade and towers. In some way the soft but tough Deccan volcanic rock has played an important role in establishing strategic forts which defended kingdoms and helped consolidate empires. The forts made by the Maratha king, Chhatrapati Shivaji, are perched on steep peaks and were made from basalt that was extracted from the slopes of the hills. The Deccan rocks enabled Aurangzeb to create efficient roadways, bridges and aqueducts in and around Aurangabad, which kept him connected to his vast empire. Pune's oldest monument is a rock cut temple made by the Rashtrakutas in the eighth century. Even the Yerwada Central Jail in Pune is made from local lava rocks! The Deccan rocks also laid the foundations that made Mumbai the economic centre and a symbol of architectural cosmopolitanism. In 1784 the first major attempt to link the seven islands of Mumbai (with causeways) was undertaken by a British engineer named Hornby Vellard. These causeways were made from volcanic rocks. Vellard had also planned the reclamation of Worli and Mahalaxmi. The British built textile mills, warehouses and factories on Parel and Elphinstone Road; the piers and berths in Cotton Green; the famous colleges—Wilsons, Elphinstone and St Xavier's—and the Bombay University; several hospitals and prominent schools—all using these grey volcanic rocks. Basaltic outcrops within the city of Mumbai like the Gilbert Hill and Sewri Mound have been quarried to make forts and landmarks of Mumbai. The dargah (shrine) of Haji Ali built in 1431 lies above a basalt island and is linked by volcanic rocks to the Worli causeway. The San Miguel Church (or St Michael's Church as the local BEST buses call it!) in Mahim and forts made by the Portuguese, all used the dark Deccan basalt. If you happen to be travelling from the airport to Santa Cruz (East), spare a moment and see the ruins of the St John the Baptist Church (in the busy commercial district of SEEPZ) which has a beautiful but rapidly crumbling facade made entirely of pahoehoe.

9. Humble Beginnings

1: One of the oldest mammals to emerge just before the demise of dinosaurs in India is from the Cretaceous outcrops near Bacharam village, Warangal, Telangana, which is described in S. Anantharaman et al., 'A possible Late Cretaceous "haramiyidan" from India', *Journal of Vertebrate Paleontology* 26 (2006): 488-90. The mammal was named *Avashishta* (meaning 'one that survives' in Sanskrit), and the researchers who discovered its teeth, skull and bones believe they show that the first Indian mammals (from around 70 million years ago or so) were of, both, Laurasian and Gondwanan origins. This discovery, that dietary diversification was a major factor in driving the evolution of early mammals, helped confirm the lessons drawn from discoveries made in Europe, China and Africa.

2: The relation between mammal-like reptiles found in the Gondwana is simply presented in Fernando Abdala and Ana Maria Ribeiro, 'Distribution and diversity patterns of Triassic cynodonts (Therapsida, Cynodontia) in Gondwana', *Palaeogeography, Palaeoclimatology, Palaeoecology* 286 (2010): 202-17, and Anthony D. Barnosky and Brian P. Kraatz, 'The role of climatic change in the evolution of mammals', *BioScience* 57 (2007): 523-32.

3: *Deccanolestes* is perhaps the most studied (and discovered) early-mammal fossil from the subcontinent. *Deccanolestes* has been discovered in at least three different sites, and it is now extensively reported in journals that at least two species (*D. hislopi* and *D. robustus*) existed. These are described in: G.V.R. Prasad and A. Sahni, 'First Cretaceous mammal from India', *Nature* (1988): 332, 638-40; A. Khosla et al., 'Discovery of a micromammal-yielding Deccan intertrappean site near Kisalpuri, Dindori district, Madhya Pradesh', *Current Science* 87 (2004): 380-83. Another mammal closely related to *Deccanolestes*, called

Sahnitherium rangapurensis (see R.S. Rana and G.P. Wilson, 'New Late Cretaceous mammals from the Intertrappean beds of Rangapur, India and paleobiogeographic framework', *Acta Palaeontologica Polonica* 48 (2003): 331–48) has been discovered in the sedimentary rocks between the Deccan lava flows, suggesting that several small and related species of mammals existed in and around the last days of the dinosaurs. A more comprehensive review on the *Deccanolestes* is provided by G.V.R. Prasad et al., 'First mammal evidence from the Late Cretaceous of India for biotic dispersal between India and Africa at the KT transition', *Comptes Rendus Palevol* 9 (2010): 63–71. The paper describes the discovery of *Deccanolestes* from the mammal-yielding Deccan intertrappean site near Kisalpuri village, Dindori district,, Madhya Pradesh. The fossil teeth of this mammal reveal an affinity to mammals from India and Africa (*Afrodon*) and Europe.

4: Reconciling evolution vis-à-vis early mammals is tricky, but most palaeontologists agree that an exchange between India, Africa and Europe, creating a 'ghost lineage' for the early ancestors (called the 'Gondwanan eutherians') between 80 to 45 million years ago, thus creating the ancestors of all modern mammalian families. It is likely that the Indian subcontinent, Eurasia and Africa are the regions where many mammalian families have evolved; see the discussion presented in D.M. Boyer et al., 'New postcrania of Deccanolestes from the Late Cretaceous of India and their bearing on the evolutionary and biogeographic history of euarchontan mammals', *Naturwissenschaften* 97 (2010): 365–77, and Mark S. Springer et al., 'The historical biogeography of Mammalia', *Philosophical Transactions of the Royal Society B* 336 (2011): 2478–502. The discovery of *Avashishta* by S. Anantharaman et al. (2006) strengthens the claim of this hypothesis.

5: On the evolution and dominance of early birds, see Alan Feduccia, *The Origin and Evolution of Birds* (New Haven: Yale University Press, 1999), and Larry G. Marshall, 'The Terror Birds of South America', *Scientific American* 14 (2004): 82–9. The rivals to these giant birds were the long-limbed land crocodiles, and these are discussed in Gordon Grigg and David Kirshner, *Biology and Evolution of Crocodylians* (New York: Cornell University Press and CSIRO Publications, 2015).

6: For nearly 58 million years (i.e., from 61 to 2 million years ago), this feathered beast dominated the Americas. This was the only time when birds competed and dominated, if not ruled, the land as the top predator. The earliest giant bird (*Titanis*) lived during the dinosaur era and was distantly related to *Tyrannosaurus rex*, and this perhaps explains the predatory instinct of the terror birds. The *Titanis* weighed more than 120 kilograms and stood nearly 5 feet tall. A heavy, razor-sharp beak, coupled with massive talons, made it the most ferocious predator around. It had the ability to swallow whole prey that was about the size of a small dog. In the case of larger prey, it would rip away its hide and fur, just as birds of prey do today. The *Titanis*, like terror birds in other land masses, perhaps died out earlier because it lived a solitary life and was eventually outsmarted by social mammals like wolves, and their eggs were preyed on by mammals.

7: Big, flightless ground birds (called 'ratites', and which include the emu, rhea, ostrich and cassowary) are true southern-hemisphere specialities. The heart of Gondwana, for the evolution of these birds, was Antarctica. Around 140 million years ago, a moist and warm climate prevailed, and tropical vegetation covered much of Antarctica and Gondwana. The early birds, although small, began to evolve in the shadow of the dinosaurs and reptiles. Around 80 million years ago, all the major Gondwanan land masses had gone their separate ways, carrying with them their cargo of life forms, including birds. India and Madagascar remained connected to Antarctica, until 72 million years ago, through an umbilical land bridge called the Kerguelen Plateau, now submerged in the southern Indian Ocean. India carried with it the ostriches, while the elephant birds remained in Madagascar. After India collided with Eurasia, these giant birds spread north to the Gobi Desert, and west into Iran, Arabia, the Levant, and then entered Somalia, before colonizing Africa. A study of the eggshells of ostriches from 10 million years ago show that they had spread through corridors and land bridges and the intensification of the monsoon and the subsequent cooling, which caused the forests to shrink and the grasslands to spread, enabling the ostrich to flourish (see Rajeev Patnaik et al., 'Ostrich-like eggshells from a 10.1 million-yr-old Miocene ape locality, Haritalyangar, Himachal Pradesh, India', *Current Science* 96 (2009): 1485–95. The rock paintings of a large bird found in Katothiya (about 25 kilometres south of Bhopal) and in the Bazaar Cave of Pachmarhi are, according to some archæologists, those of an ostrich, though many anthropologists disagree with this interpretation. However, ornaments made from ostrich eggshell have been excavated from several sites across India, suggest that ostriches may have died out as recently as 13,000 and 11,000 years ago.

8: The first occurrence of the (dyrosaurid) crocodile from the intertrappean beds of India, from Kisalpuri in Dindori district, Madhya Pradesh, is mentioned in: Ashu Khosla et al., 'Dyrosaurid remains from the Intertrappean Beds of India and the Late Cretaceous Distribution of Dyrosauridae', *Journal of Vertebrate Paleontology* 29 (2009): 1321–26.

9: For the most comprehensive review of the evolution of mammals see: Kenneth David Rose, *The Beginning of the Age of Mammals* (Baltimore: Johns Hopkins University Press, 2006).

10: The discovery of coconuts is commonplace in fuller's earth mines—see Anumeha Shukla, Rakesh C. Mehrotra and Jaswant S. Guleria, 'Cocos Sahnii Kaul: A Cocos Nucifera L.-like Fruit from the Early Eocene Rainforest of Rajasthan, Western India', *Journal of Biosciences* 37 (2012): 769–76. The same is the case with the discovery of shrimps, crabs and other marine creatures—see, for example, Krishna Kant Tiwari, 'Lower Tertiary Penaeid shrimps from Kapurdi Barmer district, Rajasthan, India', *Crustaceana* 5 (1963): 205–12. These papers suggest that the western margin of India was once covered by wet tropical forests and was close to marine and freshwater environments.

11: Lignite mines in western Rajasthan, and all the way to Cambay in Gujarat (like Vastan), hold within them a bounty of plant fossils and amber from different periods of the Eocene age (56 to 33.9 million years ago), which not only reveal several new animals (many of them with no descendants left) but also the habitat they lived in. Among all of these places, Vastan is the most intensely studied site. The full faunal list of discoveries made in Vastan are as follows: several insects, fish, frogs, lizards, ten kinds of snakes, at least two birds, and twenty-five kinds of mammals, including eight bat species, four primates, a small hyena-like mammal, two rodent-like extinct mammals (or tillodonts), a few insectivores, a small chevrotain-like creature, two rodents, an ancestor of the rabbit, and two ungulates, including large tapir-like creatures named *Cambaytherium*. Of particular interest is a marsupial that lived here from 56 to 38 million years ago (see Sunil Bajpai et al., 'First fossil marsupial from India: Early Eocene Indodelphis n.gen and Jaegeria n.gen from Vastan lignite mine, District Surat, Gujarat', *Journal of the Paleontological Society of India* 50 (2005): 147–51. A large body of work comprising the analyses of the Vastan ambers (and, to a lesser extent, the clay and lignite horizons) by researchers is listed at http://www.cambayamber.com/Cambay%20 Amber%20Research/index.html.

12: For how frogs and other amphibians evolved after India broke off from the rest of the Eastern Gondwana, see Kim Roelants, et al., 'Global Patterns of Diversification in the History of Modern Amphibians', *Proceedings of the National Academy of Sciences* 104 (2007): 887–92.

13: George A. Boulenger, *The Fauna of British India, Including Ceylon and Burma: Reptilia and Batrachia* (London: Taylor and Francis, 1890): p. 541.

14: On the evolution and distribution of early flowering plants, I benefited from reading the seminal works of Peter Crane (https://environment.yale.edu/profile/crane/pubs) and David

Dilcher. In addition, the following papers were useful: Bruce H. Tiffney, 'Seed size, dispersal syndromes, and the rise of the angiosperms: Evidence and hypothesis', *Annals of the Missouri Botanical Garden* 71 (1984): 551–76; E. Conti et al., 'Early Tertiary out-of-India dispersal of Crypteroniaceae: Evidence from phylogeny and molecular dating', *Evolution* 56 (2002): 1931–942; Daniel I. Axelrod, 'Edaphic aridity as a factor in angiosperm evolution', *American Naturalist* 106 (1972): 311–20; and Stephen McLoughlin, 'The breakup history of Gondwana and its impact on pre-Cenozoic floristic provincialism', *Australian Journal of Botany* 49 (2001):271–300. All these papers present a broad geological framework of Gondwana and its break-up and shed light on the dispersion of the flora. For how mammals responded to changing flora, see Margaret E. Collinson et al., 'Fossil Evidence of Interactions between Plants and Plant-Eating Mammals [and Discussion]', *Philosophical Transactions: Biological Sciences* 333 (1991): 197–208.

15: To understand the possible modes of dispersal of mammals, the following papers are helpful: G.V.R. Prasad and A. Sahni, 'Were there size constraints on biotic exchanges during the northward drift of the Indian plate?' *Proceedings of the Indian National Science Academy A* 65 (1999): 377–96. Another thought-provoking paper that reconstructs the potential of dispersal routes in southern Gondwana and the current constraints in understanding is by Anne D. Yoder and Michael D. Nowak, 'Has Vicariance or Dispersal Been the Predominant Biogeographic Force in Madagascar? Only Time Will Tell', *Annual Review of Ecology and Evolutionary Systems* 37 (2006): 405–31.

16: For details on the regional environment that led to isolation and influenced the evolution of mammals into different lineages, see C. Barry Cox, 'Plate Tectonics, Seaways and Climate in the Historical Biogeography of Mammals', *Memórias do Instituto Oswaldo Cruz* 95 (2000): 509–16. The paper presents possible models that show how mountain ranges and oceans acted as both barriers and land bridges for mammals between 90 and 40 million years ago.

17: E. Eizirik et al., 'Molecular Dating and Biogeography of the early placental mammal radiation', *Journal of Heredity* 92 (2001): 212–19.

18: A paper that presents the current gaps in the understanding of the Gondwanan origins of flightless birds and fish, and suggests possible solutions for reconciling their timing, is John C. Briggs, 'Fishes and Birds: Gondwana Life Rafts Reconsidered', *Systematic Biology* 52 (2003): 548–53.

19: Using molecular-sequencing methods, one paper estimates that grasses emerged between 70 to 55 million years ago (Elizabeth A. Kellogg, 'Evolutionary History of the Grasses', *Plant Physiology* 125 [2001]: 1198–205.) and suggested dinosaurs may not have eaten grasses at all, until coprolite from intratrappeans of Maharashtra proved this otherwise (Vandana Prasad et al., 2005). For a slightly revised timeline on the origin of grass species, see Sylvain Glémin and Thomas Bataillon, 'A comparative view of the evolution of grasses under domestication', *New Phytologist - Tansley Review* 183 (2009): 273–90.

20: For information on how flowering plants exploded on to the scene, see David Dilcher, 'Toward a new synthesis', 2000, and William L. Crepet, 'The Fossil Record of Angiosperms: Requiem or Renaissance', *Annals of the Missouri Botanical Garden* 95 (2008): 3–33. Crepet discusses in detail the two periods of rapid radiation of angiosperms—the Turonian (93.9 to 89.9 million years ago) and the Early Tertiary—and explores the possible reasons for what favoured the emergence of a new genus and species. In addition, the paper by Timothy J. Bradley et al., 'Episodes in insect evolution', *Integrative and Comparative Biology* 49 (2009): 590–606, reviews the evolutionary history of the insects, their role in the evolution and diversification of flowering plants, and the origins of flight. Using the evolutionary history of two major flowering species, this paper suggests that migration between Gondwana and Laurasia in the Late Cretaceous to the Early Tertiary period became possible as the Tethys gradually narrowed as Greater India approached Eurasia, and the approaching land mass acted as a raft that carried on it plants species from Madagascar to Asia (see J.E. Richardson et al., 'Historical biogeography of two cosmopolitan families of flowering plants: Annonaceae and Rhamnaceae', *Philosophical Transactions of the Royal Society of London B* 359 (2004): 1495–508).

10. Moving Heaven and Earth

1: Perhaps the one book, though quite technical to the lay reader, on the making of the Himalaya and Tibet (and other landforms) and the geological processes which created them, is Michael Searle, *Colliding Continents* (Oxford: Oxford University Press, 2012). I also recommend a short journal article and a lecture by D.N. Wadia; see D.N. Wadia, 'Geology of the Himalaya Mountains: Some Unsolved Problems', *Proceedings of the Indian National Science Association* 29 (1963): 378–87; and D.N. Wadia, 'The Himalaya Mountains: Their Age, Origin and Subcrustal Relations', delivered during the Meghnad Saha Memorial Lecture, 22 October 1965, Calcutta. These provide a good overview on the Himalaya, although the dates have been revised since.

2: On the process of how the Kohistan-Ladakh arc was formed and subsumed, see Wang-Chun Xu et al., 'Rapid Eocene erosion, sedimentation and burial in the eastern Himalayan syntaxis and its geodynamic significance', *Gondwana Research* 23 (2013): 715–25. Using zircon as an indicator, this study determined the age of the collision of Kohistan with India to be around 65 to 61 million years ago at a latitude far south of its present position. Shuhab D. Khan et al., 'Did the Kohistan-Ladakh island arc collide first with India?' *Geological Society of America Bulletin* 121 (2009): 366—84. 3: The subduction of Greater India under Eurasia is discussed in Guillaume Dupont-Nivet et al., 'Palaeolatitude and age of the Indo-Asia collision: palaeomagnetic constraints', *Geophysical Journal International* 182 (2010): 1189–198. For a detailed account of the impact of India's collision with Asia, and the deformation that took place beyond the Himalayas, see An Yin, 'Cenozoic tectonic evolution of Asia: A preliminary synthesis', *Tectonophysics* 488 (2010): 293–325. A more recent paper that reconciles the timing is Stephane Guillot et al., 'Importance of continental subductions for the growth of the Tibetan plateau', *Geophysical Research Abstracts* 15 (2013), 2013–5760.

4: The Andaman is one of the most seismically active and youngest areas on the face of Earth. The Andaman arc extends from Myanmar in the north to Indonesia in the south. It is the result of the subduction of the north-east-moving Indian Plate, beneath the Burmese Plate. Barren Island is the only active volcano located in the Andaman Sea and in India, and it marks the north-westernmost limit of this volatile region. The timing of the initiation of the Andaman subduction zone has been always in debate and there are two schools of thought that exist in this regard. The first school (S.K. Acharyya, 'Stratigraphy and tectonic history reconstruction of the Indo-Burma-Andaman mobile belt', *Indian Journal of Geology* 69 [1997]: 211–34; S. Sengupta et al., 'Nature of ophiolite occurrences along the eastern margin of the Indian plate and their tectonic significance', *Geology* 18 [1990]: 439–42), argues that the present subduction, all around the western Sunda arc, began 20 to 15 million years ago, and that a second trench, further east of the present-day Andaman trench with a protocontinent was subsumed. The second school of thought is based on the presence of melange and the highly deformed sediments that were raised after melting under the oceanic crust. Subscribers to this theory suggest that the onset of subduction took place during Late Cretaceous (between 75 to 65 million years ago), triggered, perhaps, by the separation of the Gondwanan land masses (J.R. Curray et al., 'Structure, tectonics and geological history of the north-east Indian Ocean', in *The Ocean Basins and Margins: The Indian Ocean, Vol. 6*, ed. A.E.M. Nairn and F.G. Stehli (New York: Plenum Press, 1982), pp. 399–450). A survey comparing the rocks formed due to the merging of the India and Burma plates is presented in D.K. Roy and S.K. Acharyya, 'Palaeontology, Tectonostratigraphy and Emplacement History of the Andaman Ophiolite Belt and its Correlation with the Naga Hills Ophiolites: Progress Report for Field Seasons 1987 to 1990', Palaeontology & Stratigraphy Division, Geological Survey of India, 1991.

5: Ammonites like *Ophoceras* and *Claraia* and conodonts are good indicators for calibrating the time when the Tethys closed; see H. Kozur and H. Mostler, 'Beiträge zur Mikrofauna permotriadischer Schichtfolgen. Teil I: Conodonten aus der Tibetzone des Niederen Himalaya (Dolpogebiet, Westnepal)', *Geol. Paläont. Mitt. Innsbruck Bd.* 3, 9 (Innsbruck, 1973): S.1-23. However, a trans-Tethyan survey of this kind has not been done, especially with regard to discoveries made since the time of this publication.

6: For an excellent account on Fuller's earth, see Ashok Sahni and Nagendra K. Choudhary, 'Lower Eocene fishes from Barmer, South Western Rajasthan, India', *Proceedings of the Indian National Science Academy - Part A: Physical Sciences* 38 (1972): 97-102.

7: For a summary of the economic and cultural history of food crops introduced by the Portuguese and other colonists into India, see Kenneth F. Kiple and Kriemhild Coneè, ed., *The Cambridge World History of Food, Volume 1* (Cambridge: Cambridge University Press, 2000), p. 2153; and Utsa Ray, *Culinary Culture in Colonial India* (Cambridge: Cambridge University Press, 2015), p. 280.

8: The rise of Tibet is covered eloquently in Andreas Mulch and C. Page Chamberlain, 'The rise and growth of Tibet', *Nature* 439 (2006): 670-71. Another paper that explains the processes that raised Tibet with special reference to Namche Barwa is Lewis A. Owen, 'How Tibet might keep its edge,' *Nature* 455 (2008): 748-49.
9: The arrangement of the Himalaya and the introduction to Himalayan rivers is lucidly provided in M.S. Krishnan, *Geology of India and Burma* (Madras: Higginbothams, 1968).

10: For a description on the age of rocks in high Himalayan peaks see Myrow, P.M., Hughes, N.C., Searle, M.P., Fanning, C.M., Peng, S.-C., Parcha, S.K., 2009. Stratigraphic correlation of Cambrian-Ordovician deposits along the Himalaya: implications for the age and nature of rocks in the Mt. Everest region. Geological Society of America Bulletin 120, 323-332.

11: The Gandak is also known as the Saligrami in Nepal because it brings down large quantities of Saligrams (ammonite fossils).

12: The garnet and kyanite we find embedded in the gneiss were formed when the first upheaval was taking place between 41 and 36 million years ago. These minerals give us a clear indication that these rocks were formed at great depths and pressure-cooked at very high temperatures. Geologists use gems in gneiss to mark the point from where the lighter Indian plate subducted, melted and re-emerged as gneiss encrusted with these glassy gems; for more information on this, see Rodolfo Carosi et al., 'Eocene partial melting recorded in peritectic garnets from kyanite-gneiss, Greater Himalayan Sequence, central Nepal', *Geological Society London Special Publications* 412 (2014): 5-17. Most geologists use different dating methods and apply these to the different type of rocks (ophiolites) or sediments to estimate the timing of the upheavals. But the general consensus is that the Himalayas were raised in three phases lasting more than 4 million years each.

13: For a good account on the Indus, its origins, course and fate over the past 50 million years, see Avijit Gupta 'The geographic, geological and oceanographic setting of the Indus river', in *Large River: Geomorphology and Management* (New York: John Wiley & Sons, 2008).

14: The process of cutting off of the Tsangpo (Tibet) and Irrawaddy (Myanmar) rivers 18 million years ago is described in Ruth A.J. Robinson et al., 'Large rivers and orogens: The evolution of the Yarlung Tsangpo-Irrawaddy system and the eastern Himalayan syntaxis', *Gondwana Research* 26 (2014): 112-21. The course carved by the Brahmaputra (more specifically the Tsangpo in Tibet before it enters India) is described in Jonathan C. Aitchison and Aileen M. Davis, 'Evidence for the multiphase nature of the India-Asia collision from the Yarlung Tsangpo suture zone, Tibet', in *Aspects of the Tectonic Evolution of China*, ed. J. Malpas, et al., *Geological Society London Special Publications* 226 (2004): 217—33. For details on the Brahmaputra's course, see Laura Bracciali et al., 'The Brahmaputra: tale of tectonics and erosion: Early Miocene river capture in the Eastern Himalaya', *Earth and Planetary Science Letters* 415 (2015): 25-37.

15: The Kosi floods of 2008 are reviewed in D.V. Reddy et al., 'The 18 August 2008 Kosi river breach: an evaluation', *Current Science* 95 (2008): 1668-69.

16: The events following 15 August 1950 in Assam and Pasighat in Arunachal Pradesh had far-reaching consequences for Assam and north-east India. For an elegant body of research, see Bérénice Guyot-Réchard and, in particular, her paper, 'Reordering a Border Space: Relief, rehabilitation, and nation-building in north-eastern India after the 1950 Assam earthquake', *Modern Asian Studies* 49 (2015): 931-62.

17: On the tectonic, glacial and weather processes that caused the Kedarnath tragedy, see Alan D. Ziegler et al., 'Pilgrims, progress, and the political economy of disaster preparedness: the example of the 2013 Uttarakhand flood and Kedarnath disaster', *Hydrological Processes* 28 (2014): 5985-990; and S.K. Allen et al., 'Lake outburst and debris flow disaster at Kedarnath, June 2013: hydrometeorological triggering and topographic predisposition', *Landslides* (2015): 1-13.

18: Joseph Dalton Hooker's travels in the Sikkim Himalayas in 1848 are described in Joseph Dalton Hooker, *Himalayan journals; or, Notes of a naturalist in Bengal, the Sikkim and Nepal Himalayas, the Khasia mountains, etc.* (London: J. Murray, 1854).

19: On the origins and nature of the Indian monsoons, see Peter D. Clift et al., 'Monsoon evolution and tectonics-climate linkage in Asia: an introduction', *Geological Society London* 342: 1-4. One comprehensive paper that presents the sporadic rise of the Himalaya and Tibet and how they influenced the monsoon is An Zhisheng et al., 'Evolution of Asian monsoons and phased uplift of the Himalaya-Tibetan plateau since Late Miocene times', *Nature* 411 (2001): 62-66. I found two other papers to be useful, especially in terms of the timing of geological and climatic events; these are Nigel Harris, 'The elevation history of the Tibetan Plateau and its implications for the Asian monsoon', *Palaeogeography, Palaeoclimatology, Palaeoecology* 241 (2006): 4-15; and Liu XiaoDong and Dong Bu Wen, 'Influence of the Tibetan Plateau uplift on the Asian monsoon-arid environment evolution', *Chinese Science Bulletin* 58 (2013): 4277-91. In the context of the shift of plates, see a paper on the April 2015 Nepal earthquake (referred to as the 'Gorkha earthquake'), which caused a 1-metre uplift of the Kathmandu Basin, while the high Himalaya, which is further north, subsided by about 0.6 metres; see J.R. Elliott et al., 'Himalayan megathrust geometry and relation to topography revealed by the Gorkha earthquake', *Nature Geoscience* 9 (2016): 174-80.

20: The drying of rivers, with special reference to the mythical river Saraswati, is explained by K.S. Valdiya, *Saraswati: The River that Disappeared* (Hyderabad: University Press/ISRO, 2002). The desiccation of grasslands during the last ice age that caused the expansion of deserts and the drying of rivers is discussed in P.C. Bakliwal and S.K. Wadhawan, 'Geological evolution of Thar Desert in India: Issues and prospects', *Proceedings of the Indian National Science Academy A* 69 (2003): 151-65; and B.B. Roy and S. Pandey, 'Expansion or contraction of the Great Indian Desert', *Proceedings of the Indian National Science Academy B* 36 (1971): 331-44.

21: The temperature over the Antarctic has a profound impact on the world's climate, especially the Asian monsoon; for more information on this, see An Zhisheng et al., 'Glacial-Interglacial Indian Summer Monsoon Dynamics', *Science* 333 (2011): 719–23. Another good overview of Antarctica is by Eelco J. Rohling et al., 'Antarctic temperature and global sea level closely coupled over the past five glacial cycles", *Nature Geoscience* 2 (2009): 500–04. Sunlight plays a crucial role on the surface temperature of Antarctica and the seas. It impacts the climate of the southern hemisphere and has a profound impact on the Indian monsoons. An interesting website that shows the sunlight and its effects in Antarctica, see http://www.antarctica.gov.au/about-antarctica/environment/weather/sunlight-hours. The impact of Antarctica on the Indian monsoons is simply explained in S.S. Dugam and S.B. Kakade, 'Antarctica sea-ice and monsoon variability', *Indian Journal of Radio & Space Physics* 33 (2004): 306–09. A simple account of late glacial events is described in Lewis Owen et al., 'The Quaternary Glacial History of The Himalaya', *Quaternary Proceedings* 6 (1998): 91–120.

22: For understanding how the Himalayas and Tibet influenced the onset of the mini ice-age, see Maureen E. Raymo and Peter Huybers, 'Unlocking the mysteries of the ice ages', *Nature* 451 (2008): 284-85.

23: On the slopes of the Himalaya and in the River Ganga and its tributaries (but less so in the Brahmaputra) is where an important geochemical reaction occurs. This has a profound impact in terms of regulation of regional and global climate. Atmospheric carbon dioxide is absorbed by sediments made from young rocks that make the mountain-face of the Himalaya. The simple process of moving carbon dioxide out of the atmosphere into long-term reservoirs (sinks like deep ocean-beds) was formulated by Harold Urey (see H.C. Urey, *The Planets: Their Origin and Development* [New Haven: Yale University Press, 1952], p. 245). Urey suggested that the amount of carbon dioxide in our atmosphere was governed by an equation: $CO_2 + CaSiO_4 \rightarrow CaCO_3 + SiO_2$. The weathering process occurs when slightly acidic rainwater brings dissolved carbon dioxide to the surface of fresh igneous rocks, which contain calcium-bearing silicate minerals (or $CaSiO_4$), and readily erodes. In Urey's equation, the calcium in the rocks and carbon dioxide in the water combine to make $CaCO_3$ (like limestone), while the silicate is released to make siliceous mineral (like opal and chert minerals). Urey argued that the concentration of carbon dioxide in Earth's atmosphere corresponds to the equilibrium expected for this reaction. In simple terms, the amount of CO_2 in the atmosphere gets regulated by the action of precipitation that falls on the slopes of the young mountain range and gets buried in deep sea—from the Sunderbans till the deep trenches of the Indian Ocean. Two papers need to be read in tandem to understand what occurs in the rock and river sediments, and after they get subsumed in the depths of the Bay of Bengal and the Andaman Sea. The chemical reactions—from erosion to carbon burial in the Himalaya are given in C. France-Lanord et al., 'Continental erosion and CO2 uptake- inferences from the Himalayan system', *Mineralogical Magazine* 62a (1998): 466–67; and V. Galy et al., 'Efficient organic carbon burial in the Bengal fan sustained, by the Himalayan erosional system,' *Nature* 450 (2007): 407–11. According to this paper 'the amount of organic carbon deposited in the Bengal basin represents about 10 to 20 per cent of the total terrestrial organic carbon buried in oceanic sediments'. Currently, few climate change experts are concerned about the impact of growing pollution and urbanization along the slopes and within the Ganga and its tributaries, and how this can impair the delicate Urey's reaction.

24: For how the Ganga-Brahmaputra (or the Bengal) fans formed, see V. Galy et al., 'Efficient organic carbon burial' (2007); and more recently J.M. Coleman, 'Brahmaputra River: Channel processes and sedimentation', *Sedimentary Geology* 3 (1969): 129–239, estimates that the sediment load during floods in the Brahmaputra reaches as high as 13 million tons per day; for the Ganga, it may have been higher during the same period.

For the marine process of transport and the roles of sediment and deltas, see J.A. Covault, 'Submarine Fans and Canyon-Channel Systems: A Review of Processes, Products, and Models', *Nature Education Knowledge* 3 (2011): 893-94. An excellent visual of the thickness of river sediments is aided by the National Geophysical Data Center (NGDC), the National Oceanic and Atmospheric Administration (NOAA) and the Department of Commerce of the US Government; for more details, see 'Total Sediment Thickness of the World's Oceans & Marginal Seas', available at www.ngdc.noaa.gov/mgg/sedthick.html (as viewed on 1 October 2015). One will be able to see how widespread the sediment of the Ganga-Brahmaputra and Indus is, and the thickness, which is more than 10,000 to 20,000 metres deep.

25: Geologists and, specifically, geodesist are interested in finding out how far India has subducted, and how large the boundary of India was before it merged with Eurasia. Two papers have tried to find the answer to this. The first (T.S. Balakrishnan and P. Unnikrishnan, 'Deep Structure of Himalaya', *Journal of Indian Geophysical Union* 13 [2009]: 173-82) studied the geophysical deformation and how it its impacts gravity and seismic nature. The second paper (M.N. Qureshy, 'A Geologic analysis of Bouguer anomaly map of Peninsular India', *Proceeding International Science Congress India*, 30 (1964): 675-88) uses gravity studies to find the structural relation of the Indian subcontinent. The data from this paper helps delineate the outer boundary of Greater India, and is marked by the Indus in the west and the alluvial boundary of Indo-Gangetic plain. The Indo-Gangetic alluvium boundary starts roughly from Ludhiana, travelling east, south of Shimla, towards Dehradun, cutting through north Nepal, Sikkim, Arunachal Pradesh and turning south, east of Kohima, towards the eastern border with Bangladesh. This is also validated by more recent research (see Douwe J.J. van Hinsbergen et al., 'Greater India Basin hypothesis and a two-stage Cenozoic collision between India and Asia', *Proceedings of the National Academy of Sciences* 109 [2012]: 7659-64, and the ensuing letters and comments on this article in the journal, which discuss the current limitations and further research that is needed). This paper describes the size of the Indian land mass before the collision and the far-reaching impact of India-Eurasia on Tibet, Mongolia and Southeast Asia.

11. The Carnival of Mammals

1: For the earliest records of the fossils of large mammals that were found in the Siwalik see Hugh Falconer and Proby T. Cautley, *Fauna antiqua sivalensis, being the fossil zoology of the Sewalik Hills, in the north of India*', (London: Smith, Elder and Co.,1846); and 'Description of the plates of the Fauna *antiqua sivalensis* from notes and memoranda by Hugh Falconer', ed. Charles Murchison (London: R. Hardwicke, 1868). The latter provides copious details on the fossils and their locations. The collected papers from the Barnum Brown India Expedition (1921-25) have been published in the *American Museum Novitates*, a publication of the American Museum of Natural History, New York. After Barnum Brown, the fossils were tirelessly studied by Edwin H. Colbert, Charles Craig Mook and others. For a more updated catalogue, see Rajeev Patnaik, 'Indian Neogene Siwalik Biostratigraphy: An Overview', in *Fossil Mammals of Asia: Neogene Biostratigraphy and Chronology*, ed. Wang et al. (New York: Columbia University Press, 2012), pp. 423-44.

2: Mammals that descended from trees evolved into the hooved browser (like the *Protoreodon*, which lived in western North America), and they moved in small groups, eating foliage, fruits and nuts at dawn. Such mammals were about the size of a medium-sized dog. Stalking them is a distant cousin (like *Harpagolestes*), which emerged a few million years ago from a common ancestor, which, in turn, emerged around 40 and 35 million years ago. On the emergence of early mammals, see Kenneth David Rose, *The Beginning of the Age of Mammals* (Baltimore: Johns Hopkins University Press, 2006). For an overview of how mammals diversified into modern families, see David Macdonald, *The Encyclopaedia of Mammals* (Oxford: Oxford University Press, 2006).

3: There are roughly equal numbers of mammals that eat grass or vegetation on the land to those who eat them, numbering about 270. Among the grazers are the 220 even-toed ungulates or artiodactyl species, which include pigs, hippos (including whales), camels, deer, antelopes, sheep, goats and cattle, as well as seventeen species including horses, rhinos and tapirs. There are nearly 270 carnivores who feed on these grazers. There are 5416 known species of mammals, of which rodents make for the largest number of species (2277) (see Don E. Wilson and Dee Ann M. Reeder, ed., *Mammal Species of the World: A Taxonomic and Geographic Reference* [Baltimore: Johns Hopkins University Press, 2005], p. 2142). The taxonomy and reference to all currently known mammals is also available on the website of Bucknell University: http://www.departments.bucknell.edu/biology/resources/msw3/ (as viewed on 12 June 2016).

4: It made evolutionary sense for all animals to become omnivores first before they fully evolved into meat eaters; see Roland M. Novak, *Walker's Carnivores of the World* (Baltimore: John Hopkins University Press, 2005).

5: On the life and times of *Paraceratherium* and their competition with elephants, like the *Gomphotherium* and *Deinotherium*, and perhaps other rhinos, I recommend a thoroughly enjoyable book by Donald R. Prothero, *Rhinoceros Giants: The Paleobiology of Indricotheres* (Bloomington: Indiana University Press, 2013).

6: Large herbivores too, like elephants, were spread far and wide and grew in size as they crossed from Eurasia into India and Africa. A 13-foot-long thigh bone of *Palaeoloxodon namadicus* (the old elephant of Narmada) was discovered in Narsinghpur, Madhya Pradesh, which suggests that some elephants grew to be very large, perhaps ever larger than the giant rhino, *Paraceratherium*! A recent paper has re-examined the Narsinghpur *Palaeoloxodon* and found that it would have been nearly a foot taller and a ton heavier than the biggest known *Paraceratherium*. However, more bones or, better still, complete skeletons will be needed to confirm which of the two would have been the largest land mammals; see Asier Larramendi, 'Proboscideans: Shoulder Height, Body Mass and Shape', *Acta Palaeontologica Polonica* 61 (2015): 537–74. Another paper (Mikael Fortelius, 'The largest land mammal ever imagined', *Zoological Journal of the Linnean Society* 107 [1993]: 85–101) describes the history and the science (and mis-science) behind estimating the size of *Paraceratherium*.

7: One study has computed that it took a minimum of 1.6 million, 5.1 million and 10 million generations for terrestrial mammal mass to increase 100-, 1000- and 5000-fold, respectively. On the other hand, significant decreases (for example, due to isolation on islands) can occur at more than ten times the rates of increase. The relationship is valid for terrestrial and marine mammals; see Alistair R. Evans et al., 'The maximum rate of mammal evolution', *Proceedings of the National Academy of Sciences* 109 (2012): 4187–190.

8: Why couldn't mammals become as large as dinosaurs? One reason for this is on account of the difference in the way animals regulate their body heat. Mammals are endotherms (meaning, they regulate their own body temperature) and can't evolve bodies as large as the largest dinosaurs, because they need to use so much of their physical energy—provided by the food they eat—towards keeping their bodies warm. A mammal of a given size uses ten times more energy than a reptile or a dinosaur of the same size does. About 11 million years ago, mammals had peaked in their body sizes. Two factors explain how mammals broke the existing size constraints. First, as the Tethys closed, a colder climate emerged and this permitted mammals to increase their body sizes with the intent to conserve more heat. Second, as continents merged (India with Europe, Africa got closer to Europe, and North and South America became connected), land area grew and land productivity increased, raising the available nutrients on both land and sea, and together these factors were able to support the populations of really large mammals—see Felisa A. Smith et al., 'The Evolution of Maximum Body Size of Terrestrial Mammals', *Science* 330 (2010): 1216. For those inclined to understand why herbivores become so large, a good overview from the physiological perspective is presented by M. Clauss et al., 'The maximum attainable body size of herbivorous mammals: Morphophysiological constraints on foregut, and adaptations of hindgut fermenters', *Oecologia* 136 (2003): 14–27.

9: Chota Nagpur is a corruption of the original name of a village Chutia (literally meaning 'small mouse') which lies on the outskirts of Ranchi. Chutia was the residence of the former kings of Khuhras, which was the capital of Lohardaga, Jharkhand. The rajas of Khukhra believed that they descended from the nagas (a mythical race, part human, part cobra), and the region (comprising the districts of Lohardaga, Hazaribagh, Palamau, Singhbhum and Manbhum) as a whole was called Nagpur or Chutia Nagpur. The name was gradually corrupted and eventually accepted when the British passed the Chota Nagpur Tenancy Act 1908 (under the Bengal Act VI of 1908); see Dominic Bara, *Glimpses of Life and Milieu in 19th-century Jharkhand* (Ranchi: Vikas Maitri Publishers, 2012).

10: Using deep-ocean foraminiferal $d_{18}O$ isotope ratios over the last 4 million years, one paper studied the global sea-level fluctuations (see Lorraine E. Lisiecki and Maureen E. Raymo, 'A Pliocene-Pleistocene stack of 57 globally distributed benthic $d_{18}O$ records', *Paleoceanography* 20 (2005): 1–17, and this has helped to correlate the impact of seasonal (monsoonal) climate developed after the rise of the Tibetan plateau (about 30 million years ago), the closure of the seaway between the Australian and Asian plates (around 15 million years ago) and the intensification of the monsoon approximately 10 million years ago, with the emergence of modern land masses and the flora and fauna they hold. For a more updated picture on how forests and large animals colonized South and South East Asia, see W.H. Berger, *Ocean* (Berkeley: University of California Press, 2009); and David S. Woodruff, 'Biogeography and conservation in South East Asia: How 2.7 million years of repeated environmental fluctuations affect today's patterns and the future of the remaining refugial-phase biodiversity', *Biodiversity Conservation* 19 (2010): 919–41.

11: The red panda was once considered a relative of the giant panda of China, but this is no longer the case. In fact, it is closely related to skunks and otters and emerged around 20 million years ago. The giant panda, on the other hand, belongs to the family of bears, and its ancestors evolved closer to 40 million years ago; see Roland M. Novak, *Walker's Carnivores of the World* (Baltimore: John Hopkins University Press, 2005).

12: On the geographic and genetic process of how grey langurs were isolated by the rivers Sunkosh in Bhutan and Maanas in Assam to create two distinct species of langurs, see Tashi Wangchuk, 'The Evolution, Phylogeography, and Conservation of the Golden Langur (*Trachypithecus geei*) in Bhutan', (DPhil diss., University of Maryland, 2005).

13: Soil carbon burial is the biggest source of carbon burial on land. Over the last 40 million years, greater carbon burial has led to greater cooling and the drying of global climate, which resulted in the expansion of grasses, creating large grasslands; see Gregory J. Retallack, 'Cenozoic Expansion of Grasslands and Climatic Cooling', *Journal of Geology* 109 (2001): 407–26.

14: A paper that explores how fossils became a part of myth, especially in one locality (Kurukshetra, at the base of the Siwalik range), is Adrienne Mayor, 'Place names describing fossils in oral traditions,' in *Myth and Geology*, eds. L. Piccardi, and W.B. Masse (London: Geological Society [Special Publication no. 273], 2007), pp. 245–61. '[The] mythic battle of the great Indian epic Mahabharata was said to have taken place in the rich Pliocene fossil-bone beds of the Siwalik Hills. Asthipura, "Town of Bones", was named because of the remains, thought to be those of giant heroes and war elephants, slaughtered during the legendary war.'

Also see Alexandra van der Geer et al., 'Fossil Folklore from India: The Siwalik Hills and the Mahabharata,' *Folklore* 119 (2008): 71-92.

15: Campbell Dodgson, 'The Story of Dürer's Ganda', in *The Romance of Fine Prints* (Kansas City: The Print Society, 1938), pp. 44-56. Europe's fascination with rhinos is described eloquently in L.C. Rookmaaker, 'Captive rhinoceroses in Europe from 1500 until 1810', *Bijdragen tot de Dierkunde* 43 (1973): 41.

16: As the Himalaya and Tibet rose, they became effective barriers that impacted the evolution of mammals. This is presented eloquently using an analyses of rodent fossils in Raquel López-Antoñanzas et al., 'Causal evidence between monsoon and evolution of rhizomyine rodents', *Nature Scientific Reports* 5 (2015): 9008. The rising of the Himalaya and other smaller mountains beyond it isolated species and created several distinct ones from them. In Jing Che et al., 'Spiny frogs (Paini) illuminate the history of the Himalayan region and Southeast Asia', *Proceedings of the National Academy of Sciences* 107 (2010): 13765-770, the researchers have used the genes of spiny frogs to tell the story of their evolution. According to their genetic analyses of the spiny frogs, as the Himalaya rose (between 27 and 8 million years ago), they isolated and created new species of frogs, some which became endemic within ranges in the Himalaya, while others, which dispersed eastwards into South East Asia, are widespread.

17: A common prey for smaller cats and dogs was a giant hare (*Caprolagus*), which survives today in its miniaturized version—the hispid hare, an animal which was once found from the *terai* of western Uttar Pradesh all the way to Assam, but is now rare. The ancestors of the hispid hare were spread all over, from the Levant to China, but as the Himalaya rose, the populations of the hispid hare became isolated and survived only in the lower hills of the Himalaya and the terai region. However, unlike rodents and spiny frogs, they did not speciate.

18: On how goats and their relatives evolved as the Himalaya rose, see Anne Ropiquet and Alexandre Hassanin, 'Hybrid origin of the Pliocene ancestor of wild goats', *Molecular Phylogenetics and Evolution* 41 (2006): 395-404.

19: The genetics of the big cats is complex. I relied on the following papers: Olga Phyrkina, 'Phylogenetics, genome diversity and origin of modern leopard, *Panthera pardus*', *Molecular Ecology* 10 (2001): 2617-633. This paper has estimated that the origin of modern leopard lineages took place between 470,000-825,000 years ago in Africa, which was followed by their migration into and across Asia around 170,000-300,000 years ago. The date of emergence of leopards in Asia is refined to 169,000-400,000 years ago (see Stephen J. O'Brien and W.E. Johnson, 'Big Cat Genomics', *Annual Reviews of Genomics Human Genetics* 6 (2005): 407-29) and this paper also estimated the time when intrinsic genetic variation emerged in modern cats.

21: Two papers that suggest that the size of leopards in Sri Lanka was reduced due to 'island effect', are S. Miththapala et al., 'Genetic Variation in Sri Lankan Leopards', *Zoo Biology* 10 (1991): 139-46; and S. Miththapala et al., 'Phylogeographic Subspecies Recognition in Leopards (*Panthera pardus*): Molecular Genetic Variation', *Conservation Biology* 10 (1996): 1115-132.

22: A 2016 paper by Rajeev Patnaik calibrates the evolution of modern mammals to the tectonic upheavals in the Himalaya and beyond, and the ensuing climate, sea level and vegetation changes. These, Patnaik argues, occurred in seven pulses that occurred between 23-22, 20-18, 15-14, 11-10, 9-8, 3-2 and around 1 million years ago. See Rajeev Patnaik, 'Neogene-Quaternary Mammalian Paleobiogeography of the Indian Subcontinent: An Appraisal', *Comptes Rendus Palevol* 919 (2016): 1-14. Among the mammals that followed the footsteps of the giant elephants (through the *Gomphotherium* corridor 22 million years ago) were the creodonts (a group of vicious carnivores that is now extinct) and smaller mammals, like rodents, were the first to take this trail. Rhinos, *chalicotheres* (the odd-looking large herbivores that sat on their rear legs and munched leaves from trees that are now extinct), the bear-dog (like the *Amphicyon*) and a variety of pigs left Eurasia and entered Africa between 20 and 18 million years ago, perhaps through this corridor. From the east, smaller mammals like squirrels arrived. Over the next 2 million years or so, several such exchanges took place mostly from the west. Around 7 million years ago, two elephant species, the giant-tusked *Anancus* and *Stegodon*, evolved in the Siwalik and migrated into Africa, Europe and North Asia. Gradually, between 5 and 3.5 million years ago, there was a decline in the variety of plant eaters (elephants, giraffes, deer, horses and a few primates) and a second major extinction was recorded of the giant rhinos and other large mammals, most of whom left behind no descendants. These drier conditions were, however, favourable times for rodents like rats, mouse, and other mammals like the hyrax.

12. Birth of the Whales

1: For one of the earliest records of an ancestor of a whale from India, see Kishør Kumar and Ashok Sahni, '*Remingtonocetus harudiensis*; New Combination, a Middle Eocene Archaeocete (Mammalia, Cetacea) from Western Kutch, India', *Journal of Vertebrate Paleontology* 6 (1986): 326–49.

2: Three papers provide a good introduction on the evolution of whales. These are: Mark D. Uhen, 'Evolution of Marine Mammals: Back to the Sea after 300 Million Years', *The Anatomical Record* 290 (2007): 514–22; J.G.M. Thewissen et al., 'From Land to Water: The Origin of Whales, Dolphins, and Porpoises', *Evolution Education Outreach* 2 (2009): 272–88; Philip D. Gingerich, 'Evolution of Whales from Land to Sea', *Proceedings of the American Philosophical Society* 156 (2012): 309–23. Gingerich's paper is a state-of-the-art research on the evolution of whales.

3: Whales and hippos are first cousins. The ancestor of hippos evolved in Africa 16 million years ago and exploded in number around 8 million years ago, while whales emerged from the Tethys in the Indian subcontinent from a common ancestor. Over millions of years, this common ancestor, an even-toed ungulate (or artiodactyl), evolved into vastly different varieties of mammals—the terrestrial cows and hippos on one end and the aquatic cetaceans on the other. A good introduction to how whales evolved from hoofed mammals can be found in J.G.M. Thewissen et al., 'Whales Originated from Aquatic Artiodactyls in the Eocene Epoch of India', *Nature* 450 (2007): 1190–95. The position of hippos is presented in Jean-Renaud Boisserie, Fabrice Lihoreau and Michel Brunet, 'The Position of Hippopotamidae within Cetartiodactyla', *Proceedings of the National Academy of Sciences* 102 (2005): 1537–41. Another enjoyable overview is by Donald R. Prothero and F.E. Foss, eds, *The Evolution of Artiodactyls* (Baltimore: Johns Hopkins University Press, 2007).

4: For each species, I consulted several papers of which I list just one or two. For *Indohyus*, a simple description is presented in Lisa Noelle Cooper and J.G.M. Thewissen, 'Indohyus and the Origin of Whales', *McGraw-Hill Yearbook of Science and Technology, 2010*, (New York: McGraw-Hill, 2010). To read about the discovery and first description of *Himalayacetus*, see: Sunil Bajpai and Phillip D. Gingerich, 'A New Eocene archaeocete (Mammalia, Cetacea) from India and the Time of Origin of Whales', *Proceedings of the National Academy of Sciences* 95 (1998): 15464–68. For *Nalacetus*, see Lisa Noelle Cooper, J.G.M. Thewissen and S.T. Hussain, 'New Middle Eocene archaeocetes (Cetacea: Mammalia) from the Kuldana Formation of Northern Pakistan', *Journal of Vertebrate Paleontology* 104 (2009): 1289–99. The paper which describes the discovery of *Indocetus* is Ashok Sahni and Vijay Prakash Mishra, 'Lower Tertiary Vertebrates from Western India', *Monograph of the Paleontological Society of India* 3 (1975): 1–48. Greater elaboration on new fossil discoveries made in Pakistan and India was presented by Sunil Bajpai and J.G.M. Thewissen, 'Middle Eocene Cetaceans from the Harudi and Subathu Formations of India', in *The Emergence of Whales: Evolutionary*

Patterns in the Origin of Cetacea—Advances in Vertebrate Paleobiology, ed., J.G.M. Thewissen (New York: Plenum Press, 1998), pp. 213–33.

5: A vertebra similar to that of *Remingtonocetus* has been discovered in the limestone quarry owned by Cement Corporation of India in Karbi Anglong district of Assam, a four-hour drive from Guwahati on the road to Dimapur (K. Whiso et al., 'A Fossil Mammal from Marine Eocene Strata [Jaintia Group] of the Mikir Hills, Assam, Northeastern India', *Journal of the Palaeontological Society of India* 54 (2009): 111–14).

6: Ashok Sahni and V.P. Mishra, 'Lower Tertiary Vertebrates from Western India', *Palaeontological Society of India*, Monograph 3 (1975): 1–48. This paper reported the discovery of *Remingtonocetus*, which was elaborated in a later paper (K. Kumar and A. Sahni, 'Remingtonocetus harudiensis, New Combination, from Western Kutch, India', 1986).

7: For a systematic review on *Basilosaurus*, see the introduction in Philip D. Gingerich et al, '*Basilosaurus drazindai* and *Basiloterus hussaini*, New Archaeoceti, (Mammalia, Cetacea) from the Middle Eocene Drazinda Formation, with a Revised Interpretation of Ages of Whale-Bearing Strata in the Kirthar Group of the Sulaiman Range, Punjab (Pakistan)', *Contributions from the Museum of Paleontology* 30 (1997): 55–81 (Ann Arbor, MI: University of Michigan).

8: The oldest anthropoid primate from India discovered so far (named *Anthrasimias* or coal, monkey) is from Vastan shale in Gujarat, and is dated to be about 45 million years old. This discovery suggests that India was an important centre for the diversification of the earliest primates. This mouse-sized primate, which possibly resembled a lemur, weighed about 75 grams and consumed fruits and insects. The paper also discusses other early primates discovered in this locality and other provenances (Sunil Bajpai et al., 'The Oldest Asian Record of *Anthropoidea*', *Proceedings of the National Academy of Sciences* 105 [2008]: 11093).

9: An earlier paper which describes the discoveries made and habits of early primates of South Asia is Philip D. Gingerich and Ashok Sahni, '*Indraloris* and *Sivaladapis*: Miocene Adapid Primates from the Siwaliks of India and Pakistan', *Nature* 279 (1979): 415–16.

10: Two summaries of the process to adaptation to life in freshwater and then in marine environments that I found useful are: J.G.M. Thewissen et al., 'Evolution of Cetacean Osmoregulation', *Nature* 381 (1996): 379–80. On morphological changes in bone density which enabled mammals to take to water, see Noel-Marie Gray et al., 'Sink or Swim? Bone Density as a Mechanism for Buoyancy Control in Early Cetaceans', *The Anatomical Record* 290 (2007): 638–53.

11: *Megaselachus* and other sharks from Kutch are described, along with mammals and crocodiles in (Rajeev Patnaik, 'Additional Vertebrate Remains from the Early Miocene of Kutch, Gujarat', *Special Publication of the Palaeontological Society of India* 5 [2014]: 353–65). To know more about the discovery of sharks and rays in the marine bed near Baripada in Odisha see K. Milankumar and R. Patnaik, 'Additional Fossil Batoids (Skates and Rays) from the Miocene Deposits of Baripada Beds, Mayurbhanj District, Orissa', *Earth Science India* 6 (2013): 160–84.

12: The only large carnivorous sharks that thrive in cold water are the five mackerel sharks (*Lamnidae*) which maintain a high body temperature to hunt in cooler waters. Of these, three (the White, Salmon and Porbeagle sharks) can thrive in cold waters, although they spend winters in lower latitudes (see: Leonard Compagno, Marc Dando and Sarah Fowler, eds, *A Field Guide to the Sharks of the World* [London: HarperCollins, 2005], pp. 181–85).

13: An enjoyable summary on how some land mammals increasingly took a liking to water can be found in Mark D. Uhen, 'Evolution of Marine Mammals: Back to the Sea after 300 Million Years', *The Anatomical Record* 290 (2007): 514–22.

14: After South America (which started breaking away 41 million years ago and came to its present resting place almost 23 million years ago) and Australia (which initiated the breakup 60 million years ago and reached its present position 30 million years ago) parted ways from Antarctica, there were no more exchanges of land mammals across Southern Gondwana. This is why there are no polar bears in the South Pole, but aquatic mammals like seals are found in both the northern and southern hemispheres. The ancestors of seals had evolved around 46 million years ago and were cosmopolitan. When the Southern Gondwanan land masses were still connected with land bridges (Antarctica, southern tips of Africa, South America, Australia and New Zealand), seals colonized any region that offered a cool to frigid climate. Penguins evolved in the southern hemisphere, perhaps in New Zealand around 62 million years ago or so and were restricted to areas where there were few predators, especially mammals. Penguins, therefore, were restricted to the southern hemisphere given their constraints (being flightless and threatened by predators) and lived in a few pockets before they colonized the southern, cooler margins of Gondwanan land masses. Just one species of penguin was able to cross the equator and colonize the Galapagos Islands. We don't yet know when it reached the Galapagos, given that the islands formed just 3 million years ago. It perhaps did so in cooler temperatures and travelled along the coast of Chile to reach Galapagos. This is one reason why you are unlikely to find a photograph of a polar bear (which is found in the Arctic region) with a penguin, unless it is taken in a zoo!

13. How to Make a Man

1: Among the mammals, those that stayed on trees emerged later as the ancestors of the earliest primates. Primates emerged from placental mammals, which are the most commonly occurring of all mammals. Placental mammals differ from other mammals in that the foetus is nourished during gestation via a thick spongy tissue called the placenta. The other class—called 'marsupial' (Greek *marsupion* and modern Latin *marsupialis*, for pouch), which has the kangaroo and koala as its members—differs from placental mammals. Placental mammals and marsupials diverged from a common ancestor around 110 million years ago. A handful of mammals, called '*monotremes*' (or 'single hole') retained their primitive, reptile-mammal traits such as egg-laying. This group has two members—the enigmatic platypus and the hedgehog-like echidna. Monotremes have remained unchanged since they emerged 220 million years ago. Like reptiles, they have a single orifice from which they urinate, defecate and lay eggs. For a simple yet comprehensive introduction to the origin of modern mammals, see David Macdonald, *The Encyclopaedia of Mammals* (Oxford: Oxford University Press, 2006). For a basic understanding on the origin and relationship of early primates which gives an updated estimate of the structure, relationships and tempo of primate evolutionary history, see Helen J. Chatterjee et al., 'Estimating the Phylogeny and Divergence Times of Primates Using a Supermatrix Approach', *BMC Evolutionary Biology* 9 (2009): 259. For the timing of the divergence and classes of mammals, see H. Nishihara et al., 'Retroposon Analysis and Recent Geological Data Suggest Near-Simultaneous Divergence of the Three Superorders of Mammals', *Proceedings of the National Academy of Sciences* 106 (2009): 5235–40.

2: Modern Indian primates can be divided into four groups (other than humans). In the first group are the small, goggle-eyed lorises which are found in the forests of southern and northeast India. The second group constitutes the eight species of macaques, which thrive in virtually all climates except the frigid mountains of Kashmir. The third group—langurs—has ten species as members. The fourth group comprises two species of the tailless primates known as gibbons. Only in gibbons are males and females distinct with differently coloured fur. In all other primates, the male and the female resemble each other superficially. S.H.

Prater, *The Book of Indian Animals* (Mumbai: Bombay Natural History Society, 1990) provides a good overview and illustrations of members from these groups.

3: For the habits and behaviour of *Gigantopithecus*, I referred to David W. Frayer, 'Gigantopithecus and its Relationship to Australopithecus', *American Journal of Physical Anthropology* 39 (1973): 413–26.

4: *Sivapitehcus/Ramapithecus* was thought to be one of the earliest hominins until the 1970s and this is why so much research was done in the Siwaliks by Indian and global palæontologists. On *Sivapithecus*, see Anek Ram Sankhyan, 'Late Occurrence of Sivapithecus in Indian Siwaliks', *Journal of Human Evolution* 14 (1985): 573–78; Jay Kelley, 'A New Large Species of Sivapithecus from the Siwaliks of Pakistan', *Journal of Human Evolution* 17 (1988): 305–24.

5: The European (or Eurasian) origin of African apes is described in David R. Begun, 'Dryopithecins, Darwin, de Bonis, and the European origin of the African apes and human clade', *Geodiversitas* 31 (2009): 789–816; David R. Begun, 'Sivapithecus Is East and Dryopithecus Is West, and Never the Twain Shall Meet', *The Anthropological Society of Nippon* 113 (2005): 53–64. Also see a summary by David R. Begun, 'Planet of the Apes', *Scientific American* 289 (2003): 74–84, with excellent illustrations by John Gurche.

6: There are three potential candidates of the early ape which could become the progenitor of *Homo*. Traditionally, the similarities in the skull of *Dryopithecus* and African great apes suggest that the former, which originated in eastern Europe, migrated into Africa and gave rise to the African great apes and *Homo*. A second school of thought emerged stating that Dryopithecine facial bones discovered in Spain are similar to those of the orangutan and therefore related to the *Ponginae* and not to the African apes. According to a group of Spanish palæoanthropologists, *Dryopithecus* is a better candidate for being a common ancestor of all *Hominidae* (see Salvador Moya-Sola and Meike Köhler, 'Recent Discoveries of *Dryopithecus* Shed New Light on Evolution of Great Apes', *Nature* 365 [1993]: 543–45). A third view is that they were stem homonids, which diverged creating distinct lineages of orangutans, African apes and ancestors of *Homo*.

7: The species earlier described as *Sivapithecus occidentalis* (because its enamel folding pattern was similar to *S. sivalensis* that lived in the Siwaliks) was renamed *Hispanopithecus laietanus* in 1944. There is considerable debate on the organization and nomenclature of early apes, and only more discoveries of fossils and better dating can resolve this—something not expected to occur anytime soon.

8: Prasad (see K.N. Prasad, 'Review of Miocene Anthropoidea from India and Adjacent Countries', Jurij Alxeandrovich Orlov Memorial Lecture, *Journal of the Palaeontological Society of India* 20 [1977]: 382–90) provides a comprehensive survey of major hominin discoveries and specially Dryopithecines from India and around. An earlier species discovered, which was named *Ramapithecus* (see: G.E. Lewis, 'Preliminary Notice of Man-Like Apes from India', *American Journal of Science* 27 [1934]: 161–79), is attributed to be the African Dryopithecines which may have arrived in southern Asia during the Early Miocene (David R. Begun et al., 'Dispersal Patterns of Eurasian Hominoids: Implications from Turkey', *Distribution and Migration of Tertiary Mammals in Eurasia* 10 [2003]: 23–39); Elwyn L. Simons, 'On the Mandible of Ramapithecus', *Proceedings of the National Academy of Sciences* 51 (1964): 528–35.

9: The term hominin is used for all bipedal species most closely related to *Homo sapiens* that evolved after humans and chimpanzees split. The term hominid includes all great apes, encompassing chimpanzees, gorillas, orangutans and humans. A hominin, therefore, is any bipedal species closely related to humans. Simply put, hominin are a subset of hominid. The relationship of primates, apes and early hominin is explained simply in R. Patnaik and P. Chauhan, 'India at the Cross-Roads of Human Evolution', *Journal of Bioscience* 5 (2009): 729–47.

10: To understand the interrelation of body size, metabolism, ecological dominance, sociality, life history, diet and other factors with increasing brain size in primates and early hominins, see Carol V. Ward et al., 'Body Size and Intelligence in Hominoid Evolution', in *The Evolution of Thought: Evolutionary Origins of Great Ape Intelligence*, eds Anne Russon and David Begun (Cambridge: Cambridge University Press, 2004), pp. 335–49.

11: The causes which possibly led to anatomical changes in early apes and enabled bipedalism have been reviewed in Brian G. Richmond et al., 'Origin of Human Bipedalism: The Knuckle-Walking Hypothesis Revisited', *Yearbook of Physical Anthropology* 44 (2001): 70–105.

12: A good collection of essays which presents plausible evolutionary models of early hominins, especially from the perspective of diet, can be found in Jean-Jacques Hublin and Michael P. Richards, eds, *The Evolution of Hominin Diets—Integrating Approaches to the Study of Palaeolithic Subsistence* (Netherlands: Springer, 2009).

13: At least three intestinal worms that are common to antelopes, hyenas, hunting dogs and early humans were transmitted to modern humans and their domesticated animals (dogs were the first animals that modern humans domesticated, somewhere between 135,000 and 90,000 years ago). Two papers review the crossover: Pat Shipman, 'A Worm's View of Human Evolution', *American Scientist* 90 (2002): 508; Eric Hoberg et al., 'Out of Africa: Origins of the Taenia Tapeworms in Humans', *Proceedings of the Royal Society of London B* 268 (2001): 781–87.

14: For the importance of fat and breasts, see A. Zee, 'On Fat Deposits around the Mammary Glands in the Females of Homo sapiens', *New Literary History—Views and Interviews* 32 (2001): 201–16. To know more about the emergence and 'social relevance' of buttocks, see Bobbi S. Low et al., 'Human Hips, Breasts and Buttocks: Is Fat Deceptive?' *Ethology and Sociobiology* 8 (1987): 249–57.

15: An interesting paper which reviews the evolution of bipedal locomotion with a particular emphasis on the evolution of the design of the foot is W.E.H. Harcourt-Smith and L.C. Aiello, 'Fossils, Feet and the Evolution of Human Bipedal Locomotion', *Journal of Anatomy* 204 (2004): 403–16.

16: A good review on the evolution of human birth can be found in Kate R. Rosenberg and Wenda R. Trevathan, 'The Evolution of Human Birth', *Scientific American* 13 (2003): 80–85. Apes that descended from trees and began to walk on two feet faced a challenge. They had to reconcile adapting to walk on two legs (which supported the development of biomechanically efficient, narrow pelvis) with the need to produce babies with large brains who would mature more rapidly (and therefore the need for wide pelvises). This is called Washburn's obstetrical dilemma. In 1960, Sherwood Washburn, an American anthropologist, proposed that the dilemma could be resolved by the birth of the 'foetus' at an earlier stage (compared to other great apes). The various theories proposed that counter or support Washburn have been reviewed in H. Dunsworth and L. Eccleston, 'The Evolution of Difficult Childbirth and Helpless Hominin Infants', *Annual Review of Anthropology* 44 (2015): 55–69. An April 2016 study suggests that the female pelvis becomes obstetrically suitable at the time of maximum fertility, after which it again becomes more like that of males, by reducing the size of the birth canal, which is driven by hormonal changes from the onset of puberty till the beginning of menopause. Researchers contend that the female body can modulate its pelvic dimensions 'on demand', and is not dependent on genetically fixed developmental patterns (see Alik Huseynov et al., 'Developmental Evidence for Obstetric Adaptation of the Human Female Pelvis, *Proceedings of the National Academy of Sciences* 113 [2016]: 5227–32).

17: On the role of alloparenting in *sapiens*, see an eminently readable and authoritative account in Sarah Blaffer Hrdy, *Mothers and Others: The Evolutionary Origins of Mutual Understanding* (Cambridge, MA: Harvard University Press, 2011).

18: Leopards and large cats were *Erectus*'s main nemesis. The menacing, sabre-toothed cats died out around 120,000 years ago in South Asia and Africa, and were replaced by smaller cats in South East Asia and Eurasia. This perhaps may have helped the migration of *Erectus* into Asia. Sabre-toothed cats survived in the Americas until 90,000 years ago.

19: How early *Homo* discovered and used fire and made sophisticated tools has been discussed in Kirkpatrick Sale, *After Eden: The Evolution of Human Domination* (Durham, NC: Duke University Press, 2006).

20: Our ape ancestors evolved in relatively cooler tropical forests, but as the climate warmed up, their fur became a liability and attracted more parasites. It also heated the body more, affecting the brain. With the discovery of fire that could keep them warm and the ability to wear the fur of other mammals, hair on the human body became less useful for the early *Homo*. As a result, the hair thinned and also reduced in number. The timing of when exactly hair reduced has remained elusive. Hair doesn't fossilize, and this leaves scientists to use genetics to estimate approximately when the genes that code for specific aspects of fur or hair began to change. A most promising lead for geneticists has been to study the mutation on MC1R (Melanocortin 1 Receptor) that leads to the development of blond hair, which geneticists have estimated evolved about 11,000 years ago. A combination of research on the evolution of mutation that causes reduction of hair or change in colour of hair has offered functional explanations and evolutionary correlates to different aspects of fur/hair loss. Scientists now agree that several factors worked together towards the reduction of the numbers and thickness of hair. These include environmental reasons (increased contact with water for ablution and fishing), diets (long phases of hunger, low nutrient foods and low micronutrient intake), lifestyle (use of cloth during the Ice Age, and the adaptation of a new species of lice (*Pediculus*) and infectious diseases (especially hookworm, *Taenia* and hepatitis), and these together may have led to the reduction in 'fur' and its transformation to hair. It should also be noted that much of the hair on human body is so short and flimsy that it is practically invisible to us, and several follicles that lie on the surface may not throw up any hair at all. For a good overview on why hair began to reduce in ancestors of humans, see G.D. Ruxton and D.M. Wilkinson, 'Avoidance of Overheating and Selection for Both Hair Loss and Bipedality in Hominins', *Proceedings of the National Academy of Sciences* 108 (2011): 20965–69.

21: We don't know that early apes breathed this way, but research by Takeshi Nishimura on laryngeal descent (https://www.pri.kyoto-u.ac.jp/shinka/keitou/nishimura-HP/tn_res-e.html) and the earlier works of Marc Verhaegen suggests that as the body cooled, panting reduced and the tongue and voice box began to descend, making it versatile enough to produce a variety of sounds. Another study on vocalization in early humans is: Philip Lieberman and Robert McCarthy, 'Tracking the Evolution of Language and Speech, Comparing Vocal Tracts to Identify Speech Capabilities', *Expedition* 49 (2007): 15–20.

22: The evidence that *erectus* probably lived in rocky shelters, their use of fire which they used for cooking (assumed from discovery of charred bone remains) comes from the 700,000-year-old Kao Poh Nam outcrop in Thailand. There is no evidence whether *Erectus* had knowledge to create fires and this may be hard to prove. We know that perhaps the *Erectus* or *Sapiens* that created the first flints could do so, although one can make fire without rocks and sparks also. A natural branch caught on fire can also be used to light a 'hominin fire', but there is no evidence of when controlled fire emerged. The earliest evidence of fire being harnessed was found in Swartkrans (about 1.5 million years ago) and other sites in South Africa, such as the Cave of Hearths (various hearths dated from 700,000 to 200,000 years ago), the Wonderwerk caves (about 1 million years ago), the Montagu Cave (200,000 to 58,000 years ago) and at the Klasies River Mouth (130,000 to 120,000 years ago). Zhoukoudian, China (dated 460,000 to 230,000 years ago) and two sites in Spain (Torralba and Ambrona, dated 500,000 to 300,000 years ago) also show signs of fire, tools and charred bones in cave systems.

23: The argument that cooking made us human was first made by Richard W. Wrangham, *Catching Fire: How Cooking Made Us Human* (London: Profile Books, 2010). Although many biologists who study human evolution have challenged this, Wrangham's theory has received support from other fronts as well. One study addressed the quandary of why humans developed a larger brain size per body size that sets them apart from other primates. A study published in March 2016 by Wrangham's colleagues at Harvard University suggests that even before the discovery of fire, *Erectus* may have used stone tools to pound, dice and slice meats and vegetables. This pre-processing would have also aided digestion and improved the retention of nutrients from food. Their research has found that a diet of one-third sliced meat and two thirds pounded vegetables would ensure that *Erectus* would spend 27 per cent less effort in chewing when compared to unprocessed meat and vegetables. This, the authors say, translates to at least 2.5 million fewer chewing actions, which possibly led to the shrinking of the jaw and teeth in later *Erectus* and other species that evolved. The consequence of this on the digestive system and human physiology is a matter of further research (Katherine D. Zink and Daniel E. Lieberman, 'Impact of Meat and Lower Palaeolithic Food Processing Techniques on Chewing in Humans', *Nature* 531 [2016]: 500–3).

24: A mathematical study analysed the metabolic costs and trade-offs of supporting a human-sized brain (the third most energy-consuming organ after muscle and the liver). Raw food alone cannot provide so much energy to the body, and cooked food alone allowed *Erectus* to grow larger brains. The authors suggest that eating cooked food had other advantages—*Erectus*, and later *Sapiens*, spent more time together around fire, and developed social and cognitive skills like language, art and social order (Karina Fonseca-Azevedo and Suzana Herculano-Houzel, 'Metabolic Constraint Imposes Tradeoff between Body Size and Number of Brain Neurons in Human Evolution', *Proceedings of the National Academy of Sciences* 109 (2012): 18571–76. Support for this theory comes from another unexpected quarter. Mammals sustain larger bodies and brains by spending more time feeding and sleeping. In researching brain chemistry in rats and humans, one study found that increasing numbers of cortical neurons led to decreased sleep requirement in evolution that allowed for more hours of feeding and increased body mass. This led to an increase in the number of brain neurons which mammals, especially humans, compensate through intensive feeding. Higher numbers of neurons linked to declining sleep requirements and more time spent on feeding perhaps allowed mammals to increase their size of brain and body mass, and proved to be a critical driver in the evolution of mammals (see Suzana Herculano-Houzel, 'Decreasing Sleep Requirement with Increasing Numbers of Neurons as a Driver for Bigger Brains and Bodies in Mammalian Evolution', *Proceedings of the Royal Society of London B: Biological Sciences* 282 [2015]: 41–51).

25: One of the most complete fossils of a bear-dog discovered from the subcontinent has been described in Suvi Viranta et al., 'The Anatomical Characteristics of a Giant Miocene Amphicyonid (Carnivora) Humerus from Pakistan', *Pakistan Journal of Zoology* 36 (2004): 1–6.

26: Rajeev Patnaik, 'Fossil Murine Rodents as Ancient Monsoon Indicators of the Indian Subcontinent', *Quaternary International* 229 (2011): 94–104. This paper compares the shape of teeth (dentition) of extinct and extant (or modern) rodents from different sites of the Indian subcontinent, and shows that around 4–5 million years ago, the Siwalik region received annual rainfall (between 1000 and 2000 mm), and as the monsoon intensified (between 3.6 and 2.6 and 2 to 1.8 million years ago), the rainfall doubled and became a constant feature. More recent evidence comes from the study of fossils of extinct rabbits, which suggest that they had spread along flat grasslands that evolved in Europe, but over a period of 8 million years or so (23 to 14 million years ago) they spread across Asia once the Tethys Sea closed, and became extinct around 5 million years ago (see Margarita A. Erbajeva et al., 'A New Species of the Genus *Amphilagus* [*Lagomorpha, Mammalia*] from the Middle Miocene of South-Eastern Siberia', *Historical Biology* 28 [2016]: 199–207).

14. Citius, Altius, Fortius

1: The geography of the Siwalik mountains and early fauna and flora is described in Edwin H. Colbert, 'Siwalik Mammals in the American Museum of Natural History', *Transactions of the American Philosophical Society*, new series, 26 (1935): 1–20. This paper also reviews the fossil collection in the American Museum of Natural History collected during the 1925 Siwalik Hills Indian Expedition of the American Museum of Natural History. An updated review of the Siwalik mammals before and during the time of the arrival of *H. erectus* is done by Rajeev Patnaik. See Rajeev Patnaik, 'Siwalik Palaeoecology and Early Hominin Dispersals', in *Asian Perspectives on Human Evolution*, ed., Anek R. Sankhyan (New Delhi: Serials Publications, 2009), pp. 43–51. There is now consensus that the Siwalik served as a highway for the exchange of species from Eurasia and South East Asia, especially during the middle Pleistocene (780,000 to 126,000 years ago).

2: An excellent overview of the relationship between *Australopithecus* and *Homo* is presented on the Tree of Life website (http://tolweb.org/treehouses/?treehouse_id=4438, as viewed on 12 January 2016). The website states that *Australopithecus anamensis* found in Kenya and Ethiopia was perhaps the first to evolve, and lived between 4.17 to 3.9 million years ago.

3: In July 2011, a team of archaeologists from Stony Brook University, New York accidentally discovered a site (which they labelled Lomekwi 3) with stone artifacts which are believed to be 3.3 million years old and made by *Australopithecus* or *Kenyanthropus*. These are the oldest stone tools discovered thus far and appear about half a million year older than tools made the earliest Homo, confirming that stone tool making had arisen much earlier than previously believed (see Sonia Harmand et al. (2015). 3.3-million-year-old stone tools from Lomekwi 3, West Turkana, Kenya; Nature. 521 (7552): 310–315.).

4: For a good account on early settlers, sites and tools found in India, see Irfan Habib, *People's History of India—Part 1: Prehistory* (Chennai: Aligarh Historians Society and Tulika Books, 2001). In addition, see A.P. Khatri, '"Mahadevian": An Oldowan Pebble Culture of India', *Asian Perspectives* 6 (1962): 186–97, which builds on the earliest discovery of the Oldowan tools from the Mahadev-Piparia region between Hoshangabad and Pachmarhi in Madhya Pradesh, India, but this is rather contentious. More recently, Oldowan-Acheulean intermediate tools have been discovered in the Chambal region (see: Robert G. Bednarik and Giriraj Kumar, 'Typological Context of the Lower Palaeolithic Lithics from Daraki-Chattan Cave, India' (paper presented at the symposium on Pleistocene art of Asia, IFRAO Congress, France, 6–11 September 2010, available at http://www.ifrao.com/wp-content/uploads/2015/08/12Asia3.pdf). But here is a word of caution. Oldowan generally precedes *Erectus*, and mostly Acheulean tools are associated with *Erectus*. Some palaeontologists in India believe that Oldowan tools have been found at Riwat, southeast of Rawalpindi, which lies in the Soan valley in Western Punjab of Pakistan, and in the Siwalik rocks of Himachal Pradesh. See a good overview on discoveries made in the Indian subcontinent by Parth Chauhan, 'An Overview of the Siwalik Acheulian & Reconsidering Its Chronological Relationship with the Soanian—A Theoretical Perspective', *Assemblage* 7 (2003), available at http://www.assemblage.group.shef.ac.uk/issue7/chauhan.html.

5: Tools found in South Delhi. See Nayanjot Lahiri and D.K. Chakrabarti, 'A Preliminary Report on the Stone Age of the Union Territory of Delhi and Haryana', *Man and Environment* 11 (1987): 109–16. There is an interesting incident about how a professor accidentally found a stone tool at a bus stop at the Jawaharlal Nehru University in New Delhi, which is described in Mala Dayal, *Celebrating Delhi* (New Delhi: Penguin Books India, 2010) p. 17.

6: A leading Indian anthropologist, Ravi Korisettar, has studied the nature of distribution of archaeological sites discovered in peninsular India and has suggested that the journey of *Erectus* does not support the coastal model across the subcontinent and into South East

Asia. Instead, Korisettar argues that the first settlements were made around the western and central Indian river basins. While this is evidence-based, it still needs to be validated further because the subcontinent's coasts are under-explored, and also, early sites could have submerged with the rise of the sea level.

7: It may premature to label the stone tools of Dhansi as Oldowan, the significance of this site is that it is perhaps the oldest evidence of human occupation in central India, older than 780,000 years. Archaeological Survey of India had first proposed the Pahalgam tool to be between 700,000 and 500,000 years old (see B.B. Lal ed., *Indian Archaeology 1969-70: A Review* (New Delhi: Archaeological Survey of India, 1973), p. 13, available at http://asi.nic.in/nmma_reviews/indian%20archaeology%201969-70%20a%20review.pdf), but these dates need to be re-evaluated. The earlier method of estimating dates was based on glacial cycles associated with terraces where tools were found, but this is generally not accepted widely.

8: The general route and details are provided by the National Geographic website: National Geographic, 'Global Human Journey', editor Caryl-Sue Micalizio, producer Sean P. O'Connor, *Media Spotlight*, video documentary, 2013, http://education.nationalgeographic.org/media/global-human-journey/. The overall consensus among anthropologists is that *Sapiens* had used the coastal route along Arabia and Persia and continued travelling along the coast of Gujarat, occasionally making inroads along major rivers of the west coast like the Indus, Narmada and Tapti.

9: The discovery of a skull cap in 1982 in Hathnora is described in Arun Sonakia, 'A Skull-Cap of Early Man and Associated Mammalian Fauna from Narmada Valley Alluvium, Hoshangabad Area, Madhya Pradesh (India)', *Records of the Geological Survey of India* 113 (1984): 159-72; and Arun Sonakia and Kenneth A.R. Kennedy, 'Skull Cap of an Early Man from the Narmada Valley Alluvium (Pleistocene) of Central India', *American Anthropologist*, new series 87 (1985): 612-16. The account of the discovery of the 'Narmada man' (Kenneth A.R. Kennedy, 'Prehistoric Skeletal Record of Man in South Asia', *Annual Review of Anthropology* 9 [1980]: 391-432) presents several attempts made to find hominid fossils in South Asia.

10: Since the discovery in Hathnora, a few other discoveries were reported but remain understudied and unconfirmed. The discovery of a calvarium (skull cap) from another site near Hathnora has been reported in A.R. Sankhyan et al., 'New Human Fossils and Associated Findings from the Central Narmada Valley, India', *Current Science* 103 (2012): 1461-69.

11: A diagnosis of the Hathnora skull confirms that it is closer to *Sapiens* although it bears a few characteristics of a late *Erectus* and *H. neanderthalensis*. The dating was for bovine bones and equid teeth using ESR and U-series, which show that the deposits are as young as 50,000 years (see D. Cameron et al., 'The Phylogenetic Significance of the Middle Pleistocene Narmada Hominin Cranium from Central India', *International Journal of Osteoarchaeology* 14 [2004]: 419-47) The Narmada woman, therefore, can also be younger. Also see Kenneth A.R. Kennedy et al., 'Is the Narmada Hominid an Indian Homo erectus?', *American Journal of Physical Anthropology* 86 (1991): 475-96, which raised initial doubts before arriving at the current accepted term—anatomically modern human.

12: Briana Lee Pobiner, 'Hominin-Carnivore Interactions: Evidence from Modern Carnivore Bone Modification and Early Pleistocene Archaeofaunas (Koobi Fora, Kenya; Olduvai Gorge, Tanzania)' (PhD diss., submitted to Rutgers, 2007). The dissertation presents the interactions between Oldowan hominins and larger carnivores and how they possibly shaped hominin adaptation, including morphology, foraging patterns, habitat preferences and social behaviour.

13: The Attirampakkam site in Tiruvallur district, the northern-most district of Tamil Nadu, has been studied since 1863. See the website of The Sharma Centre for Heritage Education, Chennai: http://www.sharmaheritage.com/index.php/research/attirampakkam

15. The Promised Land

1: For evidence that the early *Homo* species coexisted between 2 and 1.5 million years ago, see M.G. Leakey et al., 'New Fossils from Koobi Fora in Northern Kenya Confirm Taxonomic Diversity in Early Homo', *Nature* 488 (2012): 201–04.

2: The 'Out of Africa' event occurred twice, first with *Erectus* in several waves, and then with *Sapiens*, who left in small bands. *Erectus* left Africa around 1.9 million years ago and migrated around the tropical latitudes of Asia and Southern Europe, which, till this time, possessed the remnant of tropical forests. For a good account of competing theories proposed for the timings and routes taken by *Erectus* and *Sapiens*, see Amanuel Beyin, 'Upper Pleistocene Human Dispersals Out of Africa', *International Journal of Evolutionary Biology* (2011), Article ID 615094, 1–17.

3: Arabia was a possible bridge for the dispersal of anatomically modern humans (*Sapiens*) out of Africa. This land mass offered plenty of fresh water and food between 75,000 and 105,000 years ago, following which it rapidly turned arid and the lakes dried out (T.M. Rosenberg et al., 'Humid Periods in Southern Arabia: Windows of Opportunity for Modern Human Dispersal', *Geology* 39 [2011]: 1115–18).

4: A review of stone tools suggests that migration of *Sapiens* was along rivers and lakes in the Sahara, suggesting that a possible was taken out of Africa to enter Europe. This has been proposed by Eleanor M.L. Scerri et al., 'Earliest evidence for the structure of *Homo sapiens*: populations in Africa, *Quaternary Science Reviews* 101 (2014): 207–16.

5: A mathematic and genetic study rules out the Sinai route for out of Africa and provides new evidence that the *sapiens* may have taken the Bab-el-Mandeb strait into Asia. See Marta Melé et al. and the Genographic Consortium, 'Recombination Gives a New Insight in the Effective Population Size and the History of the Old World Human Populations', *Molecular Biology and Evolution* 29 (2012): 25–30.

6: The global dispersal of modern humans occurred through two major waves: an initial eastern dispersal around the Indian Ocean starting 75,000–62,000 years ago, and a later dispersal into Eurasia 38,000–25,000 years ago (Morten Rasmussen et al., 'An Aboriginal Australian Genome Reveals Separate Human Dispersals into Asia', *Science* 334 [2011]: 94–98). This time frame of 75,000 to 70,000 years ago is again confirmed through the use of genetic analyses that map the co-evolution of the tuberculosis bacterium in *Homo* population that migrated and grew. The genetic analyses of the bacterium's mutations in specific sets of genes confirm when humans left Africa and suggest patterns of dispersion and possible population size (Iñaki Comas et al., 'Out-of-Africa Migration and Neolithic Co-Expansion of Mycobacterium Tuberculosis with Modern Humans', *Nature Genetics* 45 [2013]: 1176–82).

7: There were at least three eruptions from Toba, and the last one, called the Youngest Toba Tephra or YTT, which occurred about 74,000 years ago, was the largest and most intense. (When the magma reaches the surface, rapidly expanding gas bubbles get released in a massive explosion, and the molten fragments of volcanic rock that disperse far and wide are called tephra). The three eruptions from Toba have near similar chemical fingerprints, hence it is difficult to distinguish these except through direct dating from the deposit in ash beds. No Toba ash has been dated older than 74,000 in India, and therefore, it is the last Toba event which is of interest to us from the perspective of the evolution of the *Sapiens*. The role of the Toba volcano in Sumatra, Indonesia, in the dispersal of *Sapiens* has been a topic of immense interest, but has attracted few researchers. The cooling effects of the Toba volcanic eruption caused the forest to be replaced by open grassland in central India, which potentially created a population bottleneck for *Homo* and other mammals (see Martin A.J. Williams et al., 'Environmental Impact of the 73 Ka Toba Super-Eruption in South Asia',

Palaeogeography, Palaeoclimatology, Palaeoecology 284 [2009]: 295–314). Michael Petraglia of Oxford University and Ravi Korisettar of Andhra University have worked for over two decades on studying the impact of Toba in India and Asia. The complete work of Petraglia et al. can be found at www.researchgate.net/profile/Michael_Petraglia. In particular, the following publications are useful: Michael D. Petraglia and Bridget Allchin, eds, *The Evolution and History of Human Populations in South Asia: Inter-Disciplinary Studies in Archaeology, Biological Anthropology, Linguistics and Genetics* (Dordrecht: Springer, 2007), and M. Petraglia et al., 'Middle Palaeolithic Assemblages from the Indian Subcontinent before and after the Toba Super-Eruption', *Science* 317 (2007): 114–16.

8: Tigers are less closely related to lions, leopards and jaguars and closer to the snow leopard. In August 2010, a team of scientists discovered a skull of what they believe was the ancestor to all modern cats in southwest Tibet. They called this *Panthera blytheae*, and the fossil has been dated to be between 5.95 and 4.1 million years old (see Z. Jack Tseng et al., 'Himalayan Fossils of the Oldest Pantherine Establish Ancient Origin of Big Cats', *Proceedings of the Royal Society of London B: Biological Sciences* 28 [2014]: 24–42). Another study (Brian Davis et al., 'Supermatrix and Species Tree Methods Resolve the Phylogenetic Relationship within Big Cats, Panthera (Carnivora: Felidae)', *Molecular, Phylogenetics and Evolution* 56 (2010): 64–76) conducted genetic analysis of all large cats and found that the tiger began evolving 3.2 million years ago, from an ancestor that was related to the ancestor of the endangered snow leopard. Although all of these cats are grouped together under one genus (*Panthera*), they are different from the other smaller cats. The relationship between large cats is entangled with perhaps several crossovers among cousins before they became distinct species. Their ancestor split from other cats around 4.3 to 3.8 million years ago, and the ancestors of tigers and snow leopards were the first to branch off, around 3.9 million years ago, while lions and leopards split from one other about 3.1 to 1.95 million years ago.

However, the Toba event affected the population of large mammals, like tigers, elephants and even langurs, in South Asia, and this is seen through genetic studies (see: Shu-Jin Luo, *What Is a Tiger? Genetics and Phylogeography, Tigers of the World*, second edition [Amsterdam: Elsevier, 2009]). Luo has analysed the genetic diversity of the modern tiger and finds that this was constrained when the Toba volcano super-eruption in Sumatra, Indonesia, reduced their range and broke up the population of the tigers (called a demographic bottleneck). Changes in the ecology and isolation led to the evolution of a distinct tiger subspecies. The timing of this constraint matches with the occurrence of the Toba super-eruption event that occurred approximately 72,500 years ago and led to the recent coalescence time for extant tiger populations. Tigers subsequently recolonized India and the rest of Asia with the return of a suitable climate and habitat (see also Shu-Jin Luo et al., 'Phylogeography and Genetic Ancestry of Tigers [Panthera tigris]', *PLoS Biology* 2 [2004]: 2275–93). Geneticists have used sophisticated tools and have found that lions had established in east and southern Africa between 324,000 and 169,000 years ago, and expanded about 100,000 years ago into Central and North Africa and then into Asia (see Agostinho Antunes, 'The Evolutionary Dynamics of the Lion Panthera leo Revealed by Host and Viral Population Genomics', *PLoS Genetics* 4 [2008]; available online at http://journals.plos.org/plosgenetics/article?id=10.1371/journal.pgen.1000251). Lions came to India from Eurasia 100,000 years before present, while tigers entered India from South East Asia around 72,000 years before present.

9: To understand the behaviour of Neanderthals, I consulted Kirkpatrick Sale, *After Eden: The Evolution of Human Domination* (Durham, NC: Duke University Press, 2006). Molecular studies suggest that Neanderthal and *Sapiens* DNA indicate that the latter left Africa much earlier. *Sapiens* mixed with Neanderthals around 100,000 years ago outside Europe and then again around 50,000 years ago. The genetic study which found disease-causing genes in Neanderthals which may have been transmitted to *sapiens* is Corrine N. Simonti et al., 'The Phenotypic Legacy of Admixture between Modern Humans and Neanderthals', *Science* 351 [2016]: 737–41).

10: Genetic studies suggest that the Jarawa of the Andamans are 'relict' populations of those who migrated out of Africa and have since survived in (genetic) isolation, perhaps as long as 65,000 years ago. In contrast, the Shompen of the Nicobar are closely related to populations from South East Asia and are more recent residents of the Islands, and perhaps reached there about 10,000 years ago. See Kumarasamy Thangaraj et al., 'Reconstructing the Origin of Andaman Islanders', *Science* 308 (2005): 996; and Kumarasamy Thangaraj et al., 'Response to Comment on "Reconstructing the Origin of Andaman Islanders"', *Science* 311 (2006): 470. Also see Hua-Wei Wang et al., 'Mitochondrial DNA Evidence Supports Northeast Indian Origin of the Aboriginal Andamanese in the Late Paleolithic', *Journal of Genetics and Genomics* 38 (2011): 117–22. It is now widely accepted that the Jarawa bear close affinity to ancestral South Indian groups (D. Reich et al., 'Reconstructing Indian Population History', *Nature* 461 [2009]: 489–94.). For a comprehensive review of the genetic origins of the people of India, see Rakesh Tamang and Kumarasamy Thangaraj, 'Genomic View on the Peopling of India', *Investigative Genetics* 3 (2012): 20.

11: Excavations in the sediments (some of which are about 25 metres thick) in the Kurnool Caves have revealed several diverse mammals and suggest that these were rock shelters of early human settlers. A part of a jaw and a tooth has been found and has been attributed to *Homo sapiens*, but these have not been investigated or dated (K.N. Prasad, 'Pleistocene Cave Fauna from Peninsular India', *Journal of Caves and Karst Studies* 58 [1996]: 30–34; Paul S.C. Tacon et al., 'Mid-Holocene Age Obtained for Nested Diamond Pattern Petroglyph in the Billasurgam Cave Complex, Kurnool District, Southern India', *Journal of Archaeological Science* 40 [2013]: 1787–96). The caves could have been occupied by successive occupants, but determining the age remains challenging and controversial. Also see Michael Petraglia et al., 'Human Occupation, Adaptation and Behavioral Change in the Pleistocene and Holocene of South India: Recent Investigations in the Kurnool District, Andhra Pradesh', *Eurasian Prehistory* 6 (2009): 119–66, which presents the findings from archaeological studies done the Kurnool caves.

12: Although numerous *Sapiens* fossils and bones have been found in South Asia like in Badatomba Lena in Sri Lanka, and Bhimbetka, Jwalapuram and several sites in north India and Gujarat, none of these are over 40,000 years old. The earliest-dated fossil of anatomically modern *Homo sapiens* from South Asia was found in Sri Lanka. See Kenneth A.R. Kennedy and Siran U. Deraniyagala, 'Fossil Remains of 28,000-Year-Old Hominids from Sri Lanka', *Current Anthropology* 30 (1989): 394–99.

13: Two reviews describe the early quest of *Homo* in India. The first, by A.P. Khatri, presents a bibliography of studies, reports and museum records covering 100 years (1863–1963) of discovery of tools and artefacts from across India (A.P. Khatri, 'A Century of Prehistoric Research in India', *Asian Perspectives* 6 [1962]: 169–85). The second, by Kennedy, presents a scathing critique of the state of anthropology in South Asia and more importantly India, and criticises the means and methods, institutions and the people behind them, and their collective failures (see: Kenneth A.R. Kennedy, 'The Uninvited Skeleton at the Archaeological Table: The Crisis of Paleoanthropology in South Asia in the Twenty-First Century', *Asian Perspectives* 42 [2003]: 352–67). See also Jay Kelley, 'The Hominoid Radiation in Asia', in *The Primate Fossil Record*, ed., Walter Carl Hartwig, (Cambridge: Cambridge University Press, 2002), which presents a short history of discoveries of hominids made in the Indian subcontinent.

14: Irfan Habib (*People's History*, 2001) presents an overview of stone tools and industry in the Indian subcontinent. More recent discoveries in Atbarapur can be found in Claire Gaillard et al., 'Technological Analysis of the Acheulian Assemblage from Atbarapur in the Siwalik Range (Hoshiarpur District, Punjab)', *Man and Environment* XXXIII (2008): 1–14; and Claire Gaillard et al., 'Lower and Early Middle Pleistocene Acheulian in the Indian Sub-Continent', *Quaternary International* 223–224 (2010): 234–41.

15: Stone artefacts from Attirampakkam date the earliest Acheulian levels as no younger than 1.07 million years ago (1.51–0.07 million years ago). The results were surprising and suggest that India was already occupied by hominins with the ability to make Acheulian tools like hand-axes and cleavers, indicating that the spread of bifacial technologies across Asia occurred earlier than previously known (Shanti Pappu et al., 'Early Pleistocene Presence of Acheulian Hominins in South India', *Science* 331 [2011]: 1596–99).

16: August G. Costa, 'First Late Pleistocene Fossil Fauna from the Dry Coastal Zone of Tropical Southwestern Asia: Implications for Early Human Dispersals along Inundated Coastal Corridors' (poster presented at Paleoanthropology Society Annual Meeting, [NESPOS], Department of Anthropology, Indiana University, 2013). This poster describes fossil discoveries near the seaside village of Gopnath in Bhavnagar district of Gujarat, which was perhaps a coastal oasis which was formed when the onset of glaciation reduced sea levels, and helps chart the path of human dispersals out of Africa and into India.

17: A collection of palæontological, anatomical and anthropological researches of the changes in human brains with changes in diet and environment, and how this led to the evolution of human behaviour and society is described in Stephen C. Cunnane and Kathlyn M. Stewart, eds, *Human Brain Evolution, The Influence of Freshwater and Marine Food Resources* (Hoboken, NJ: Wiley, 2010). On the evolution of human bipedal locomotion with a particular emphasis on the evolution of the foot, see W.E.H. Harcourt-Smith and L.C. Aiello, 'Fossils, feet and the evolution of human bipedal locomotion', (2004).

18: Two papers that explain the role of parents and grandparents among *Sapiens* are James E. Coxworth et al., 'Grandmothering Life Histories and Human Pair Bonding', *Proceedings of the National Academy of Sciences* 112 (2015): 11806–11; F. Marlowe, 'Paternal Investment and the Human Mating System', *Behavioural Processes* 51 (2000): 45–61.

19: A recent summary of possible modes of dispersal can be found in the research by Huw S. Groucutt et al., 'Rethinking the Dispersal of Homo sapiens Out of Africa', *Evolutionary Anthropology* 24 (2015): 149–64. The authors conclude: 'A difficulty in inferring the routes and chronology of *Homo sapiens*' dispersal out of Africa is that it requires integration of many different sources of information, each with its own ambiguities and assumptions.'

20: A review of analyses of tools (specifically something called point technology, which studies how stones are shaped into more sophisticated tools) from the Thar Desert suggests that these are perhaps contemporaneous with those found in North Africa and Arabia, which indicates that the dispersal of *Sapiens* took place around 71,000 years before present (James Blinkhorn et al., 'Middle Palaeolithic Point Technologies in the Thar Desert, India', *Quaternary International* 382 [2015]: 237–49). When did *Sapiens* leave Africa and under what conditions remains a prickly issue among anthropologists and divides them into camps. One set of adherents believes that *Sapiens* had already arrived in India before the Toba event took place. Stone tools found above and below the ash layer, they believe, belong to *Sapiens*. Others differ and believe that the cooling forced *Sapiens* to leave Africa, when sea levels receded. This helped them to hop across land bridges (see Stephen Oppenheimer, 'A Single Southern Exit of Modern Humans from Africa: Before or after Toba?', *Quaternary International* 258 [2012]: 88–99), and reach India between 70,000 and 60,000 years before present (see James Blinkhorn et al., 2015). A third school of anthropologists, like Paul Mellars of the University of Cambridge, UK, have studied the archaeological and cultural evidence of anatomically modern humans and believes that the *Sapiens* could not have left Africa before the Toba event and left only around 60,000 years ago, based on the earliest art outside of Africa which is similar to that found in Africa (Paul Mellars et al., 'Genetic and Archaeological Perspectives on the Initial Modern Human Colonization of Southern Asia', *Proceedings of the National Academy of Sciences* 110 [2013]: 10699–704). Recent genetic studies of diseases (human mitochondrial diseases as well as infectious diseases that have co-evolved with humans, like tuberculosis and malaria, among others) and the timing when they crossed over and spread across the world suggest the likely time to be before 60,000 years ago and after 72,000 years ago (or post-Toba).

21: Other than the fossils of *Sapiens* found in Israel, there have been recent discoveries in eastern China, including numerous teeth of a similar age, and these are summarized in Song Xing et al., 'Hominin Teeth from the Early Late Pleistocene Site of Xujiayao, Northern China', *American Journal of Physical Anthropology* 156 (2015): 224–40.

22: The most accessible book on how genetics is used to trace the history of the ancestors of *Homo sapiens* and their journey out of Africa is Spencer Wells, *Journey of Man—A Genetic Odyssey* (Princeton, NJ: Princeton University Press, 2002). It remains, to date, the most enjoyable work on the subject. More recent studies of the complete human mitochondrial DNA (mtDNA) sequence data from aboriginal populations like those in Malaysia (see V. Macaulay et al., 'Single, Rapid Coastal Settlement of Asia Revealed by Analysis of Complete Mitochondrial Genomes', *Science* 308 [2005]: 1034–36), and the Andaman Islands (K. Thangaraj et al., 'Reconstructing', 2005) indicate that these populations on the Indian Ocean coast contain some of the oldest mtDNA lineages outside Africa. This evidence supports an expansion along the southern route. The proposed timing of a southern migratory path was between 65,000 and 60,000 years ago, when the sea level was at a local minimum, which would have made crossing the Red Sea possible, see P. Forster, 'Ice Ages and the Mitochondrial DNA Chronology of Human Dispersals: A Review', *Philosophical Transaction of the Royal Society of London B* 359 (2004): 255–64.

23: Although the argument that the Chinese are descendants of *Erectus* (or Peking man as the *Erectus* fossil was called) has been debunked by Western science, it remains acceptable among Chinese paleoanthropologists. Chinese anthropologists have claimed until recently that the facial features of the Chinese were present in the *Erectus* (as interpreted by them through fossils). Barry Sautman, a social scientist at the Hong Kong University of Science and Technology, has argued that China has used palaeoanthropology as means of propaganda for racial nationalism to project superiority of the race and Chinese-ness. See B. Sautman, 'Peking Man and the Politics of Paleoanthropological Nationalism in China', *Journal of Asian Studies* 60 (2001): 95–124. The quote of Chinese as being 'earth's original inhabitants' is from Lin Yan, *Zhonggup minzu de youlai (Beginnings of the Chinese Race)* (Shanghai: Yongxiang yinshuguan, 1947), pp. 27–29. A 'Sino-centric worldview' about being a distinct race manifests in Chinese society; see an excellent social commentary by Jiayang Fan, 'Lessons about China and Race from a Detergent Ad', 9 June 2016, *The New Yorker*, available at http://www.newyorker.com/news/daily-comment/lessons-about-china-and-race-from-a-detergent-ad.

24: The oldest primate skeleton (*Archicebus*), which is about 55 million years old, has been discovered in China. The fossil is important and suggests that primates had diverged from tarsiers and anthropoids from around this time or perhaps ever earlier (X. Ni et al., 'The Oldest Known Primate Skeleton and Early Haplorhine Evolution', *Nature* 498 (2013): 60–64).

25: The current controversies, reconciliations and agreements among some paleo-anthropologists who work in the subcontinent have been discussed in P.R. Chauhan and R. Patnaik, 'Inter-Disciplinary Perspectives on Indian Paleoanthropology and Prehistory', *Quaternary International* 269 (2012): 1–8.

26: Dates obtained on calcrete from the Acheulian site of Nevasa in Maharashtra and Yedurwadi in Karnataka are more than 350,000 years old (S. Mishra, 'The Age of the Acheulian in India—New Evidence', *Current Anthropology* 33 (1992): 325–28), and these are not for the tools. Also, a more recent dating study of the calcrete has found that these are no older than

200,000 years. There is no definitive age for Nevasa artefacts yet.

27: A lifetime's work of K. Paddayya's research has contributed to the current understanding of hand tools, especially in three sites—Hunsgi-Baichbal Basins, Budihal and Isampur. The dating of the Isampur tools as being 1.2 million years old was done by Paddayya (K. Paddayya et al., 'Recent Findings on the Acheulian of the Hunsgi and Baichabl Valleys, Karnataka, with Special Reference to the Isampur Excavation and Its Dating', *Current Science* 83 [2002]: 641–47), but has been challenged (see S.K. Acharyya's letter and the response by Paddayya in *Current Science*, http://tejas.serc.iisc.ernet.in/~currsci/jan252003/127.pdf). Recent studies, however, have assigned a date of about 700,000 years to the Isampur tools. The importance of the dating of the tools and the site is discussed in Robin Dennell, *The Palaeolithic Settlement of Asia*, Cambridge World Archaeology Series (Cambridge: Cambridge University Press, 2009), pp. 375–77.

28: In 1944, a Harvard palæontologist named Hallum Movius drew something called the Movius line, dividing the area where Acheulean tools where found and beyond which there were none. The line runs southeast from across Europe into the Middle East and end east of Kolkata and cuts through parts of Bangladesh. East of the Movius line, across the whole of southeast Asia and into China, only the older, simpler Oldowan tools have been found (see HL Movius 1948. The Lower Palaeolithic cultures of Southern and Eastern Asia. Transactions of the American Philosophical Society, 38: 329–426.). This suggests that East Asian bifaces may be independent Acheulean-like innovations.

29: The first reported find of hominin remains in India was mentioned form the Narmada valley by W. Theobald in 1860 (see: W. Theobald, 'On the Tertiary and Alluvial Deposits of the Central Portion of the Nerbudda Valley', *Memoirs of the Geological Survey of India*, [Calcutta: Government Press, 1860). Tragically, the skull was lost in the Asiatic Society of Bengal Museum. An overview of discoveries is presented in Rajeev Patnaik et al., 'New Geochronological, Palaeoclimatological and Palaeolithic Data from the Narmada Valley Hominin Locality, Central India', *Journal of Human Evolution* 6 (2009): 114–33.

30: The first inhabitants of Papua New Guinea hopped across the islands of the Indonesian archipelago and arrived about 50,000 years ago in several waves. The land they encountered had a remarkable effect on their cultural development. Papua New Guinea's terrain is marked by imposing mountains, vast forests and extremely rugged territory, and these populations became isolated. Over time, they developed distinct cultures and languages, but possess very few differences in their genes. This is one reason why Papua New Guinea is one of the world's most diverse and fascinating cultural landscapes. When the Portuguese explorer Jorge de Meneses (or Menezes) landed here in the early sixteenth century, he named it 'Ilhas dos Papuas' or the Island of the Fuzzy-Haired People.

31: Evolutionary thinker Stephen Jay Gould (1941–2002) marvelled at the diversity that followed the big freeze 550 million years ago (or the Cambrian explosion) and wondered at the improbability of re-creating the same life forms once again if it were to commence again. 'Wind back the tape of life to the early days of the Burgess Shale; let it play again from an identical starting point, and the chance becomes vanishingly small that anything like human intelligence would grace the replay.' (Stephen Jay Gould, *Wonderful Life*, Chapter 1, [New York: W.W. Norton and Company, 1989]).

LIST OF ILLUSTRATIONS

Grateful acknowledgement is made to the following for permission to carry copyright material:

List of Illustrations

INDEX

INDEX

INDEX